T0146211

Sovereign Skies

HAGLEY LIBRARY STUDIES IN BUSINESS, TECHNOLOGY,
AND POLITICS

Richard R. John, Series Editor

Sovereign Skies

The Origins of American

Civil Aviation Policy

Sean Seyer

JOHNS HOPKINS UNIVERSITY PRESS   BALTIMORE

© 2021 Johns Hopkins University Press
All rights reserved. Published 2021
Printed in the United States of America on acid-free paper
9 8 7 6 5 4 3 2 1

Johns Hopkins University Press
2715 North Charles Street
Baltimore, Maryland 21218-4363
www.press.jhu.edu

Library of Congress Cataloging-in-Publication Data

Names: Seyer, Sean, 1979– author.
Title: Sovereign skies : the origins of American civil aviation policy / Sean Seyer.
Description: Baltimore, Maryland : Johns Hopkins University Press, [2021] | Series:
    Hagley library studies in business, technology, and politics | Includes
    bibliographical references and index.
Identifiers: LCCN 2020022454 | ISBN 9781421440538 (hardcover) | ISBN
    9781421440545 (ebook)
Subjects: LCSH: Aeronautics, Commercial—Law and legislation—United States—
    History. | Aeronautics, Commercial—Government policy—United States—
    History. | United States. Air Commerce Act. | Convention Relating to the
    Regulation of Aerial Navigation (1919 October 13) | Aeronautics and state—United
    States.
Classification: LCC KF2439 .S49 2021 | DDC 343.7309/7—dc23
LC record available at https://lccn.loc.gov/2020022454

A catalog record for this book is available from the British Library.

*Special discounts are available for bulk purchases of this book. For more information, please
contact Special Sales at specialsales@jh.edu.*

Johns Hopkins University Press uses environmentally friendly book materials,
including recycled text paper that is composed of at least 30 percent post-consumer
waste, whenever possible.

Contents

Preface

This book began with a question. While studying aviation history as a graduate student at Auburn University, I came across an aeronautical treaty crafted in the aftermath of World War I that the United States signed but did not ratify: the Convention Relating to the Regulation of Aerial Navigation. I was struck by the complete lack of discussion concerning this convention's impact on the United States; as far as the existing literature was concerned, US aviation regulation occurred completely detached from the rest of the world. But how could an international agreement that delineated the proper relationship between the state and the airplane *and* provided a set of aeronautical rules and standards have had absolutely no effect on the American experience? The more I sifted through numerous archival collections, the more it became apparent that the origins of US foreign and domestic civil aviation policy had not occurred in isolation but instead arose within an intricate web of geographic and historical interconnectivity. As I dove further into the material, this project morphed from a legislative history into a broader "US in the world" story that offered an understanding of how new technologies can prompt a reevaluation of domestic political power and facilitate the diffusion of international standard practices and norms.

The book before you would not have come into existence without the assistance and support of a multitude of individuals. William Trimble shepherded this project through its early stages as a dissertation. Discussions with Dominic Pisano, Roger Launius, Robert van der Linden, and others at the Smithsonian Air and Space Museum helped to reconceptualize the dissertation into a more refined book project. Elizabeth Demers, formerly of Johns Hopkins University Press, saw the potential in this project early on, and current Johns Hopkins University Press senior acquisitions editor Matt McAdam has offered continued encouragement and support. I am deeply indebted to series editor Richard John, whose constant urging to expand the scope of the book undoubtedly led to a better finished product.

Grants and fellowships from the Herbert Hoover Presidential Association, the Smithsonian Institution, the Pan Am Historical Foundation, and the University of Kansas provided valuable funding for research and writing. Alan Meyer, Andrew Russell, Dale Urie, Chris Forth, and Brian Frehner provided useful feedback on the draft (either in whole or in part), as did participants in two Linda Hall Library works-in-progress seminars under the leadership of the unflappable Benjamin Gross. I am eternally appreciative for the assistance of numerous archivists who, continually asked to do more with less, cheerfully helped me comb through a seemingly endless amount of material. My colleagues in the Humanities Program at the University of Kansas have provided an intellectually engaging interdisciplinary environment as well as the institutional and emotional support necessary to see this project through to its conclusion. Various members of the Society for the History of Technology and the Society for Historians of American Foreign Relations provided valuable feedback on this project at numerous conference panels, and I am especially grateful to John Krige, Angelina and Jason Callahan, David Lucsko, Emily Gibson, Lee Vinsel, and Laurence Burke for their assistance and friendship.

My parents, Dennis and Ellen Seyer, have provided unequivocal love and support over the years. And I could not ask for a better partner to share this life with than Carol Woods, who has learned more than she ever wanted to know about aviation history.

Sovereign Skies

Introduction

With the "war to end all wars" fresh in the nation's collective consciousness, Americans across the country gathered on May 30, 1922, to honor those who had fought and died in the service of their country. In Washington DC, this Memorial Day observance took on special significance. Over the past eight years, the Lincoln Memorial—a shrine to the national memory of an earlier conflict—had slowly taken shape on the far west side of the National Mall, and its afternoon dedication now drew thousands in their Sunday best. Dignitaries in attendance included members of Congress, the US Supreme Court, the diplomatic corps, foreign representatives (including the recently arrived German ambassador), Tuskegee Institute president Robert R. Morton, Vice President Calvin Coolidge, surviving members of the Union Army, and President Abraham Lincoln's eighty-three-year-old son, Robert Todd Lincoln.[1]

President Warren G. Harding personally accepted the memorial on behalf of the nation from Supreme Court Chief Justice William Howard Taft, president of the Lincoln Memorial Commission. As the president delivered his prepared remarks, a Curtiss JN-4 "Jenny" circled above the crowd several times. Although officers at nearby Bolling Field had personally requested that local pilot Herbert Fahy not fly over the area that afternoon, the allure of quick cash from a "New York newspaper photographer" determined to get the perfect shot proved too hard to resist.[2] As he circled the crowd well below the ninety-nine-foot height of the Lincoln Memorial, Fahy's engine drowned out President Harding and greatly frightened those below.[3]

Those present at the Lincoln Memorial dedication had just cause for concern at the sight of Fahy's aircraft overhead. Other than extremely lim-

A film reel captured the appearance of Herbert Fahy's JN-4 "Jenny" above the Lincoln Memorial Dedication on May 30, 1922. Lincoln Memorial Dedication, 1922, Records of the Office of the Chief Signal Officer, RG 111, NARA II. Online version, https://www.youtube.com/watch?v=DVUbzOk8mCc, accessed Oct. 16, 2018.

ited provisions in a select few states and municipalities, aviation remained a lawless and unregulated frontier nearly nineteen years after Orville Wright's first flight. In this aerial Wild West, anyone could take off in any machine at any moment, and sensational stories of aircraft accidents—accompanied by vivid postcrash images—appeared regularly in newspapers. Without a mandatory inspection of the airplane or examination of the pilot, those in attendance could not take for granted the structural integrity of the JN-4 flying overhead or Fahy's ability to operate it. In fact, less than two months later the very same aircraft took an unexpected nosedive one hundred feet above the west bank of the Potomac River, severely injuring Fahy and killing his passenger Lewis Swan.[4] Had this unfortunate incident occurred during the Lincoln Memorial dedication, the death of key government officials and civilians could have been an unprecedented national tragedy. Fahy's reckless actions "tremendously annoyed" President Harding, incensed Secretary of War John W. Weeks, and drew the ire of the press.[5] But Fahy

had broken no laws. With Harding's full support, Weeks took the only punitive action open to him and revoked Fahy's commission in the Air Service
Reserve Corps for "disrespect to the President, interference with a great sol-
emn and national occasion, and endangering the lives of many people."[6]

The inability of the federal government to prevent and punish Fahy's
flight stands in stark contrast to the current situation in the United States.
Today, the federal government possesses "exclusive sovereignty of airspace"
and as such exercises ultimate authority over aviation.[7] With a budget of
more than $17.45 billion and nearly forty-five thousand employees, the Fed-
eral Aviation Administration (FAA) manages access to this ethereal domain
through the National Airspace System (NAS), a network of 518 air traffic
control towers and 154 Terminal Radar Approach Control facilities that carried more than a billion passengers in fiscal year 2018.[8] Prior to boarding,
each passenger underwent a security screening by employees of the Transportation Safety Administration (TSA), a branch of the Department of
Homeland Security that was established in the wake of the September 1,
2001, terrorist attacks. Passengers then boarded an FAA-approved aircraft,
where FAA-licensed crews followed detailed FAA-crafted procedures as the
aircraft traveled along a predetermined FAA-issued flight plan. US airspace is
divided into five classes with their own distinct operational guidelines, and
numerous permanent and temporary restricted areas exist throughout the
country.[9] But no area is more tightly regulated than the Flight Restricted
Zone within the Washington, DC, Special Flight Rules Area, a thirty-nautical-
mile radius around Ronald Reagan Washington National Airport (DCA) that
extends to an altitude of eighteen thousand feet. Only "Approved Carriers"
can land at DCA, and civilian flights within the Flight Restricted Zone are
limited to those operating under instrument flight rules that have received
a waiver from the FAA and TSA. Because of national security concerns, the
FAA recommends that pilots operating under visual flight rules (predominantly used in noncommercial general aviation) "think of the Flight Restricted Zone as a 'no-fly' area."[10] Whereas a complete absence of regulations
and means of enforcement meant Fahy faced minimal ramifications for his
1922 flight over the Lincoln Memorial, any modern reenactment would be
met with harsh penalties and lethal force. In response to the airplane, the
atmosphere—a previously open and lawless frontier—has become a primary arena for the modern regulatory state.[11]

This book is about how Americans initially grappled with the proper re-

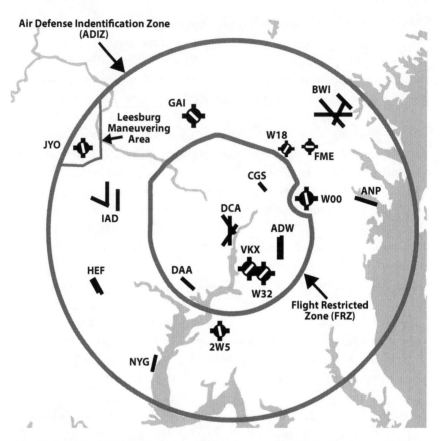

The Federal Restricted Zone around Washington, DC, within the larger Air Defense Identification Zone (renamed the Special Flight Rules Area in 2009), which includes Andrews Air Force Base and Davison Army Airfield. Map from Federal Aviation Administration Advisory, effective Aug. 30, 2007, https://www.faa.gov/news/updates/adiz_frz/media/DC_ADIZ-FRZ_Advisory.pdf.

lationship between the airplane and government in the decade after World War I, a process that laid the foundation for the steady expansion of the federal government's aerial dominance over the next one hundred years. It focuses not on invention or innovation but on how individual actors and relevant interest groups sought to develop rules and structures for civil aviation in both domestic and international contexts.[12] Domestically, aviation policy took the form of the 1926 Air Commerce Act and subsequent Air Commerce Regulations, while at the international level it manifested within the 1928 Pan American Convention on Commercial Aviation. In this book,

I argue that US aviation policy developed in response to three interrelated and mutually reinforcing factors: the airplane's unprecedented ability to provide controlled movement within three-dimensional space, shifting notions of constitutional federalism, and the need to maintain compatibility with the provisions of the Convention Relating to the Regulation of Aerial Navigation crafted as part of the Versailles Peace Conference in 1919. The creation of US aviation policy was not solely technologically determined or socially determined but instead arose out of a complex interaction among the "demands" of the airplane, the values of American society in the early twentieth century, and international considerations. By focusing on what Thomas Misa and Daniel Carpenter dubbed the "middle"- or "mezzo"-level activities of career civil servants, industry representatives, and "standard-setting bodies," this book demonstrates how a relatively small group of bureaucrats, military officers, industry leaders, and engineers drew upon previous regulatory schemes and international principles in their struggle to define government's relationship to the airplane.[13] At its most fundamental level, this study addresses how regulatory ideas initially develop, morph as new players adopt and adapt them, crystallize into a consensus among relevant interest groups, and become institutionalized within either formal or informal processes.[14]

Orville Wright's 120-foot flight on December 17, 1903, over the dunes of Kitty Hawk, North Carolina, marked a clear departure from millennia of human experience.[15] Although human beings had been ascending in lighter-than-air craft since the Montgolfier brothers' demonstrated their hot air balloon in 1783, such devices remained susceptible to the whims of the wind. The airplane, on the other hand, transformed the atmosphere into a seemingly limitless avenue for human conveyance subject only to the will of the pilot. The airplane's ability to provide powered and controlled flight—its very technological essence—constituted a radical change from what had come before, which, in turn, prompted new ways of thinking about spatial arrangements, warfare, empire, trade, communication, and society itself.[16] Viewed as "an instrument of reform, regeneration, and salvation, a substitute for politics, revolution, or even religion" that promised an untold level of global interconnectivity, the airplane also brought with it unique challenges.[17] Unlike radio, another emerging technology of the period, the airplane constituted a very real and immediate physical threat that could, and did, come crashing down on people and property below.[18] The airplane's freedom of movement also constituted a profound challenge to traditional notions of sovereignty, the belief that states possess full autonomy in their affairs and "some reasonable degree of control over both their borders and their territory."[19] Louis Blériot's 1909 flight from Calais, France, to Dover, England, showed that "no nation could afford to ignore a new machine capable of soaring over traditional barriers and borders."[20] Airplanes could carry mail and commercial goods, but they could also carry contraband liquor to thirsty Prohibition Era Americans. The same airplane that dropped advertisements over a crowd could easily be converted to drop a bomb. And the very aerial perspective that provided breathtaking panoramas and inspired works of artistic beauty also offered potential enemies knowledge of fortifications and troop concentrations.[21] The experience of World War I only heightened the airplane's potential danger, transforming it from "a peacetime marvel . . . to a fearsome tool of war, a bringer of destruction and death."[22] For all its benefits, the dual-use technology of the airplane constituted a clear threat to lives, property, and national security.[23]

Government had a duty to protect its citizens and constrain the destabilizing aspects of powered flight.[24] But where did the regulatory authority for aviation lie in the United States, a nation composed of individual sovereign states and a federal government of enumerated powers? This question constituted a new chapter in the larger historical debate over the nature of domestic sovereignty within the Constitution's federalist system, "the overriding issue Americans confronted before and after independence" that continues to shape American life to this day.[25] With their ability to move rapidly among municipal and state jurisdictions, new industrial means of transportation unknown at the time of the nation's founding became primary flashpoints for this debate. Steamboats, railroads, and the automobile each prompted a reevaluation of the border between state and federal authority that influenced regulatory discussions concerning the airplane. Advocates of uniform aviation regulation pointed to the precedents of federal steamboat and railroad laws while they fought vehemently against state legislation that could precipitate a repeat of the automobile's state-centered system, one that they viewed as fundamentally incompatible with the airplane's speed and unique capacity for powered flight.[26]

But Americans did not address the airplane's place within constitutional government and the proper regulations for powered flight in isolation. Rather, representatives of the nascent aircraft industry, civilian and military departments within government, and the engineering profession did so within the context of a new postwar international civil aviation regime that recognized the nation-state as the primary regulatory authority for aviation. Drafted as a means to balance the need for security with the desire to facilitate aerial commerce, the 1919 Convention Relating to the Regulation of Aerial Navigation established a set of "principles, norms, rules, and decision-making procedures" for international aviation.[27] Although the United States signed but did not ratify this first multilateral air convention, it nonetheless profoundly shaped the origins of US civil aviation policy. Facilitated by increased networks of exchange as a result of wartime Allied cooperation and compelled by Canada's adoption of the convention's provisions and norms, American policymakers simply could not ignore this new postwar international civil aviation regime as they worked to craft a domestic response to powered flight. When the United States sought to create an air convention for the Western Hemisphere specifically tailored to its economic and security concerns, its Latin American neighbors drew upon established international aeronautical principles and norms to successfully constrain Washington's attempt to secure complete aerial hegemony.

Aviation presented a unique policy challenge for the United States. The airplane's ability to disregard terrestrial checkpoints and geographical barriers heralded an age of increased connectivity that simultaneously undermined the territorial integrity of political boundaries. Faced with a device that threatened to weaken this central tenet of state sovereignty after World War I, Allied representatives reaffirmed the sanctity of national borders and drafted a system of rules and regulations for international air travel. But within the United States, uncertainty over the delineation between state and federal authority under constitutional federalism further complicated matters. As military and civil officials in Washington and industry representatives debated the proper regulatory response to this new device, they drew upon historical analogies as well as ideas and solutions that—just like the airplane—also crossed borders. Despite the historical rhetoric of a "city on a hill" and contemporary tirades against globalism, the United States has never been an island unto itself, and this was as true in the first decades of the twentieth century as it is in the first decades of the twenty-first.[28] By placing the creation of US civil aviation policy in a "wider view," this book calls attention to the ways that certain technologies prompt a reconfiguration of the contours of power within constitutional federalism, the role of international regimes on domestic policy-making, and the contentious nature of sovereignty.[29]

1

Where Does the Regulatory Power Lie?

Transportation and Federalism before
World War I

Assistant Legislative Counsel for the House of Representatives Frederick P. Lee was on a mission. For almost a year he had worked with William P. MacCracken Jr., a Chicago lawyer and the chair of the American Bar Association's Special Committee on the Law of Aviation, on legislation to place "all air navigation in the United States" under federal control through a new Bureau of Aeronautics in the Commerce Department.[1] The two lawyers had come to believe that "uniform regulation . . . is . . . absolutely indispensable," and only federal control could "avoid that diversity of State regulation which has arisen in respect to motor-vehicle traffic."[2] In a legal brief to the House Interstate and Foreign Commerce Committee in support of their bill—known as the Civil Aeronautics Act of 1923—they argued that Congress's constitutional authority to regulate interstate and foreign commerce under the commerce clause "keeps pace with new developments of time and circumstance." They looked to the precedents of federal steamboat and railroad regulation to support their call for federal control over aviation and chided Congress's failure to prevent the creation of diverse state automobile laws.[3] But there was one overriding problem. Drafted a mere five years after the Montgolfier brothers first ascended in a hot air balloon, the US Constitution did not explicitly grant the federal government jurisdiction over the atmosphere or objects traveling within it. In fact, the Tenth Amendment's declaration that "any powers that are not the exclusive powers of the United States government, or that are not specifically denied the states' governments by the Constitution, are the powers of the states, or of the people" would seem to give regulatory authority over aviation to the states.[4]

Lee's memo in support of the 1923 Civil Aeronautics Act vividly shows that the origins of aviation regulation in the United States were intertwined with fundamental questions about the division of sovereign power within the Constitution's federal system, a tension originating with the Federalist and Anti-Federalist debates of 1787 that remains central to the American experience today. At the heart of the matter lay the discrepancy between a state's limited territorial jurisdiction and the rapid movement made possible by new industrial modes of conveyance. Steamboats traveled along major rivers that meandered through or constituted the borders of several states, railroads spanned multiple jurisdictions, and automobile owners traversed thoroughfares that crossed state lines. With each advancement in transportation, the divergence between an individual state's regulatory authority and the practical use of these new devices prompted a reevaluation of the very contours of domestic sovereignty under the Constitution's federal system.[5]

These new industrial modes of transportation shaped Americans' perceptions of the extent of federal regulatory authority that, in turn, set the parameters for discussions on aviation after World War I. Over the span of one hundred years, individuals such as US Supreme Court Chief Justice John Marshall, regulatory pioneer Charles Francis Adams, lawyer Charles Thaddeus Terry, and prominent jurist Simeon Baldwin called for an expansion of federal authority to properly regulate these new machines. In the case of steamboats and railroads, the inability of states to adequately address threats to public safety and consumer welfare compelled Congress to pass regulatory legislation based on a reinterpretation of the commerce clause. But this was not the case with the automobile. Although the automobile clearly allowed for interstate commerce, and motorists beseeched Congress for uniform national laws, regulation of this new individual means of motorized transit became locked at the state level. In their struggle to place aviation under federal control, government officials, industry leaders, and members of the legal profession drew upon the lessons of steamboats and railroads while they adamantly fought against a repeat of the path-dependent process that had led to the state-centered regulatory system for the automobile, one they viewed as fundamentally incompatible with powered and controlled flight.[6] In order to fully understand the post–World War I debate over aviation, we must look first at the broader historical context of transportation regulation in the United States.

Full Steam Ahead

Aaron Ogden found himself in an unenviable position in early February 1824. For the previous ten years, he had operated a steamboat service between New York City and Elizabethtown, New Jersey, under a New York State–granted monopoly first awarded to Robert Fulton and his partner Robert Livingston after their successful 1807 steamboat demonstration. Now Ogden was embroiled in a Supreme Court case against his former partner Thomas Gibbons, whose competing steamboat service operated under a federal license issued under the Federal Coastal Licensing Act of 1793. Gibbons had turned to the courts, and, unsurprisingly, the New York Court of Errors upheld Ogden's state monopoly. In response, Gibbons and his lawyer, the famed orator Daniel Webster, appealed the decision.[7] Ogden's lawyers argued that the federal law applied only to the types of vessels that existed at the time of its passage, therefore placing the new technology of the steamboat outside of the federal law. It now fell on the Supreme Court to determine which of the two operating licenses reigned supreme. Although ostensibly a case about steamboats, *Gibbons v. Ogden* addressed the very nature of federal power under the commerce clause.

Gibbons came before a Supreme Court under the direction of Chief Justice John Marshall. A staunch Federalist from Virginia who had served in the Continental Army during the Revolution, Marshall remained acutely aware of the economic fragmentation that had precipitated the Constitution's creation and the need for a strong federal authority to serve as a bulwark against future political disintegration.[8] Under the weak central government of the prior Articles of Confederation, Rhode Island, Massachusetts, Pennsylvania, New Hampshire, and New York had passed protective tariffs against their neighbors to foster their own domestic markets. In May 1787, state delegates met in Philadelphia to address the rapidly deteriorating situation, and they quickly agreed on the need for a stronger central authority to provide a counterforce against the centrifugal tendencies of confederation.[9] To prevent the future erection of tariff barriers among states, the new Constitution granted Congress the power "to regulate commerce with foreign Nations, and among the several States, and with the Indian Tribes," and to enact "all laws which shall be necessary and proper" in the course of exercising that power.[10] Under this authority, the newly established Congress issued licenses under the aforementioned Federal Coastal Licensing Act, set import duties on a wide range of consumer goods, fixed graduated tonnage duties on all ships entering designated US ports, and undertook the "support, maintenance and repairs" of state-erected navigational aids to coastal trade.[11] Clearly, members of Congress believed they possessed the authority to foster and administer coastal trade from the very beginning of constitutional government.[12] But the full extent of Congress's constitutional authority to regulate commerce remained uncertain. Did the term "commerce" refer solely to the actual act of exchange or to "all gainful activities intended for the marketplace?"[13] Could states exercise a concurrent regulatory power? If so, how would conflicting state and federal statutes be reconciled? *Gibbons* provided Marshall the perfect opportunity to define the extent of federal power under the commerce clause.

In a unanimous decision written and delivered by the chief justice, the Supreme Court declared that the New York law clearly conflicted with the Federal Coastal Licensing Act of 1793 and invalidated Ogden's monopoly based on its recently elucidated doctrine of federal supremacy.[14] But the court did not stop there. Determined not to return to the fractious days of confederation, Marshall delineated a broad sphere of federal authority based on "a theory of the Constitution as a constituent act of the people of the United States, not a compact among sovereign states."[15] After defining commerce broadly as "the commercial intercourse between nations" and not simply the discrete exchange of goods, Marshall declared that Congress's ability to regulate commerce must therefore include an implied power to regulate navigation. He further argued that because "commerce among the States cannot stop at the external boundary line of each State, but may be introduced into the interior," this implied power to regulate navigation extended into the political jurisdiction of individual states.[16]

The Supreme Court's ruling in *Gibbons* would have a profound impact on subsequent debates over the relationship between federal and state power, particularly in the realm of transportation. It would not be difficult to argue that an implied power to regulate navigation also applied to the movement of goods and services through other means. In fact, Marshall's explicit rejection of the defense's argument that steamboats differed from traditional sailing vessels and therefore fell outside of Congress's current regulatory power supported future arguments for the federal regulation of other modes of conveyance unknown at the nation's founding. But *Gibbons* did not permanently settle the tension between state and federal authority as Marshall also recognized that states possessed sovereign authority to pass laws for the general welfare and well-being of their citizens through their traditional police powers.[17] Under his successor Roger Taney—the author of the infamous 1857 Dred Scott decision—the Supreme Court undermined the doctrine of supremacy that undergirded *Gibbons* when it declared that the federal government's constitutional right to regulate both commerce and navigation "did not imply a prohibition on the States to exercise the same power."[18] Although the extent of federal power waxed and waned as the domestic sovereignty debate continued, Marshall's broad definition of commerce in *Gibbons* has allowed the commerce clause to serve as a primary justification for the extension of federal authority in such diverse areas as Progressive Era antitrust law, New Deal recovery programs, the post–World War II civil rights struggle, and the Affordable Care Act of the early twenty-first century.[19]

Public safety concerns would compel Congress to regulate steamboat navigation in the decades after *Gibbons*. Steamboats, with their ability to navigate upstream, transformed the nation's major rivers "from continental drains into two-way arteries of commerce," and smokestacks and paddle-wheels forging up the mighty and tumultuous waterways of the frontier came to symbolize American western expansion just as much as covered wagons.[20] The increased speed and subsequent reduction in freight costs precipitated a "market revolution" that transformed Cincinnati, St. Louis, Louisville, and other Midwest river cities from small trading outposts to major urban and industrial centers.[21] While steamboats provided clear benefits, the harnessing of atmospheric pressure differentials within "large wood, copper, or cast-iron containers" made steam engines susceptible to explosion. By 1830, an estimated 350 lives had been lost in sixty-four separate steam boiler explosions.[22] As it became apparent that the enlightened self-interest of steamship owners would not correct the situation, Alabama, Louisiana, Kentucky, Illinois, Wisconsin, and Missouri passed safety legislation that included boiler inspections, the licensing of engineers, and navigational requirements.[23] But these laws applied only within each state's jurisdictional boundaries, while major rivers crossed multiple states. The limits of an individual state's sovereign authority simply did not align with actual steamboat use. Deaths continued to mount because—as the Mississippi legislature informed Congress—the ability to fully address the issue existed "beyond the power" of states.[24] Advocates of federal aviation regulation nearly a century later would adopt the very same position.

Federal steamboat regulation manifested itself within two separate pieces of legislation: an initial act in 1838 and a revised one in 1852. The first empowered district judges to appoint "skilled and competent" inspectors to certify the hulls and boilers of all steamships engaged in the transportation of goods and passengers along all "navigable waters of the United States" prior to the awarding of a federal license but suffered from the absence of an effective enforcement mechanism and uniform inspection criteria.[25] Despite these practical shortcomings, the 1838 act is significant for two reasons. First, the overwhelming consensus surrounding the bill illustrates the extent to which the domestic sovereignty debate in relation to steamboats had shifted since *Gibbons*. Guided by a technical report from the Franklin Institute—a precursor to the engineering societies of the late nineteenth and early twentieth centuries—even fierce political rivals Andrew Jackson and Daniel Webster could agree on the need for Congress to act and its constitutional right to do so.[26] Second, it represented the first direct federal regulation of a means of transportation that had not existed at the time of the nation's founding, and it occurred through legislation rather than a constitutional amendment. With *Gibbons* as precedent, Congress subsumed the steamboat under its recognized regulatory authority despite technological differences between it and traditional vessels. In 1852, Congress addressed the shortcomings of the 1838 act by empowering federally employed inspectors with the ability to compel testimony and creating a presidentially appointed nine-member administrative board within the Treasury Department— later known as the Steamboat Inspection Service—to promulgate proper inspection procedures and navigation regulations.[27] This stricter form of federal oversight demonstrably reduced steamboat accident deaths from 1,155 between 1848 and 1852 in the west alone to a total of 224 on both eastern and western waterways from 1854 to 1858.[28] As the nation lurched closer to civil war, it had been clearly demonstrated that a well-defined and enforceable system of federal regulation could mitigate the negatives of a new means of transportation.

The steamboat's introduction sparked a reevaluation of the relationship between state and federal authority in the new Republic that would have an enormous impact on discussions of transportation regulation for decades to come. Marshall's decision in *Gibbons* provided an expansive definition of commerce that serves as an important mechanism for the enlargement of federal power to this day. After *Gibbons*, Congress's ability to regulate navigation in the service of foreign and interstate commerce did not stop at a state's jurisdictional boundary. Perhaps even more important for the early twentieth-century debate over aviation regulation, Marshall's rejection of a regulatory distinction between sailing vessels and steamboats offered a way to apply the federal commerce power to transportation technologies unknown at the Constitution's creation. And the establishment of a system of steamboat inspection and licensing under the Steamboat Inspection Service—transferred along with the Bureau of Navigation to the newly established Department of Commerce and Labor in 1903—showed how a centralized administration could adequately address issues of transportation safety that transcended the limited jurisdictions of the various states.[29] But as a consensus was developing around the proper level of sovereign authority that could best address the dangers of steamboats, another application of steam power generated new questions about the boundary between state and federal power.

Reining in the Iron Horse

Perhaps no single technical advance has had a greater influence on the course of American political and economic development than the railroad. As it expanded from a single three-mile horse-drawn track in 1827 to "almost 250,000 miles of line" less than ninety years later, this "combination of steam locomotive, fixed rails, and a train of carriages or wagons" captured the American imagination, united the nation, and propelled unprecedented growth.[30] The proliferation of railroads affected every aspect of American life as private railroad companies pioneered new corporate management structures and competitive practices to accumulate previously unheard-of wealth, political influence, and economic power. After the Civil War, various constituencies called for government action to address perceived mismanagement, rate discrimination, and corruption. Regulatory solutions to the so-called railroad problem began at the state level, but, as with steamboats, it became apparent that the scale of the problem required a federal solution.[31] In response to calls from western farmers, eastern merchants, and regulatory advocates, Congress took action and established the Interstate Commerce Commission (ICC) in 1887 to regulate interstate railroad rates. Over the following decades, the ICC's operational limitations prompted a further reimagining of government's relationship with the railroads that contributed to and gained momentum from emerging Progressive ideology.

Whereas the federal government's regulatory relationship with the railroads was practically nonexistent before the Civil War, by the time the United States entered World War I railroads had been declared a matter of national interest and the ICC had been empowered to regulate those aspects of intrastate rail transportation that affected interstate commerce.

As a capital-intensive industry that required access to large stretches of land, railroad companies were intimately connected with state governments from their inception.[32] In the hope of competing with New York's widely successful Erie Canal, states issued charters that exempted newly founded private railroad companies from taxation and granted broad powers, including eminent domain.[33] In exchange for this broad delegation of state authority, charters sought to apply prevailing cultural assumptions about roadway access to railroading. Traditionally, ownership of the avenue of conveyance (the canal or roadway) and the means of conveyance (the ship or wagon) remained separate—anyone who paid the requisite toll could access the thoroughfare. Under common law, carriers for hire that utilized public roads and waterways were expected to charge all customers a reasonable fee as determined by a mixture of market competition, judicial review, and state regulation. As common carriers, railroads thus remained "subject to the common law obligations to make only reasonable charges," but—unlike roadways and waterways—railways themselves were privately owned thoroughfares that "combined the building and maintaining of a road . . . with the operation of traffic on that road."[34] To uphold this traditional distinction between the avenue and the means, state legislatures embedded within charters provisions for maximum toll rates (for *access* to the railway) and maximum transportation rates (for the *carriage* of goods).[35]

But such provisions, however, proved increasingly difficult to enforce as operational track mileage mushroomed from 2,818 to 70,268 between 1840 and the Panic of 1873.[36] As railroad companies stitched the nation into an interconnected national market, they extended beyond the jurisdictional boundaries of their state charters. The Pennsylvania Railroad provides as excellent example of this divergent trend. Initially established by the Pennsylvania legislature in 1846 to create and operate a two-hundred-mile line wholly within the state between Harrisburg and Pittsburgh, a wave of post–Civil War growth saw the Pennsylvania Railroad extend its tentacles into New Jersey, Delaware, Maryland, New York, Ohio, Indiana, and Illinois.[37] In their struggle to compete with rival rail companies and waterborne transportation, railroads increasingly adopted survival strategies such as preferential rates, secret rebates, and voluntary pooling agreements that both western raw material producers and eastern manufacturers viewed as unfair and corrupt. Could states adequately respond to the situation?

Charles Francis Adams Jr., the descendent of two American presidents, believed not only that they could, but should. Operating under the belief that the unprecedented power of railroads represented the defining issue of the era, Adams rejected the idea that either pure market competition or legislators susceptible to corruption could effectively address the "railroad problem."[38] Rather, he called for the creation of expert state-level commissions to oversee the railroads and inform the public of their nefarious practices.[39] A prolific writer, Adams's ideas spread rapidly, and states responded in two primary ways. A "supervisory-advisory" commission, also known as a "sunshine commission," exposed wrongdoing and harnessed public opinion to supervise the states' railroads, investigate charter compliance, and prevent accidents. Pioneered with the 1869 establishment of the Massachusetts Railroad Commission operated under Adams's guidance, this type of commission proliferated across New England. Western states, however, took a more hands-on approach in large part due to an active lobbying campaign by the National Grange of the Order of Patrons of Husbandry in response to farmers' increased reliance on railroad connections to distant markets. These "supervisory-mandatory" commissions not only investigated the railroads but possessed the authority to take legal action against railroads for noncompliance. But as was the case with state-based steamboat safety regulations, state commissions could not effectively address the totality of the problem. When states with strong regulatory commissions like Illinois mandated rate reductions within their borders, interstate corporations such as the Pennsylvania Railroad simply raised rates in other states. The regulatory actions of individual states therefore produced "spillover effects" that influenced commerce both within and among several states.[40]

Faced with the possibility of conflicting state provisions and a torrent of litigation, railroads challenged these "supervisory-mandatory" commissions in court based on three arguments: (1) they "deprived the roads of their property without due process of law," (2) ever-changing regulation by commission went contrary to the railroad's charters and therefore violated the sanctity of contract, and (3) rail transportation affected interstate commerce and therefore fell within the regulatory purview of the federal government rather than the states.[41] Railroad lawyers and executives clearly recognized that their enterprises had outgrown a state's limited jurisdiction and only the federal government could effectively alleviate the negatives of competition through sound regulation. They were not alone in this belief. While serving aboard the Massachusetts commission, Adams himself recognized that "the railroad system has burst through State limits" and could no longer "be organized or controlled under State laws."[42] At the seventh annual session of the National Grange in 1874, chairman Dudley W. Adams professed that a true solution would require "Congress to avail itself of its constitutional right to regulate commerce between the States."[43] President of the Illinois State Farmers' Association William C. Flagg noted that "national legislation or regulation is needed" to ensure that state regulatory commissions did not restrict the ability of midwestern farmers to access burgeoning national and international markets.[44] Railroads had clearly outgrown the geographically limited jurisdictions of the very states that birthed them, a reality that prompted otherwise disparate factions to look for a federal remedy.

The Supreme Court was not yet willing to make that leap. Its decision in *Chicago, Burlington & Quincy Railroad Company v. Iowa* (1877) affirmed the power of government to regulate railroads for the public good but explicitly nested such power at the state level. Drawing on a position expounded in *Munn v. Illinois* earlier that year that "when private property is devoted to a public use, it is subject to public regulation," the Supreme Court declared that because railroads constituted "carriers for hire . . . engaged in a public employment affecting the public interest," they were "subject to legislative control as to their rates of fare and freight."[45] But by upholding both the Chicago, Burlington & Quincy Railroad's Iowa charter *and* Iowa's 1874 rate law, the Supreme Court gave its blessing to a system of state-based regulation that seemed increasingly out of step with the operational reality of railroads. The court's ruling in *Chicago* is even more perplexing in light of two flanking rulings wherein the court declared that the nature of commerce remained the same "whether . . . by land or by water," that rail transportation was inherently national in nature, and that Congress's constitutional powers "are not confined to the instrumentalities of commerce . . . when the Constitution was adopted, but . . . keep pace with the progress of the country."[46] Although the Supreme Court clearly held Marshall's expansive view of commerce enunciated in *Gibbons*, it refused to move against state regulation. Through the 1870s and 1880s, eastern shippers, midwestern producers, and the railroads themselves increasingly looked to Congress to remedy a state-centered system sanctioned by *Chicago* that appeared increasingly disconnected from the court's other rulings.

Within Congress, national security and public health concerns had prompted a slow reconceptualization of federal authority over railroads initially begun in response to the outbreak of civil war.[47] To ensure the movement of troops and material during the sectional conflict, Congress had authorized the president "to take possession of any or all the telegraph lines . . . [and] railroad lines in the United States" in the name of "public safety."[48] With the Union Pacific Railroad's 1862 charter to construct a transcontinental line, Congress took its first step into the realm of rate regulation when it included the ability to "reduce the rates of fare" if the company's net earnings exceeded 10 percent of total costs.[49] To prevent discriminatory state laws that threatened Reconstruction Era unity, in 1866 Congress declared that all steam-powered railroad companies possessed the right to transport "passengers, troops, government supplies, mails, freight, and property" across state lines.[50] In response to public safety concerns, Congress provided national minimum standards for the rail transportation of cattle in 1873 that could not be fully addressed through the jurisdictionally limited legislation of New York, Massachusetts, Maine, and Illinois.[51] This power to regulate interstate cattle transport prompted the recently appointed Senate Select Committee on Transportation Routes to the Seaboard to conclude that such authority no doubt extended to "other particulars, including the tariff on freights."[52] The Republican-dominated House of Representatives agreed and approved its first bill to regulate rates the next year, but final passage remained elusive. After President Chester A. Arthur's 1883 call for legislation to "protect the people at large in their interstate traffic against acts of injustice which the State governments are powerless to prevent," the Republican Party incorporated federal railroad legislation in its official platform.[53] By 1885, both the House and the Senate had passed legislation regulating interstate commerce that differed on the particulars.[54] Nevertheless, a consensus that Congress possessed the constitutional authority to regulate interstate commerce by rail had clearly emerged among all interested parties.[55] But just as had occurred with steamboat safety regulation and would occur in the next century with the issue of aviation regulation, the passage of federal railroad legislation ultimately manifested in response to an acute sense of crisis.

An about-face by the Supreme Court in *Wabash, St. Louis & Pacific Railway Company v. Illinois* (1886) precipitated the emergency. Confronted with nine years of evidence since *Chicago* that states simply could not adequately address the problem, the Supreme Court applied the implied power to regulate navigation from *Gibbons* to the railroads, another technology unknown at the nation's founding. With six new associate justices confirmed since its *Chicago* decision, a practically reconstituted Supreme Court invalidated an Illinois rate law, asserted that railroads possessed a "national character," and proclaimed that any such regulation "should be done by the Congress of the United States under the commerce clause of the Constitution."[56] As President Grover Cleveland rightly pointed out, *Wabash* left "an important field of control and regulation . . . entirely unoccupied," and Congress quickly responded.[57] The Interstate Commerce Act (ICA) prohibited the use of rebates, preferential rates, rate discrimination between short and long hauls, and pooling. It also established a new ICC to investigate and enforce compliance.[58] In their desire to "remove all politically divisive and potentially dangerous policy decisions from the political arena," Congress chose to create an independent administrative body with a mixture of executive, legislative, and judicial powers.[59] But the ICA constituted a starting point, not a final destination, and in the following decades Progressive presidents Theodore Roosevelt and William Howard Taft oversaw an extension of the ICC's authority over "all instrumentalities and facilities of [interstate] shipment or carriage" that eventually included rail, wired and wireless telegraphy, and the telephone.[60]

While the regulation of ever-expanding transportation and communication companies became a central pillar of Progressive policy, by the end of Taft's term federal power still did not include the ability the regulate intrastate rates that affected interstate commerce. In the years leading up to the outbreak of World War I, the Supreme Court drew upon Marshall's assertion in *Gibbons* that commerce did not "stop at the external boundary line of each State" to extend federal authority to intrastate rail transportation.[61] Associate Justice Charles Evans Hughes proved central to this process. A keen Republican legal mind, Hughes had won the governorship of New York in 1906 with the support of fellow New Yorker Theodore Roosevelt only to resign four years later to accept Taft's nomination to the Supreme Court. Prior to his assent to the governorship, Hughes had served on the state's investigative committees into the gas and insurance industries and—although he would later butt heads with Franklin Delano Roosevelt over the extent of federal power to enact New Deal programs—he firmly believed in government's ability to address social ills.[62] Channeling the spirit of Marshall, Hughes's opinions in two cases in the early years of the Wilson administration saw the Supreme Court recognize that "wherever the interstate and intrastate transactions of carriers are so related that the government of the one involves the control of the other, it is Congress, and not the State, that is entitled to prescribe the final and dominant rule."[63] When state and federal interests could not be neatly disentangled, the doctrine of supremacy elevated federal regulation over that of the states. For Hughes, the entire rail network had to be viewed as a national system, and any state regulation that could affect the larger system must give way to the federal power. Hughes's rulings not only provided a clear precedent for advocates of federal aviation regulation in the next decade, but as secretary of state in the Harding administration, his conception of federal power and "strongly" held views "in favor of international cooperation" would help shape US aviation policy.[64]

Over a period of roughly ninety years, railroads went from an experimental technology in an agrarian nation to the dominant means of transportation for a rapidly maturing industrial society. Initially bestowed with sovereign authority by state legislatures, the power of railroad corporations quickly grew to rival and then surpass the very state governments that spawned them. As rail transportation went from a convenience to a necessity and the railroad industry pioneered new corporate structures and bureaucratic procedures, citizens turned to state governments to address unfair and corrupt practices. But by the 1880s it became overwhelmingly apparent that states simply lacked the authority to effectively regulate an industry whose operations extended beyond their limited geographical borders. Only the federal government could provide the level of regulatory control necessary, and this recognition prompted a steady reconceptualization of Congress's authority under the commerce clause over the next three decades. By the time Wilson won reelection in 1916, the vision of interstate commerce that Marshall had espoused in *Gibbons* applied to railroad rate regulation, and Congress had extended the ICC's mandate to include railroad safety devices.[65] To meet the logistical demands of total war, the federal government even took direct control of the railroads under the Federal Railroad Control Act of 1918.[66] The steady expansion of federal regulatory authority over steamboats and railroads served as powerful precedents for placing aviation under federal control, but the state-centered regulatory approach for the automobile developed during the first decades of the twentieth century showed that one could not assume that aviation would naturally fall under federal jurisdiction.

The Road Not Traveled

Things were not looking good for Charles Thaddeus Terry. As Legislative Committee chairman for the recently established American Automobile Association (AAA), the forty-three-year-old Dwight Professor of Law at Columbia Law School had spent the past several years spearheading the organization's push for a federal automobile bill, one that he believed would unquestionably "become the law of the land."[67] Based on a definition of commerce rooted in *Gibbons* that placed interstate travel (i.e., navigation) under federal jurisdiction "irrespective of the purpose of the intercourse or transit," Terry's bill called for the creation of a Motor Vehicle Bureau within the Department of Commerce and Labor that would issue federal automobile registrations and license plates.[68] While Terry sought to place interstate travel by automobile under federal control, as secretary for the National Conference of Commissioners on Uniform State Laws he recognized that states retained constitutional authority over intrastate automobile travel.[69] To reconcile these two sovereign spheres, his bill called for a dual registration system wherein motorists could obtain a nationally recognized federal registration number only after they had fully complied with their home state's licensing and registration requirements.[70] But now, at a February 1910 hearing of the House Committee on Interstate and Foreign Commerce—the very same body that would approve an extension of the ICC's authority over telephone and telegraph communications a month later—representatives were raising serious objections. Cognizant of the impact that such an extension of federal authority could have on their state's ability to administer institutionalized racial segregation under Jim Crow, Southern Democrats flatly dismissed Terry's proposal. Even Republicans more amicable to federal power remained convinced that automobile licensing and registration clearly fell within a state's police power. They pointed out that the federal license of a motorist who had completed a rigorous examination in their home state would be just as valid as someone who simply paid a fee in a different state and viewed Terry's suggestion that federal pressure would compel states to establish rigorously uniform standards as an unconstitutional interference with a state's police powers.[71] Although a revised bill did squeak out of committee on the second to last day of the 61st Congress, it faced unsurmountable constitutional objections and died on the House floor.[72] Terry was unable to convince the Republican-dominated House of the need to extend federal regulatory authority over commerce to the automobile, and Democrats were more than happy to let the issue die when they took control in 1911.[73]

US drivers continue to feel the ramifications of this obscure, century-old drama to this day. The federal government has asserted regulatory control over interstate trucking as well as safety and environmental aspects of the automobile over the course of the twentieth century, but the licensing and registration of automobiles remains at the state level.[74] A system of reciprocity agreements among states allows Americans to freely drive across the country, but upon changing their state of residence they have a brief window (usually thirty days) in which to secure a driver's license and automobile registration in their new home state. Although this "patchwork of diverse policies" that subjects Americans to unnecessary hassle and widely divergent fees remains difficult to enforce, the belief that the responsibility for automobile licensing and registration rests at the state level has become ingrained within the national fabric.[75] Even the revelation that vulnerabilities in the state licensing system allowed perpetrators of the 9/11 terrorist attacks to secure driver's licenses and therefore flying lessons could not dislodge the deep-rooted conviction that such activities remained a state prerogative.[76] This suboptimal system is a direct result of a path-dependent process that began with Congress's rejection of Terry's bill and received the Supreme Court's sanction in two cases prior to America's entry into World War I. As auto registration and licensing—and their accompanying revenue—became ever more entrenched at the state level, the establishment of a national system became increasingly economically and politically unlikely. In his memoranda on the 1923 Civil Aeronautics Act to the House Committee on Interstate and Foreign Commerce, Frederick Lee chided Congress for allowing the creation and continuation of disparate and often contradictory regulations for the licensing of automobile operators, vehicle registration, and rules of the road, and warned that similar inaction in the case of the airplane would gravely endanger public safety.[77] For postwar advocates of uniform aviation regulation, a repeat of the automobile's state-based framework had to be avoided at all costs. But those concerned with the ever-encroaching extension of federal power at the expense of state autonomy, particularly congressional Democrats from the Jim Crow South, saw the automobile's state-based regulatory system as an example worth emulating.

State automobile licensing and registration arose in direct response to the perceived negatives of municipal ordinances, which had themselves emerged to address the public safety threat of the newly developed automobile. Vivid newspaper accounts of accidents and injuries increasingly appeared at the turn of the century as motorists drove their new horseless carriages on urban streets already bustling with horses, cyclists, and pedestrians.[78] In Chicago, the Welsh's usual evening carriage ride through Lincoln Park ended with Mrs. Welsh suffering a broken ankle when their carriage overturned after the clicking and stammering gasoline engine of an automobile spooked their horse.[79] A similar incident in Mineola, Long Island, left eighty-year-old farmer Valentine H. Hallock in critical condition after he "sustained a compound fracture of the collar bone," broken ribs, and a possible pierced lung.[80] Four racing automobiles caused a horse, with carriage in tow, to sprint in front of an oncoming express train in Reading, Massachusetts, and the subsequent collision killed a young boy and severely injured his brother.[81] Tragedy befell thirty-year-old Virginia Lassiter—sister of Representative Francis Lassiter of Petersburg, Virginia—when an electric automobile frightened her team of horses. As she hastily dismounted the carriage in the hope of preventing personal injury, "her dress got entangled in one of the wheels, and she fell headforemost into the street, fracturing her skull." Virginia never regained consciousness and died shortly after.[82] Tragic scenes like these repeated themselves across the country as the emerging technology of the automobile, with its potential for rapid personal mobility, entered American roadways.[83] Such a threat to public safety clearly called for a regulatory response.

Municipalities had traditionally regulated roadways within their jurisdiction, so it was only natural that they would be the first to respond to the challenges of the automobile.[84] City governments justified automobile ordinances based on their previously granted authority to regulate horse-drawn traffic and utilized the three existing tools of speed limits, vehicle registration, and operators' licenses. Such municipal ordinances raised two important questions. The first revolved around whether a municipal government's authority to regulate roadways extended to the new technology of the automobile, a concern centered on technological distinction that echoed the defense's position regarding steamboats in *Gibbons*.[85] Galvanized into local auto clubs, early adopters of the automobile viewed local ordinances, such as Chicago's expansive 1899 law, as discriminatory taxation and challenged them in court.[86] Differences in legal interpretation initially led to conflicting rulings as judges struggled to define the automobile's relationship to municipal government in the first years of the new century. Whereas the DC Court of Appeals exempted electric automobiles for hire from an 1871 license tax on all "proprietors of hacks, cabs, omnibuses, and other vehicles" because they were not explicitly mentioned in the ordinance, the Michigan Supreme Court and the Fifth District Court of Pennsylvania took the opposite position when they upheld automobile registration ordinances as proper expressions of Detroit's and Pittsburgh's existing regulatory authority.[87] While the First District Appellate Court of Illinois recognized that speed limits and provisions for bells, lamps, and brakes fell under Chicago's established police powers, it also ruled that any automobile licensing provisions constituted an unjust tax "upon one class of citizens in the use of the streets, not imposed upon the others."[88] Despite this favorable ruling, concern for public opinion in the face of negative press led Chicago Automobile Club members to ultimately accept the entire ordinance as an "expedient and advisable" way to prevent reckless driving.[89] By the middle of the first decade of the twentieth century, drivers begrudgingly acknowledged the need for automobile regulation, and the courts increasingly viewed such measures as falling within the authority of municipal governments.

But municipal regulation did not correspond with the automobile's actual use. Automobiles traveled along roadways that passed through various city and county jurisdictions, and motorists quickly recognized that varying local regulations directly conflicted with the automobile's promise of rapid mobility. Take, for example, the case of a fictional motorist from Joliet, Illinois, in 1902. To legally drive in their hometown, they only needed to affix a light at the front of their vehicle at night. But to remain law-abiding citizens when driving through neighboring Chicago, they would have to pass that city's license examination, pay a registration fee, display a tag with their registration number, stay below the eight-mile speed limit, and follow Chicago's particular rules of the road. If they continued their journey north into Waukegan, they then had to pay *that* city's two-dollar registration fee, display a different tag with a different registration number, adhere to a six-mile-per-hour speed limit, and follow a different set of rules of the road.[90] The additional expense and legal uncertainty arising out of disparate municipal ordinances unnecessarily complicated automobile travel, and motorists looked to state governments to liberate them from the "evils of local ordinance."[91] As a result of a lobbying campaign spearheaded by the New York City–based Automobile Club of America, the New York legislature passed the nation's first state-level automobile regulation in 1901 that set statewide minimum speed limits, required a state registration, and set standards for lights and brakes, while a revised bill three years later invalidated and prohibited municipal licensing and registration laws.[92] Other states quickly followed suit. By 1910 thirty-three states required state registration and eleven required state driver's licenses.[93] In the first decade of the twentieth century, various courts affirmed a state's authority to regulate automobiles while they simultaneously recognized a municipality's ability to pass local ordinances that did not conflict with state law.[94]

Although this shift toward state supremacy addressed motorists' concerns about "multitudinous conflicting local regulations," automobiles carried passengers between states as well as between cities.[95] Instead of fully addressing the issue of uniformity, state laws merely shifted the focal point of motorists' concerns from the municipal level to the state level. As fully sovereign and independent authorities within their borders, each state could require out-of-state motorists to fully adhere to their own particular registration and licensing laws. This would not only unnecessarily complicate interstate driving—particularly among the tightly compacted New England states and the numerous urban areas around the country that sprawled beyond the jurisdiction of a single state—but it would also negatively affect the flow of commerce by automobile between states. By mid-decade, motorists who had previously viewed state supremacy as their salvation from municipal regulation increasingly recognized the need to address "the evils growing out of the many different State requirements."[96] A nationwide licensing and registration system would make "motoring easier, more pleasant and freer from annoyances," but Congress possessed no explicit authority over roadways other than the power to "establish post-roads" under Article I, Section 8, of the Constitution.[97] Interstate travel by steamboat and rail had already been equated with commerce in *Gibbons* and *Wabash* and subsequently placed under federal regulatory authority. Could this same power over interstate travel be extended to the automobile? A growing number of motorists thought so. William H. Hotchkiss, president of AAA, believed that "automobile travel from one State to another is commerce, and as such should have Federal supervision."[98] New York lawyer Xenophon P. Huddy, author of one of the first comprehensive treatise on American automobile law, argued that "when there is a need for uniformity in legislation, then action of Congress is not only desirable, but authorized" and called for legislation placing "all highways and roads leading directly from one State to or into another" as well as "all other highways, roads and streets necessary and convenient to the use of all direct interstate highways" under federal jurisdiction.[99] For Huddy and others, the *potential* use of the automobile in interstate commerce warranted an extension of federal authority to roadways, an area traditionally under state control.

Charles Thaddeus Terry thoroughly agreed with the need for congressional action and viewed disparate state regulations as burdens on interstate commerce analogous to the very state tariffs that had prompted the calling of the Constitutional Convention in 1787. Terry's first opportunity to challenge what he viewed as discriminatory state-based regulation came when, as legal counsel for the National Association of Automobile Manufacturers, he defended Harry Unwen against a ten-dollar fine for driving in New Jersey in a vehicle registered solely in his home state of New York.[100] The loss of this case on appeal to the New Jersey Supreme Court prompted Terry—now firmly convinced that motorists possessed an "absolute right" to travel freely among the states—to vigorously pursue both a federal licensing and registration bill and a uniform state law for standardized rules of the road to address the current "intolerable condition."[101] Although his proposed federal bill carried the endorsement of the AAA, the National Grange, and the American Road Makers' Association, Terry proved unable to convince a majority in Congress on the pressing need for federal automobile licensing and registration.

A Supreme Court ruling could compel Congress to take up the issue of automobile regulation as had occurred with *Wabash* and the railroads, and advocates of national uniformity pinned their hopes on two cases that came before the court in President Wilson's first term. The first concerned the 1910 arrest of District of Columbia resident John T. Hendrick for driving his automobile in Prince George's County without the requisite Maryland registration. Hendrick's lawyers argued that the Maryland law "impose[d] a direct burden on commerce and intercourse between the states" and restricted a citizen's right to "passage through and egress from the States of the Union."[102] In a decision delivered less than a year after Justice Hughes had supported an extension of federal authority over those intrastate rail operations that affected interstate commerce, the Supreme Court upheld the Maryland law on the grounds that "the movement of motor vehicles over highways" constituted "a proper subject of police regulation." By declaring that states could "prescribe uniform regulations . . . in the absence of national legislation," the Supreme Court in *Hendrick v. Maryland* affirmed a state's authority to regulate automobiles under its police power as it had with railroads in *Chicago, Burlington & Quincy Railroad Company v. Iowa*.

But the Supreme Court could reverse its position just as it had with railroads in *Wabash*, and *Kane v. New Jersey* promised such a possibility in the fall of 1916. The case centered around Frank J. Kane, who at the time of his October 1908 arrest in Paterson, New Jersey, was in possession of a New Jersey driver's license while driving a car registered in New York on his way to Pennsylvania. Because Kane was only passing through New Jersey on his way to a neighboring state, the New York–based AAA saw this as the perfect test case. Terry joined Kane's legal team on appeal to the Supreme Court. There he vigorously argued that the 1906 New Jersey law constituted a "burden and tax upon interstate commerce in violation of the Constitution," infringed upon the Fourteenth Amendment, and represented class discrimination due to its sole applicability to motorists. But there would be no *Wabash* for the automobile. In his decision, recently confirmed Justice Louis Brandeis—who had gained fame successfully defending Oregon's state labor laws in the 1908 case *Mueller v. Oregon*—drew heavily upon the court's ruling in *Hendrick* to affirm the right of states to secure both "reasonable compensation for special facilities afforded as well as reasonable provisions to insure [*sic*] safety."[103] With first *Hendrick* and then *Kane*, the Supreme Court affirmed that states possessed the authority to require licensing and registration under their police powers to ensure public safety and proper maintenance of their roadways. In doing so, it dismissed the notion that the *possible* use of automobiles in interstate commerce supported an extension of federal authority into a realm traditionally recognized as under state jurisdiction. Despite Terry's best efforts, Congress and the Supreme Court proved unwilling to extend the federal commerce power to the automobile as had occurred with steamboats and railroads.

An explosion of automobile ownership during the Wilson administration cemented this state-level system of driver's licensing and automobile registration. Spurred on in large part by the reduction in automobile prices as a result of Henry Ford's pioneering production techniques, state automobile registrations jumped from 1,190,393 to 3,367,889 between 1913 to 1916, and total gross registration revenue rose from $8,192,253 to $25,863,369, an increase of 215 percent.[104] Because licensing and registration fees supported road construction, a feedback loop emerged wherein new and improved roadways encouraged more Americans to purchase automobiles, and this in turn stimulated both new automobile purchases and owners' demands for more improved roadways. In addition, expanding revenues from licensing and registration also provided necessary financial support for a growing number of state programs, a dynamic that helped to lock this important revenue stream at the state level. The funneling of federal funds to state governments through the Federal Aid Road Act of 1916 only reinforced state control over roadways and, by extension, the automobile. By the time Republicans retook control of Congress after World War I, this state-based system had become firmly entrenched, and any attempt to transfer it to the federal level would have been political suicide. To facilitate interstate travel in the absence of a rational and centrally managed system, states developed reciprocal provisions that recognized out-of-state licenses and registrations for varying periods of time, a suboptimal approach but one that seemed to better complement contemporary notions of federalism.[105]

Although his efforts proved ultimately unsuccessful, Terry's push for a federal system of automobile licensing and registration should not be dismissed as an irrelevant regulatory dead end. His argument in favor of federal action was based on an understanding that the means of conveyance used in interstate commerce—defined by Marshall to include the act of navigation, or transit—had no impact on the federal government's constitutional ability to regulate. In an era where the precedents of *Gibbons* and *Wabash* were being used to justify federal control over intrastate rail transit and new means of communication, automobiles seemed to represent merely the latest development to require an expansion of federal power under the commerce clause. Terry's attempted solution to this thorny domestic sovereignty issue—a dual regulatory system—only raised further concerns, and he was unable to convince legislators to act on his proposal due to several unresolved questions. Although Congress clearly possessed the authority to regulate interstate commerce that "may be introduced into the interior" via steamboats and railroads by this time, how would one determine the dividing line between intrastate and interstate travel by an ever-increasing number of privately owned and individually operated automobiles? Did states retain control over automobiles until each individual driver physically crossed states lines, or did the *potential* for a driver to engage in interstate commerce support the extension of federal authority over intrastate travel as well? Because roadways constituted an interconnected network, this would arguably necessitate an expansion of federal authority over all the nation's roadways, as Huddy had proposed, something a majority in Congress simply would not support without a charge from the Supreme Court.[106] Although a national system would be more convenient for motorists and automobile travel would no doubt facilitate interstate transactions, elected representatives remained unwilling to surrender their states' authority in the absence of an overwhelmingly pressing need to do so. As a result, states passed their own laws, Congress continued to refrain from action, and an inefficient yet surprisingly persistent state-based regulatory system for the automobile took root in the United States. Just as Terry was struggling in vain to convince national legislators and the courts to act, another legal mind was grappling with sovereignty issues as they related to the recently invented airplane. Could Congress act to regulate this new means of conveyance, and under what authority? Or would a similar state-based system arise, one even more ill-suited to address the unprecedented challenges arising from a means of conveyance that traveled over political jurisdictions and terrestrial barriers?

Taking to the Air

The Wright brothers' achievement of powered, controlled flight on December 17, 1903, over the dunes of Kitty Hawk, North Carolina, marked a decisive turning point in human history. Orville's initial twelve-second, 120-foot flight inaugurated a new era, one where human beings could actively move through the air in a way previously reserved to the imagination.[107] But what goes up must come down, and while the airplane compelled a sense of both awe and reverence it also represented a clear threat to public safety.[108] On September 17, 1908, First Lieutenant Thomas Selfridge became the first casualty of an airplane crash when Orville Wright's machine plummeted to earth during a demonstration flight for the US Army at Fort Myer, Virginia. Over the next four years, more than 250 "Early Bird" aviators lost their lives as they took to the sky in rudimentary machines made of wood, cloth, and wires.[109] Wherever an airplane ascended, spectators eagerly gathered to view the wonder of flight and possibly glimpse a thrilling accident. Each time an aircraft succumbed to gravity—something that might occur at any moment—it threatened not only the life of the pilot but also the lives and property of those on the ground beneath it. Pilots also faced complications from the common law maxim *cujus est solum ejus est usque ad coelum* (he who owns the soil owns up to the sky), an archaic interpretation of aerial ownership that transformed even an innocent and uneventful flight over private property into trespassing.[110] As America's state-centered regulatory system for the automobile crystalized at the end of the twentieth century's first decade, lawyer Simeon Eben Baldwin led the charge to regulate the airplane in the public interest by couching his proposal for federal control within a decidedly international context.

With the existence of numerous nations in close proximity, the movement of aircraft across national borders took on an increased urgency in Europe. Even prior to Wilbur Wright's 1908 European flight demonstrations, European jurists were locked in a debate over the atmosphere's relationship to the sovereign state below. Proposals spanned a spectrum that ranged between two opposing positions. The first, most connected with the British lawyer John Westlake, called for the extension of a nation's sovereignty into the atmosphere. The second, argued most forcefully by French jurist Paul Fauchille, would apply Hugo Grotius's seventeenth-century doctrine that the high sea could not be owned and therefore should remain open to free navigation to the atmosphere above a certain altitude.[111] Seemingly lost in this intellectual debate, however, was the actual ability of states—particularly those with extended national frontiers—to physically prevent the entrance of aircraft into their territory or limit foreign aircraft to Fauchille's designated free navigation zone. Nevertheless, concerns over practical enforcement proved irrelevant at this time as an inability to reconcile the opposing stances of freedom of the air and aerial sovereignty undermined an attempt by legal theorists to create an international air agreement in Paris during the summer of 1910.[112]

As director of the new Comparative Law Bureau of the American Bar Association (ABA), Chief Justice of the Connecticut Supreme Court Simeon Eben Baldwin remained fully abreast of this debate underway across the Atlantic. A leading force in the ABA's creation in 1878 and a respected railroad attorney, Baldwin had been a lifelong Republican who retained his strong conservative leanings even after issues with James Blaine's 1884 nomination prompted him to join the Democratic Party.[113] For Baldwin, the unprecedented capabilities of the airplane, while wonderous, were also cause for grave concern. Because the airplane remained under the constant pull of gravity, it presented an "immediate . . . peril to all within it or below it." In addition, the airplane's ability to fly "over the borders of one sovereignty into those of another as swiftly and irresponsibly as a bird" presented unique security challenges that differentiated it from prior industrial transportation technologies.[114] Both of these potential threats necessitated a government response. But should that response occur at the municipal, state, or federal level? The passage of the nation's first air ordinance by the city council of Kissimmee, Florida, in 1908 augured a repeat of the municipality–state struggle over automobile regulation.[115]

Rather than allow a similarly haphazard response to this new aerial threat, Baldwin hoped to properly address the "airplane question" at its formative stage. He divided questions surrounding the regulation of aviation in the United States into three primary issues: the nature of aerial ownership, flight among nations, and jurisdiction over the airplane within the federal system. For Baldwin, the only way "to promote the public interest" was for each "independent nation . . . to regulate the use of the air above its territory."[116] Therefore, his solution to a new device that allowed its operator to discount political borders was not to reduce national sovereignty as Fauchille had proposed, but rather to extend it. Possessing full sovereignty over their airspace, nations would then create "adequate international agreements" to "harmonize the aeronaut's rights" and facilitate cross-border flight.[117] To address jurisdictional issues under the Constitution, Baldwin authored a draft bill that, like Terry's automobile bill, divided regulatory authority between the federal government and the states, with the former in charge of interstate and foreign flights and the latter regulating intrastate flights through the passage of a uniform state law promulgated by the National Conference of Commissioners on Uniform State Laws.[118] Like Marshall, Baldwin viewed interstate and foreign commerce as falling under the federal government's prevue regardless of the means of conveyance used or the avenue. But his strict jurisdictional division between interstate and intrastate flights did not account for the reality of the airplane, a device that could shift from a means of intrastate to interstate travel even more quickly and easily than an automobile. Despite Baldwin's wealth of experience and clout, the ABA's Committee on Jurisprudence and Law Reform chose not to support his pro-active bill based on a belief that "commerce by air has not yet attained suffi-cient growth on which to justify its regulation by Congress."[119] With aircraft still a novelty, a House Interstate and Foreign Commerce Committee that had recently expressed serious concerns over Terry's automobile bill, and no endorsement from the nation's leading legal association, a federal aviation law would have to wait.

Although federal action remained elusive, Baldwin got his chance to reg-ulate aviation in the name of public safety when he won the governorship of Connecticut in 1910. In his initial message to the legislature, the state's first Democratic governor in fourteen years called for a statute to address this new aerial threat.[120] Passed June 8, 1911, and based directly on the state's 1909 automobile law, this first state air law in the nation required the regis-tration of all aircraft within Connecticut and mandated that all pilots flying over someone else's land possess a state-issued license.[121] It even included a provision almost directly lifted from the state's automobile law that recog-nized out-of-state pilot's licenses and aircraft registrations on a reciprocal basis, a clause that—because no other state had passed a similar air law at the time—effectively closed Connecticut's airspace to all out-of-state resi-dents. With his signature on this pioneering aviation bill, Baldwin believed he had established state control over the airspace, addressed the thorny legal issue of aerial trespass within Connecticut, and secured the public safety from a dangerous new threat overhead. But the passage of laws and their enforcement are two very different things, and soon a two-time dropout of the Massachusetts Institute of Technology would vividly illustrate the sheer impracticality of Baldwin's new state law.[122]

Possessing a keen mechanical mind and a daredevil's flair, twenty-seven-year-old Bostonian Harry Atwood had learned to fly at the Wright brothers' school outside Dayton, Ohio, in the weeks before the passage of Baldwin's aviation law.[123] While serving as chief instructor at a Boston flying school operated by the Burgess Company—a licensed producer of Wright aircraft—Atwood decided to fly his Wright Model B biplane to New London, Connecti-cut, to watch the Harvard–Yale Regatta on June 30. Once there, the crowd greeted him with exuberant and prolonged cheers, and the town's mayor even braved a brief flight with the aviator after treating him to lunch.[124] Buoyed by the newfound attention that wings could bring, Atwood continued south over the rest of Connecticut and arrived in New York City. There he fought powerful air currents above the buildings of Manhattan's Wall Street District, landed at Governor's Island in New York Harbor, and both thrilled the public and unnerved military observers as he buzzed passenger ferries and circled the Statue of Liberty "not more than five feet above the tip of the torch."[125] A trained pilot for less than two months, Atwood now held the American cross-county flight record.

With his exploits front-page news, Atwood announced that he would continue on to Washington, DC, and airplane fever gripped the nation's capital as newspapers promised the imminent arrival of the daring aviator. When weather and mechanical difficulties delayed his arrival, future head of the World War II Army Air Force Lieutenant Henry "Hap" Arnold and another Army aviator—no doubt upset about all the attention Atwood received in the press—took off in a Wright plane from the Army airfield at College Park, Maryland, to give the aerial-enamored citizens of Washington an impromptu air show.[126] Their circles over downtown Washington "turned the Senate topsy turvy, sent Vice President [James] Sherman and an automobile full of senators on a 3-mile wild goose chase, broke up the work of the clerks in the government departments, and held a goodly portion of the residents of Washington down on the White House grounds until their dinners were cold."[127] When Atwood finally did appear over the National Mall in the late afternoon of July 13, he proceeded to put on a nearly two-hour spectacle. Crowds looked up in awe from the streets and rooftops below as he circled the Washington Monument multiple times, buzzed the White House, and repeatedly cut off his engine to initiate a steep dive. The next day, he presented himself to President Howard Taft in dramatic fashion. After completing a late lunch, the president stepped out onto the south portico of the White House. On the lawn below him stood various military and civilian representatives alongside Atwood's proud mother. Atwood landed on the south lawn, Taft personally awarded him a gold medal from the Washington Aero Club, and the aviator presented his mother to the president before taking off into the wild blue.[128]

While Atwood's flight from Boston to Washington, DC, made him a media sensation and national hero, his travel through Connecticut without a state-issued license and registration as required by Baldwin's new air law also made him a criminal. President of the Connecticut Aero Club Arthur Holland Forbes called for Atwood's arrest but none came, and the aviator

Harry Atwood taking off from the South Lawn of the White House, July 14, 1911. Library of Congress, http://hdl.loc.gov/loc.pnp/hec.00357.

continued to make front-page headlines in August when he crossed through seven states on the first flight from St. Louis to New York City.[129] Even in this early period of aeronautical development, Atwood's flight demonstrated the limits of state air regulation. As the editors of the legal journal *Bench and Bar* pointed out, Atwood's flight "ruthlessly burst through the invisible barriers so carefully erected by the Connecticut Legislature."[130] Aloft in his Wright flier, Atwood could pass through the airspace of several small states in New England without having to land, and, unlike with terrestrial transport, law enforcement could not erect border checkpoints in the sky. This disparity between the airplane's freedom of movement and state regulatory control would only increase as aeronautical technology continued to improve. The very same year that Congress failed to pass Charles Thaddeus Terry's federal automobile bill, Atwood's flight foreshadowed the complications that could result from a similar failure to assert federal control over the airplane.

Despite the clear lessons of Atwood's flight, the subsequent passage of a handful of state and local air laws in the name of public safety pointed to a repeat of the automobile's state-based system. In 1913, Massachusetts passed an air bill similar to Connecticut's that drew upon its own automobile law.[131] Two years later, the territorial legislature of Hawaii required operators to hold either a governor-issued license or be documented pilots of the US Army, Navy, or Hawaiian National Guard.[132] On August 22, 1916, the commissioners of Nutley, New Jersey, passed a municipal ordinance forbidding stunt flying and the dropping of objects from aircraft.[133] Although leadership of the New York–based Aero Club of America (ACA)—the premier aeronautical society of the Early Bird Era—called for placing balloons and "flying water craft moving at levels which affect shipping" under the authority of the Steamboat Inspection Service now housed in the new Department of Commerce and Labor, no such action materialized.[134] It looked like the state-based system for automobile regulation that received judicial sanction with *Hendrick* and *Kane* would soon make its way into the sky.

By the time the United States entered World War I in April 1917, a web of often conflicting precedent existed concerning transportation's place within the Constitution's federal system. Drawing on *Gibbons*, commerce via steamboat and rail had been interpreted to include the act of navigation itself, and federal regulatory power over commerce in both areas now extended into state jurisdiction. But this approach had not been extended to the automobile, an individual means of transportation that remained under state police powers despite its obvious use in interstate commerce. Should the states or federal government address safety issues arising from the airplane, a device that allowed for unrestrained movement within three-dimensional space? The borderless nature of the airplane meant that the answer to this domestic sovereignty question would not be decided within a strictly domestic context. World War I accentuated the national security implications of powered flight, and America's participation in the creation of a postwar international air convention would prompt American aviation manufacturers, civilian policymakers, and military officials to formulate domestic aviation policy within a transnational context.

World War I and the Internationalization of American Aviation Policy

Howard Coffin did not mince words in his letter to President Wilson. As vice president of the Hudson Motor Car Company with a reputation for innovation and efficiency, Coffin had brought his crusade for standardization to the cause of American aircraft production as the wartime civilian chairman of the Aircraft Board.[1] Now, based on a three-month investigation of the aeronautical situation in Europe since the armistice, Coffin warned Wilson that Great Britain, France, and Italy all planned to pursue "a definite policy of aerial expansion." The United States risked falling even further behind its current global ranking of "fourth, and probably fifth" in aviation unless it continued its wartime policy of "intimate international cooperation." Like Simeon Baldwin nearly a decade earlier, Coffin believed that the airplane's speed and freedom of movement necessitated an international agreement to fully address issues of safety and "questions of . . . national and international boundaries." Prompted by French Prime Minister Georges Clemenceau's recent call for a postwar air convention as part of the Versailles Conference, Coffin astutely recognized that Canada's membership in the British Empire practically guaranteed Ottawa would sign on to such an agreement. Canadian adherence coupled with the existence of nearly four thousand miles of contiguous border between the two North American neighbors would undoubtedly place pressure on the United States to conform to the convention's new "standardized international regulations."[2] Coffin implored Wilson to appoint a delegation composed of representatives from the military services, civilian agencies, aircraft manufacturers, and engineering societies to ensure full US participation in a proposed commission to draft the convention.[3] Only a thoughtfully chosen delegation representative of

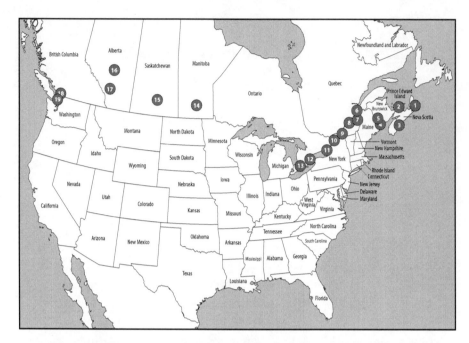

Cities affected by Canada's wartime aviation regulation were, from east to west, (1) Sydney, Nova Scotia; (2) Charlottetown, Prince Edward Island; (3) Halifax, Nova Scotia; (4) St. John, New Brunswick; (5) Fredericton, New Brunswick; (6) St. Jean, Quebec; (7) Quebec, Quebec; (8) Valcartier, Quebec; (9) Montreal, Quebec; (10) Ottawa, Ontario; (11) Kingston, Ontario; (12) Toronto, Ontario; (13) London, Ontario; (14) Winnipeg, Manitoba; (15) Regina, Saskatchewan; (16) Edmonton, Alberta; (17) Calgary, Alberta; (18) Vancouver, British Columbia; and (19) Victoria, British Columbia. Map created using Microsoft Bing Maps, found at http://www.bing.com/maps/.

the various aviation constituencies could ensure that the convention corresponded with the needs and interests of the United States.

Precedent supported Coffin's belief that Canada's postwar air policy would influence US aviation. On September 17, 1914—five weeks after the Dominion of Canada followed Great Britain into the fight against Germany and nearly three years before the United States entered the war—Prince Arthur William Patrick Albert, governor-general of Canada, issued "Orders and Regulations Respecting Aerial Navigation." Designed as a wartime security measure, these emergency regulations established a ten-mile prohibited zone in the airspace around nineteen major Canadian population centers and thirty-nine wireless stations. All aerial frontiers except that between the United States and Canada were closed. US pilots had to receive authorization prior to entering Canada and undergo a thorough pre- and post-visit aircraft inspection at one of eleven designated landing areas or face a $5,000 fine and up to five years in prison.[4] As a result of these wartime measures, the first national regulations applicable to US pilots were promulgated in Ottawa, not Washington, DC. But just as with Atwood's flight through Connecticut discussed in the previous chapter, the passage of air regulations did not magically create an aerial barrier at the border. Reports of unauthorized US aircraft entering Canadian airspace continued, and border sentries reportedly fired on them in at least one instance.[5] By March, American pilots had failed to comply with Canada's wartime regulations on over twenty occasions in British Columbia alone, and the British Embassy in Washington warned Secretary of State Robert Lansing of the "danger of regrettable incidences occurring if the practice . . . continues."[6] A round of letters from Lansing to border state governors did little to correct the situation, and unauthorized flights remained an issue until the United States officially joined the Allied cause in April 1917.[7]

Both Coffin's letter of February 14, 1919, and Canada's wartime air regulations over four years earlier vividly illustrate how World War I infused an international dimension into discussions over US aviation policy. Four years of existential conflict transformed the airplane from "a peacetime marvel . . . to a fearsome tool of war, a bringer of destruction and death" that, coupled with the airplane's permissive relationship with state sovereignty, prompted increased national security concerns.[8] British experience with a sustained bombing campaign from German Gotha and Giant bombers, Brig. Gen. William "Billy" Mitchell's use of concentrated air power for the Saint Mihiel and Meuse-Argonne offensives, and the initial development of bombing strategies that targeted noncombatants behind the front transformed H. G. Wells's earlier visions of mass aerial destruction from science fiction to apocalyptic revelation.[9] The postwar airplane constituted a clear threat to national security, but it also promised rapid communication, unprecedented commercial interconnectivity, and peaceful interchange. To reconcile the right of sovereign states to control their borders with the airplane's freedom of movement—a central tension inherent in international aviation to this very day—Allied policymakers inaugurated an international air regime based on the principle of air sovereignty. The "principles, norms, rules, and decision-making procedures" embedded within the 1919 Convention Relating to the

Regulation of Aerial Navigation would affect aviation policy for both member and nonmember nations alike for years to come.[10]

World War I propelled international concerns into the center of discussions over postwar American aviation policy. Simeon Baldwin's prewar recognition that international considerations supported federal control could be easily dismissed in 1911, but this was no longer the case eight years later. World War I thrust aerial security and sovereignty to the fore, and the international response to these pressing issues ensured that the debate over the airplane's place within the American federal system would occur within a transnational frame to an extent unknown with steamboats, railroads, and automobiles. Like their Allied counterparts, Coffin and like-minded individuals both in government and the aircraft industry recognized that the airplane not only negated state borders but also national borders. Influenced by wartime experience and the creation of a postwar air convention, Smithsonian secretary Charles Walcott, Bureau of Standards director Samuel Stratton, future chief of Air Service Maj. Gen. Mason M. Patrick, Howard Coffin, Manufacturers Aircraft Association general manager Samuel S. Bradley, W. Clayton Carpenter of the State Department, and others came to view the domestic regulation of aviation through a decidedly international lens. This melding of the domestic and international is most visible through the increased recognition of the centrality of the US–Canadian aerial relationship. As further detailed in the next chapter, postwar aerial policy entrepreneurs came to view American participation in the air convention as a way to facilitate international flight between the United States and Canada *and* justify the federal control necessary to prevent a much-feared extension of the automobile's state-based regulatory system to aviation. But postwar civil aviation was not a priority for a president intensely focused on his Fourteen Points and proposed League of Nations.[11] As a result, the United States squandered a prime opportunity to develop a sound aerial policy in consultation with various domestic aeronautical constituencies that could guide an inclusive delegation in its efforts to ensure this new international air agreement best aligned with US national interests. The convention's very existence—and a belief among relevant American interest groups that aviation regulation was inherently international in nature—meant that the development of American aerial policy over the next decade would occur within the context of an existing international regime that did not fully align with US desires.[12]

Regulating for Security

For famed aviator Ruth Law, the news that she could no longer fly must have been beyond frustrating. Her heavily publicized aerial activities over the previous two years clearly demonstrated both her high level of aeronautical skill and its usefulness to the war effort. In November 1916, she had braved mind-numbing cold to set a nonstop distance record of five hundred miles as part her pioneering flight from Chicago to New York.[13] When President Wilson presided over a ceremony to bathe the Statue of Liberty in light for the first time less than a month later, it was Law who had thrilled the "hundreds of thousands" in attendance when she circled the 305-foot monument to American freedom and opportunity.[14] Unable to enlist in the Air Service after the United States entered the war on account of her sex, she took to the air to stimulate wartime patriotism on the home front. Permitted to wear an Army uniform by the War Department, Law "bombed" Cleveland, Ohio; Joplin, Missouri; Chicago; and other midwestern cities with pamphlets to promote the first Liberty Loan and male enlistment.[15] Although lobbying efforts to secure her an Air Service commission broke at the wall of War Department policy, her flights continued to raise awareness for both the Liberty Loan Program and the Red Cross through the summer of 1918.[16] But Law's patriotic work came to an abrupt end when the Joint Army and Navy Board on Aeronautic Cognizance revoked her license "for military reasons" that August.[17] A skilled aviator, fierce advocate for gender equality, and effective promoter for the war effort, Ruth Law now joined the scores of other Americans prevented from flying during the war due to national security concerns.

The military's authority to prohibit civilian flights in the United States arose out of heightened concerns over the national security implications of the aerial gaze. Apprehension over the loyalty of the nearly eight million foreign-born and second-generation Germans in a United States now at war fed into widespread fears of sedition and subversion.[18] The godlike views offered by the airplane provided a perfect platform for foreign intelligence gathering, and even Ruth Law was forced to defend herself against slanderous allegations of spying for the enemy.[19] With the passage of the Espionage Act of June 1917, collecting information in flight for the purposes of sabotage became a federal crime, and aerial meets and exhibitions were quickly banned.[20] Nevertheless, persistent reports of unidentified aircraft flying over military installations prompted the newly established Joint Army and Navy Board on Aeronautic Cognizance to support an executive order to restrict civilian flying.[21] Wilson's approval of Presidential Proclamation No. 1432 on February 28, 1918, made the United States, its possessions, and the Panama Canal Zone areas of "military operations"; tasked the Joint Board with licensing pilots; mandated the painting of these license numbers along the upper and lower surfaces of each wing; and authorized the US military to fire on all unmarked aircraft.[22] With the stroke of a pen, the licensing of pilots—something previously limited to the ineffective laws of Connecticut, Massachusetts, and select municipalities—came to encompass the entirety of US territory. Under this authority, the Joint Board issued licenses that were "not evidence of [a] pilot's qualifications nor the reliability of [the] aircraft used" predominantly to pilots from "companies working on government contracts."[23] Requests for licenses to fly at public exhibitions, to promote civilian sales, and for pleasure were flatly denied. Ruth Law received License #100 to allow for her continued promotion of the Liberty Loan, but the overall approval rate for civilian license requests during the first months of the program stood at a mere 18 percent.[24] Once issued, licenses could be rescinded at any moment, something that both Ruth Law and Harry Atwood found out. (The Joint Board granted Atwood License #119 in December 1919 to allow experimental work for the Navy and withdrew it after he failed to complete the contract.)[25] The first federal aviation regulations in the United States, a global leader in civil aviation today, resulted in an effective ban on civilian flying.

Wilson's executive order sought to address issues of wartime aerial security, but what would happen after the war's conclusion? For those seeking to foster civil aeronautics, a few equally unpleasant scenarios could occur.[26] First, the restrictive military-dominated system that issued licenses without pilot examinations or aircraft inspections might continue. Second, the federal government could withdraw from the field of aviation regulation altogether and return to the prewar status quo. This would undoubtedly precipitate ineffective and potentially conflicting state laws that could prevent the creation of a uniform federal regulatory system through a path-dependent process similar to what had occurred with the automobile. Third, Congress could pass legislation that went too far and severely curtail civilian flying in the name of both national security and public safety. As farsighted government officials and industry leaders looked toward the postwar world, all three of these possibilities seemed increasingly inadequate and even dangerous. Wartime experience showed that a nation's future security would require the procurement of cutting-edge aircraft, but a continuation of wartime levels of investment in peacetime would not become a viable political option in the United States until the Cold War. Only a thriving peacetime commercial aviation sector would provide the industrial capacity and trained pilots necessary to facilitate rapid mobilization in a future conflict. This would require federal legislation that would simultaneously uphold national security interests, ensure public safety, promote aeronautical development, and stimulate civil and commercial aviation. As the war lurched closer to its conclusion, the National Advisory Committee for Aeronautics (NACA) and the Manufacturers Aircraft Association (MAA)—two war-forged organizations composed of distinct interest groups with overlapping regulatory priorities— became primary sites for the formulation of postwar aviation policy.

Modeled after a similar British coordinating committee, the NACA arose out of an addendum to the 1915 Naval appropriations bill and became the "coordinating body" for aeronautical discussions within the executive branch.[27] Although various government departments had experimented with the airplane in pursuit of their individual organizational missions, the creation of a permanent body composed of presidentially appointed representatives from the Army, Navy, and civilian departments alongside private aeronautical experts attested to the growing importance of aviation and "America's realization that it must keep abreast of developments."[28] Specifically charged with "the scientific study of the problems of flight . . . and their application to practical questions," circumstances soon required the NACA to reach beyond its research-focused mandate and address issues such as wartime production, interallied standardization, aviation insurance, and regulation. As chair of the NACA's wartime executive committee and chair of the postwar committee as a whole, Smithsonian secretary Charles Walcott pursued these matters with vigor. An accomplished scientist with thirteen years' experience as the director of the US Geological Survey who shared his predecessor Samuel Langley's interest in aeronautics, Walcott's commanding presence and willingness to engage the NACA in policy matters quickly elevated the fledgling committee to a preeminent position in American aeronautics.[29] In this effort Walcott could count on the unwavering support of two other NACA members. Samuel Stratton's interest in aeronautics sprung from his fascination with standardization as a means to increase productivity, reduce waste, and ensure public access to quality goods, a goal that he vigorously pursued as director of the Bureau of Standards within the newly independent Department of Commerce.[30] Johns Hopkins University physics professor Joseph S. Ames brought scientific credibility to the new committee.[31] Walcott's successor as executive committee chair in the years after the armistice, Ames's reserved yet industrious leadership style served as a fitting complement to the Smithsonian secretary.[32] Together, these three civilians fashioned the NACA into the primary incubator for the development of a postwar regulatory policy that would spread among the executive branch in the immediate postwar years.

With no direct representation on the NACA, the newly established MAA became the focal point for industry's regulatory efforts. As the possibility of direct US military involvement grew in the spring of 1917, the NACA and Howard Coffin combined forces to ensure that American industry could meet Allied demands for aircraft. By this time Coffin had been central to the growing industrial preparedness movement for nearly two years, first as chair of the Naval Consulting Board's Subcommittee on Industrial Preparedness and then as head of the president's Committee on Industrial Preparedness prior to his appointment on the congressionally mandated Council of National Defense.[33] Throughout this period, Coffin remained acutely aware of aviation's growing role in the European conflict. In February 1917, he worked with Sidney D. Waldon—a fellow member of the Society of Automotive Engineers with ties to both the Packard Motor Car Company and Cadillac—to establish the Aircraft Manufacturers Association (AMA) to promote more efficient production through voluntary cooperation to meet Allied demands for aircraft.[34] While the AMA addressed a range of issues including the standardization of materials and production practices, the patent situation remained the most pressing. The broad scope of the Wright brothers' 1906 patent and their subsequent litigation in its defense not only led to a prolonged and acrimonious dispute with Glenn Curtiss but had also cast a chilling shadow over all aeronautical development in the United States.[35] Manufacturers faced significant legal exposure unless they paid considerable licensing fees that were then passed along to aircraft purchasers, the largest being the US military. The situation clearly inhibited the nation's ability to produce the unprecedented number of aircraft many viewed as essential for victory. By mid-March, it became apparent that the AMA's voluntary nature simply could not address the patent issue, and both the Navy and War Departments

In this vivid illustration, waves of American aircraft destroy the German "sea serpent." The rapid and unprecedented expansion of the nation's fledgling aircraft industry needed to make this possible required a solution to the aeronautical patent situation. *Flying* 6, no. 6 (July 1917): 476.

directly requested that the NACA oversee a solution. At a joint meeting of the NACA and AMA, Coffin recommended the creation of a cross-licensing agreement along the lines of the automobile industry's administered by the National Automobile Chamber of Commerce.[36] Coffin brought in W. Benton Crisp, a New York attorney and counsel for the Burgess Company who had previously represented Henry Ford in the 1909 Seldon patent case that had prompted the cross-licensing agreement's creation.[37] With Crisp's assistance, the Walcott-led NACA Subcommittee on Patents ironed out a cross-license agreement for the aeronautics industry.[38] To avoid legal action arising out of wartime production, aircraft manufacturers would pay $200 per aircraft into a pool to generate a total of $50,000 in royalties to both the Wright-Martin Aircraft Corporation and Curtiss Aeroplane and Motor Corporation.[39] By mid-May, Coffin had become chair of a new Aircraft Production Board within the Council of National Defense at Walcott's suggestion, and he threw the board's full support behind the NACA's patent solution.[40] Coffin's connections, Walcott's guidance, and Crisp's legal expertise had combined to bring about what historian Alex Roland dubbed "the NACA's . . . finest hour in the Great War," the creation of a cross-license agreement for the aviation industry.[41]

On July 24, 1917, the same day that President Wilson signed a massive $640 million aircraft procurement bill, Crisp oversaw the incorporation of the MAA to coordinate the cross-license agreement under its three "impartial" voting trustees: Ames of the NACA, railroad investor Albert H. Flint, and Crisp himself.[42] At their first meeting immediately following the organization's creation, the board of directors unanimously elected Crisp as general counsel and Frank H. Russell of the Burgess Company as MAA president.[43] Russell was the perfect person to lead the new organization. From 1910 to 1911, he had served as general manager at the Wright Company before rising to become president of the Burgess Company, a Wright licensee. (When Burgess became a subsidiary of the Curtiss Aeroplane and Motor Corporation in 1920, Russell became vice president of the largest US aircraft manufacturer under its new president, Clement M. Keys.)[44] With no industry representative on the NACA, the MAA under Russell's leadership became the primary lobbying organization for aircraft manufacturers through the hard work of its new general manager Samuel S. Bradley, who would prove instrumental in the formulation and promotion of industry's postwar regulatory vision. Brought in two weeks after the MAA's creation, the former Brooklyn superintendent of parks knew little about the aviation business but worked diligently to establish and maintain "a more intimate and continuous relation between the Association and directing heads in Washington."[45] Within a year, Bradley was managing the pool's nearly two hundred patents held by the Wright-Martin Aircraft Corporation, the Curtiss Aeroplane and Motor Corporation, the Boeing Airplane Company, and others while shuttling from the MAA's New York headquarters to Washington on a regular basis.[46]

Bradley began planning for the postwar world even as conflict continued to rage in Europe. Wilson's wartime executive order on licensing could not last forever, nor would the millions of dollars' worth of military appropriations to aircraft manufacturers. A robust civilian market, buoyed by demand from ex-military pilots, could offset losses from reduced government procurement. But such a market would be difficult to cultivate without defining the rights, responsibilities, and liabilities of aviators, and Americans would remain hesitant to purchase or fly in aircraft without some assurance of safety. Just as with antebellum steam boiler explosions, legislation was needed to ensure public safety, and it needed to be a federal law to avoid conflicting state legislation as had occurred with the automobile. Aware that the future cancellation of Wilson's wartime executive order without new legislation would negatively affect the aircraft industry and therefore his job, Bradley worked to galvanize support for federal control among the MAA and NACA. By July of 1918, he had convinced Russell that the MAA needed to "co-operate with other agencies" on a bill to place civilian flight and licensing under federal control "at an early date" to avoid "hasty and ill-considered legislation."[47] Convinced of the need to address the issue of federal control "before the several States and subdivisions thereof pass legislation complicating this situation," Bradley asked Ames to bring the matter to the attention of the NACA, but the MAA trustee initially viewed legislation as outside the committee's strict scientific mandate.[48] Ames proved more responsive when the abrupt cancellation of nearly all wartime contracts soon after the armistice of November 11, 1918, raised concern that US firms would quickly fall "five years behind the rest of the world in aeronautical development."[49] After Walcott, Ames, and Director of Military Aeronautics General William L. Kenly of the NACA met with Bradley and "about ten of the manufacturers" a week later, Walcott became personally convinced of the need for federal legislation and called for it in the NACA's *Annual Report*.[50] Prodded by Bradley's anxieties over industry's postwar viability, the NACA now actively pursued a legislative agenda outside of its scientific mandate.

But what would a federal law look like? Two draft proposals quickly emerged from the NACA that would set the ideological course for the committee over the next seven years. The first arose out of an NACA subcommittee established in the wake of the meeting with MAA members. Under Walcott's chairmanship and Stratton's direction, this Interdepartmental Conference on Aerial Navigation drafted legislation to establish "a Board consisting of the Secretaries of State, War, Navy, Commerce, and Treasury, and

the Postmaster General" to "prepare rules and regulations governing aeronautical navigation" that would "have the effect of law" after presidential
approval.[51] A different organizational approach from that of the Steamboat
Inspection Service and the Interstate Commerce Commission, this cabinet-
level board represented an attempt to reconcile the military and civilian aspects of aviation under a single administrative apparatus. But the broadness
of the two-paragraph bill itself caused problems. Commerce Secretary William C. Redfield withheld support based on concerns over the delegation
of seemingly unlimited legislative authority to an executive body, and this
draft legislation never made it to Congress.[52] The second proposal emerged
in response to a perceived emergency. Unable to justify the continuation of
wartime restrictions on civilian licensing after the armistice, the Joint Army
and Navy Board on Aeronautic Cognizance lifted them in late January 1919.[53]
Legally bound to issue pilot's licenses under Wilson's executive order but
without the authority and resources to certify pilot qualifications, inspect
aircraft, or regulate air traffic, the board simply granted a license to every
applicant. As chair of the Joint Army and Navy Board on Aeronautic Cognizance, Kenly urged his fellow NACA members to take action. Now in agreement that "the duty of *administering* regulations governing aerial navigation
in the United States should be placed under the Department of Commerce,"
the NACA approved a measure drafted by Walcott "based on the rules and
regulations" of the Navy's Bureau of Navigation and the Steamboat Inspection Service. Considered temporary emergency legislation only, it called for
regulation under the Secretary of Commerce to both remedy the domestic
licensing crisis and prevent "complications . . . arising by unlicensed irre-
sponsible aircraft crossing the borders between the United States and both
Canada and Mexico."[54] Although it received Wilson's support, Walcott's bill
went nowhere in Congress, and the aeronautical situation in the United
States continued its descent toward aerial anarchy.

While these two early legislative proposals from the NACA did not become law, they remain important for several reasons. First, they foreshadow
the central role that the committee would play in the policy-making process
over the next several years. As a presidentially sanctioned body composed of
representatives from every department with a direct interest in aeronautics
(minus the Post Office), the NACA served as a central hub for regulatory
discussions, and the solutions proposed there readily diffused throughout
the executive branch. Second, both of these proposals show a clear desire

from the outset to place civilian aviation under civilian control. Global war may have accentuated the military aspects of aeronautics, but the civilian members of the NACA where unwilling to completely subsume aviation under military authority. Third, while prompted to undertake a legislative solution by Bradley and the MAA, industry did not have a role in the creation of either proposal. As shown in the next chapter, this allowed an ideological disconnect to develop between these two principal constituencies that only prolonged federal inaction. Finally, the emergence of two very different regulatory approaches in just two months—through an interdepartmental board and the Secretary of Commerce—shows that a high level of fluidity remained among NACA members as to the best framework for domestic regulation.

Perhaps most importantly, these initial legislative discussions prompted NACA members to recognize the increasingly interconnected relationship between aviation's international aspects and national legislation. Any federal legislation would, by necessity, apply to foreign aircraft operating in US airspace, and just as varying state laws within the United States would complicate domestic flights, so too would differing national laws impede international flights. To allow the airplane to achieve its full potential to connect people and places, nations needed to legislate as nodes within a larger global network rather than as isolated regulatory islands. A carefully worded international agreement could both guarantee a nation's established right to protect its territory and facilitate unencumbered international flight among member nations through the creation of uniform domestic regulations. Two months prior to Coffin's request that Wilson appoint a representative delegation to help draft the proposed air convention, Walcott informed the president of the need to draft "an international agreement on freedom of the air" at the upcoming Peace Conference to avoid "future irritation and international complications."[55] Like Baldwin seven years earlier, Walcott, Stratton, and Coffin astutely perceived that there could be no strict divide between "national and international aspects" when it came to regulating the borderless technology of the airplane; the two were mutually connected and therefore needed to be addressed in tandem.[56] Events would prove them correct as the postwar air convention profoundly shaped subsequent US regulatory discussions over the following years. But neither the NACA nor industry representatives would have any direct say in the convention's creation.

The United States and the Air Convention:
A Missed Opportunity

A week after Wilson departed for the Paris Peace Conference with Secretary of State Robert Lansing at his side, the French chargé d'affaires ad interim approached the US State Department about US participation in a postwar air convention.[57] Thrust into the acting secretary of state position in Lansing's absence, State Department counselor Frank L. Polk reached out to the American Mission in Paris for guidance.[58] When none came, both Secretary of War Newton D. Baker and Secretary of the Navy Josephus Daniels agreed that "it might be unwise to initiate an independent inquiry until after the Peace Conference adjourns," and Polk postponed his response.[59] But the French simply would not let the matter go. Convinced that "all civil aviation is more or less adaptable to military purposes," French Prime Minister Clemenceau lobbied Allied leaders relentlessly on the need for a postwar civil aviation agreement.[60] His calls found a welcome reception in Great Britain, where a Civil Aerial Transport Committee composed of representatives from the interested government offices and industry had already concluded that an international aviation agreement was "of urgent importance" and had produced both a draft convention and an imperial air navigation bill rooted in the principle of aerial sovereignty.[61] Under increased pressure to commit the United States to participate in the proposed aeronautical commission and faced with continued silence from the American Mission, Polk finally accepted the French invitation on January 23 and asked Baker and Daniels to nominate delegates.[62] Polk's cable to Paris the next day detailing his decision blindsided the American Mission.[63] President Wilson adamantly maintained that postwar aviation issues had "no pertinency to the Peace Conference," but Polk had already committed the United States.[64] Recognizing that "it would be too late to retract" this commitment "without . . . embarrassment," Wilson reluctantly affirmed Rear Adm. Harry S. Knapp and Maj. Gen. Mason M. Patrick, commander of America's wartime aviation in Europe and future postwar chief of the Air Service, as US delegates to this new aeronautical commission.[65] The United States was now committed to the creation of an international air agreement that its president personally viewed as a distraction.

As this episode illustrates, US involvement in postwar Allied aeronautical discussions came about not as the result of a systematic effort to shape international aviation but out of a comedy of errors. Wilson's decision to personally travel to Paris with Lansing, continual French pressure on the State Department in Washington, and Polk's actions in the absence of direction from the American Mission culminated in a US delegation composed solely of military representatives.[66] Although Patrick had worked with Coffin and the NACA as US representative on the wartime Inter-Allied Aviation Committee—a joint military coordinating body that convened in Paris to discuss "all questions relating to aviation which concern more than one of the Allies"—his presence was no substitute for the direct representation by various interested executive departments and industry that Coffin sought.[67] And, other than the NACA's hastily drafted emergency legislation and Walcott's recognition of the need for complementarity between an international agreement and domestic laws, postwar US aerial policy remained woefully undeveloped by the time Knapp and Patrick sat down with their counterparts in Paris to draft the international air convention. Wilson's dismissive response to a postwar air agreement stood in stark contrast to the foresighted leaders of Britain and France, who oversaw the creation of draft conventions in consultation with industry before assembling delegations comprised of both military officers and legal scholars.[68] Rather than pursue a proactive and concerted effort to influence a convention that would, in Bradley's words, "affect aerial development for all time," the United States stumbled into the postwar aeronautical conference without a clear enunciation of its aerial policy or desired outcomes.[69]

The Convention Relating to the Regulation of Aerial Navigation—the world's first multilateral air agreement—came out of the work of the Inter-Allied Aeronautical Commission (IAAC), a new body composed of two representatives from the United States, Great Britain, France, Italy, and (ultimately) Japan "to study all aeronautical questions" in relation to the Peace Conference and "draft a Convention in regard to International Air Navigation in time of peace."[70] At their first meeting on March 6, 1919, delegates divided their work among three subcommissions: the Military Subcommission would address aeronautical issues pertaining to the Peace Conference, while the Legal, Commercial, and Financial Subcommission and the Technical Subcommission would work on the details of the international convention.[71] Members readily agreed on basic principles that would guide the commission's work on the new convention over the next two months: a nation's unconditional right to absolute aerial sovereignty; the right of aircraft

A gathering of delegates from Britain, France, Italy, Japan, and the United States on the Inter-Allied Aeronautical Commission who drafted the 1919 Convention Relating to the Regulation of Aerial Navigation. Only two members of the American delegation are present in this photo, Lieut. Ralph Kiely (*ninth from the left, back row*) and Rear Adm. Harry S. Knapp (*seated, second from left*).

from member states to enter into and pass through the airspace of contracting states (innocent passage); equality of treatment and nondiscriminatory access to air fields; registration based on nationality; the compulsory and uniform licensing of pilots, airworthiness inspections for aircraft, and rules of the air; a permanent air commission; a requirement that member states pass corresponding domestic legislation; and the convention's suspension in times of war.[72] Whereas the unresolved tension between aerial sovereignty and freedom of the air had prevented an agreement in 1910, the war had burned away this theoretical dispute. Delegates recognized that a right to innocent passage should exist among convention adherence, but there was no doubt that it would remain wholly subservient to a nation's own security interests.

Despite this general consensus, the military-dominated US delegation disagreed with its allies on one important point: the universal applicability of prohibited areas. Whereas the British and Italian draft conventions simply allowed for the creation of prohibited zones and the French called for their application "without distinction of nationality," the US draft explicitly allowed a state's own military and state aircraft to enter its prohibited zones.[73] The military members of the US delegation recognized that national security concerns would require American military and state aircraft to access certain areas that were otherwise off-limits to non-US aircraft. But the US delegation found itself in the minority on this issue, and both the Military Subcommission and the Legal, Commercial, and Financial Subcommission agreed that, while each contracting state possessed the right to establish prohibited zones, any such zones should apply equally to the aircraft of *all* contracting states.[74] Outnumbered, the US delegation accepted this position, but, as detailed in chapter 6, opposition to the universal application of prohibited areas—particularly in regards to the Panama Canal Zone—would become a cornerstone of US aerial policy.

One interesting point of note was the role of American delegates in the awarding of the "N" designation to US aircraft, a seemingly random choice that has sparked numerous origin theories.[75] This far-reaching decision was the result of two Navy officers on the Technical Subcommission, Lt. Ralph Kiely and Lt. Cdr. John Lansing ("Lanny") Callan, a former representative of the Curtiss Aeroplane and Motor Corporation in Italy who oversaw naval air operations there from April to October 1918.[76] To determine the nationality of an aircraft in flight and facilitate wireless communication via radio, the Technical Subcommission adopted a system consisting of five Roman characters. The first letter proceeding a hyphen would denote the nationality of the aircraft while the four letters after it would serve as the aircraft's registration and identification number. The subcommission's initial report assigned the United States the letters "U" and "S," but the final version of the convention substituted "N." Why assign a letter with no clear connection to the country's name? The Navy had originally used the "N" designation in its pioneering radio work, and it had become institutionalized within the subsequent international radio regime. Kiely and Callan personally ensured the transfer of "the letters now assigned to the various countries by the International Radio Commission" (presumably the Bureau of the International Telegraph Union headquartered in Berne, Switzerland) to the international air agreement.[77] Since the subcommittee recommended that all aircraft transporting ten or more people "be equipped with wireless apparatus," the use of the "N" designation for aircraft and radio maintained a level of operational consistency among these two emerging technologies.

Table 2.1. Comparison of International Radio Call Signs and Those of the 1919 Air Convention

International Bureau System	Aeronautical Convention System
Great Britain: B, G, M, Y, Z	Great Britain: G
France: F	France: F
Italy: I	Italy: I
Japan: J	Japan: J
United States: N, W	United States: N

Sources: Table compiled from Department of Commerce, Bureau of Navigation, Radio Service, *Commercial and Government Radio Stations of the United States* (Washington, DC: Government Publishing Office, 1921), and International Air Convention, S. Doc. No. 66-91, at 35.

The Inter-Allied Aeronautical Commission's acceptance of the final draft convention on May 22, 1919, was a watershed event in aviation history, but it contained numerous provisions that would come to be seen as out of alignment with US national interests.[78] The largest revolved around an attempt to reconcile the principle of aerial sovereignty with a freedom of innocent passage, an inherent tension within international aviation to this day. With the horrors of war still fresh in their minds and growing concerns over the airplane's future military implications, commission members subjected any rights to innocent passage among member states to each nation's aerial sovereignty. An additional clause that declared air routes to be "subject to the consent of the State flown over" further ensured that even flights among convention adherents ultimately remained subordinate to a state's own security and economic concerns. Despite this blatant cognitive dissonance, delegates felt assured that a desire for reciprocal access would compel states to open their airspace to fellow convention adherents.[79] But what of those states whose geographic isolation made reciprocal access to their airspace unnecessary, or those states whose limited or nonexistent aeronautical development made a grant of reciprocal access irrelevant? The convention's recognition of absolute aerial sovereignty meant that overflight rights ultimately rested on a state's permission, even among its adherents. This transformed peacetime access to airspace into an important bargaining chip for nations that possessed limited economic or military tools with which to confront hegemonic and colonial power. As discussed in chapter 6, US leaders would find that "what's good for the goose is good for the gander" as Latin American nations effectively harnessed their equally sovereign aerial rights to thwart Washington's desired hemispheric aerial policy in the 1920s.

Five other clauses in the convention would also prove problematic for the United States. First, to promote the convention's rapid adoption the commission included a requirement that member states close their airspace to nonmember aircraft. This exclusionary clause meant that if either the United States or Canada ratified the convention while the other did not, there would be a legal barrier to flights between the two countries. And the same dynamic could occur in respect to other American nations. Second, the inability to confiscate aircraft based on patent infringement and a requirement that all such cases had to be adjudicated in the aircraft's home country did not provide for the strict legal protections that the MAA and its patent-holding members desired. Third, uniform customs procedures among the eight technical annexes that provided standards on operational matters such as airworthiness, licensing, log books, aircraft markings, signals, and rules of the air would require the United States to surrender its sovereign right to fully control the transfer of goods and people across its border. Fourth, the previously mentioned universal applicability of prohibited zones would prevent US aircraft from flying in areas it sought to ban other nations' aircraft from, such as the Panama Canal. With the benefit of hindsight, however, the decision to make the proposed International Commission for Air Navigation "part of the organization of the League of Nations" and place arbitration under the jurisdiction of the League of Nation's Permanent Court of International Justice presented the biggest roadblock to US adherence.[80] Membership on this new international air commission would provide the United States a "seat at the table" from which to amend the convention's questionable provisions, and the US delegation readily accepted intertwining the proposed air commission with President Wilson's beloved league. In doing so, they ensured that ratification of the air convention would become politically difficult once domestic opinion shifted against membership in the new multilateral organization.[81]

In keeping with diplomatic precedent, Knapp and Patrick submitted reservations on behalf of the United States.[82] They rejected the requirement that patent infringement claims be litigated within the aircraft's country of registration, something that no doubt aligned with the desires of the MAA. They also reserved on the requirement that all members accord each other's aircraft equal access to and treatment at airfields since it remained unclear whether the federal government possessed the constitutional authority to enforce compliance at state- and municipal-owned facilities. Half of the US delegation's reservations arose from an ardent belief that customs provisions "should not properly be a part of this Convention" since they would hinder the full freedom of the US government to establish its own aerial customs policy.[83] Inexplicably, Knapp and Patrick did not submit a reservation against the universal applicability of prohibited zones at this time, although it would soon come to dominate the aerial concerns of both military branches.

The US approach to the postwar air convention serves as a lesson on how *not* to engage in policy creation. Wilson's lack of interest and the military's monopolization of the US delegation meant that the State Department, relevant civilian executive branch agencies, Congress, the NACA, industry, and the engineering profession had no direct say in the creation of a document to establish global aeronautical standards that would—due to the borderless nature of the airplane—undoubtedly influence domestic policy.[84] Much of the subsequent legislative delay in the United States stems from disagreements among these various groups that could have been minimized with effective, foresighted leadership. Focused on multiple issues facing the postwar world and blinded by a desire to establish the League of Nations, Wilson failed to see the convention as an opportunity to bring together relevant domestic interest groups to both delineate the nation's aeronautical priorities and ensure that a postwar air agreement best aligned with those priorities. Instead, a preexisting international agreement with provisions that clearly conflicted with US interests would now shape national aerial policy.

Industry and the NACA Have Their (Belated) Say

Denied the direct input in the convention's creation that Coffin had sought, both the MAA and the NACA were forced to respond to the new agreement after the fact. They did so in similar yet decidedly different ways. Industry and by extension the engineering profession worked to shape US postwar air policy through the MAA's close relationship with the War Department. Both Coffin and Bradley fully recognized the importance of the proposed international agreement for the nascent American aircraft industry. Not only would it establish rules for international flight, but US membership could provide a rationale for uniform federal legislation that would assure the public that aviation was safe, and this, in turn, would result in a wider use of aircraft and increased sales. But there was also a patriotic element in their interest. As the nation that gave birth to the airplane, they viewed aircraft manufacture not merely as a business pursuit but as a means to strengthen the country and promote an image of an America on the cutting edge of technological development. All of these desired outcomes would be much more difficult if the United States remained officially outside of the emerging postwar air regime. Unable to shape the specific provisions of the convention alongside Knapp and Patrick, aircraft industry leaders nevertheless remained committed to influencing the Wilson administration's response to it. In the summer of 1919, Coffin and select MAA members joined Assistant Secretary of War Benedict Crowell—a Yale-educated mining engineer from Ohio whose interest in aeronautics led him to become the postwar president of the Aero Club of America—on a joint military/industry venture to Europe sanctioned by Newton D. Baker, the progressive Secretary of War who would play a primary role in laying the foundations of US aerial policy in the last two years of the Wilson administration.[85]

Both the creation and final recommendations of this American Aviation Mission (AAM) show the level in which the war fused the domestic and international aspects of aviation within the minds of the mission's members, particularly those from industry: Coffin, Bradley, president of the Wright-Martin Aircraft Corporation George H. Houston, and soon-to-be president of Curtiss Aeroplane and Motor Corporation Clement M. Keys.[86] Despite Bradley's hope, the AAM arrived in Europe too late to play a part in the convention's drafting and instead served as a fact-finding mission to gather information on the "status of aeronautical development" within Allied countries in order "to recommend certain legislation to Congress."[87] The AAM's charges—clearly crafted in consultation with Coffin—show that international considerations were at the heart of the mission from its very inception.

1. To obtain information as to the possibilities of *international cooperation* in the development of the aircraft art, particularly in its commercial phases.

2. To consult with the allied authorities with the view to establishing a basis of active cooperation in the establishment of the *international conventions* necessary for the government of the navigation of the air.

3. To consult with allied aeronautical authorities concerning the relations of the civil, military and naval branches of aeronautical development and the governmental methods best designed for their control.

4. To establish permanent and definite channels of contact insuring a free interchange of aircraft data and the adoption of *international standards*.

5.   To visit such civil and military works related to aircraft as may be of special
     interest in the future development of the air.[88]

The AAM undertook an extensive study of the postwar aeronautical situation in Europe. In Paris, they met with Supreme Allied Commander Ferdi-
nand Foch, Chief of the Air Service General Duval, and other French officials
between tours of production plants.[89] Houston, Keys, and Bradley continued

The American Aviation Mission (popularly known as the Crowell Commission)
aboard the USS *Mount Vernon* en route to Europe. *From left to right*, Samuel
Bradley, Lt. Col. James A. Blair, assistant secretary of war Benedict Crowell
(*standing center*), Howard Coffin (*seated center*), Capt. Henry Mustin of the US
Navy, president of Wright-Martin Aircraft Corporation George H. Houston, and
Curtiss Aeroplane Corporation vice president Clement Keys. Chief of Air Service Supply Col. Halsey Dunwoody joined the rest of the Mission's members
upon their arrival in Paris. At Admiral Knapp's suggestion, Lt. Cmdr. Callan, fresh
from his work on the air convention, joined the AAM as an advisor upon its
arrival in Europe and worked to ensure that its final recommendations would
"conform as closely as possible with the rules and regulations" of the proposed
convention. Crowell to Callan, June 6, 1919, folder 1919: Orders to Duty, box 6,
and Callan to Ispettorato Aeronautica, June 11, 1919, folder 1919–1920, box 4,
both in John Lansing Callan Papers, Library of Congress, Washington, DC.
Photo courtesy of the American Heritage Center, University of Wyoming, Laramie, Wyoming.

their work in France while the rest of the AAM's members investigated conditions in Italy from June 19 to June 22, and the combined mission arrived in London on June 25.[90] After interviewing Chief of Air Staff Hugh Trenchard, Secretary of State for Air Winston Churchill, Maj. Gen. William S. Brancker, and others, they boarded the RMS *Aquitania* and arrived back in the United States on July 20.[91]

A domestic policy consensus rooted in a technologically deterministic internationalism developed as AAM members hammered out various committee reports aboard ship on their return trip to the United States. In language that echoed Simeon Baldwin's presentation to the American Bar Association nearly a decade earlier, the Commercial Development Committee composed of Houston, Keys, and Bradley concluded that "aerial transportation . . . is not local in its nature, but . . . is essentially of a national and international character."[92] Bradley, as the only member of the Legal Committee, recommended that the United States ratify the convention with a sole reservation to its patent clause, pass federal legislation "in harmony with the convention," and actively discourage "concurrent legislation by the several States."[93] But it was the Organizational Committee under Coffin's chairmanship that most forcefully argued that the airplane's "great speed and range of operation" *necessitated* the creation of "international agreements governing the construction, operation and safety of aerial apparatus." Because US aircraft would have "no option but to follow these rules" for international flight, Coffin's committee asserted that the United States must ratify the convention and remain actively committed to its proposed permanent air commission, something that would be best facilitated by "a single authoritative point of contact" in Washington responsible for America's "international . . . , legal, operational, technical and political" affairs.[94] The AAM's final recommendation to establish "a single government agency" for all aeronautical matters under a "civilian secretary for air" therefore represented merely the most efficient way to bring about uniform federal regulation, promote America's continued engagement in international aeronautics, and provide a military-industrial apparatus that could quickly meet the procurement needs of the next war.[95] Of course it didn't hurt that a single department would also address interservice conflicts that negatively affected appropriations, streamline procurement, and funnel a steady stream of research and development funds to an industry heavily reliant on government purchases recently gutted by the cancellation of wartime contracts. Aviation historians remember the AAM primarily for its support of a unified Department of Air, but it is important to recognize that its members came to this position only because they placed domestic matters within a decidedly international context.

Not all of the AAM's members accepted the full implications of the Mission's final recommendations. Capt. Henry Mustin—one of the Navy's first pilots who had overseen air operations from the USS *Mississippi* in the 1914 US occupation of Veracruz prior to his wartime service—remained deeply concerned that the operational and technical differences between land-based and naval aviation would be lost in a single bureaucratic apparatus.[96] As the sole Navy representative on the AAM, he submitted reservations to its final report that would retain the Navy's control over personnel, operations, and experimental development. Mustin's reservations would preclude the creation of a single unified ministry along the lines of Great Britain, but they did not represent a complete rejection of a Department of Air. Like his fellow AAM members, Mustin still supported concentrating commercial elements and general aviation training within a cabinet-level organization and even left the possibility of a unified offensive air force open to "further investigation."[97] Most importantly, he fully concurred with the need for federal control and US engagement in the convention's postwar air regime. While Mustin's minority report did not represent a decisive break from the AAM's overall recommendations, it lent support to the NACA's pursuit of a different administrative path discussed in the next chapter.

With both the AAM's industry representatives and Walcott fully cognizant of the importance of international considerations to American aviation, it should come as no surprise that the MAA and the NACA both recommended that the United States sign on to the new air convention. But the two interested organizations greatly differed on their proposed criteria for ratification. Determined to wait until the various government departments and industry had a chance to analyze the document, Secretary of State Lansing refused to submit to pressure from his counterparts on the Supreme Council that "the American Delegation . . . accept what it did not approve."[98] At Lansing's direction, Major General Patrick personally delivered the convention and the secretary of state's request for comments to the AAM aboard the RMS *Aquitania*.[99] Convinced that "the American [military] delegates could not pretend to represent the aeronautical industry in this country," the MAA established an International Convention Committee (MAA-ICC)

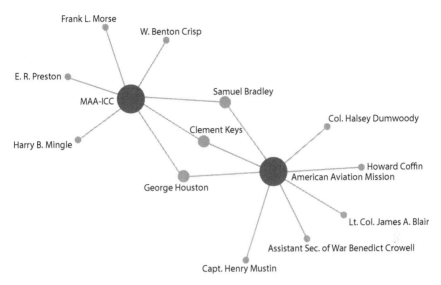

Membership on the Manufacturers Aircraft Association's International Convention Committee (MAA-ICC) included MAA counsel and trustee W. Benton Crisp, lawyer and Standard Aircraft Corporation president Harry B. Mingle, Thomas-Morse Aircraft Corporation president Frank L. Morse, E. R. Preston of the Goodyear Tire and Rubber Company, and American Aviation Mission members Bradley, Houston, and Keys. Network graphic of membership overlap between the American Aviation Mission and the Manufacturers Aircraft Association's International Convention Committee created with Palladio, http://hdlab.stanford.edu/palladio/.

to draft the AAM's response to the convention based on confidential feedback. Reactions to the convention spanned the gamut, from Director of Air Service and NACA member Maj. Gen. Charles T. Menoher's belief that it "should form the basis of our own regulations" to complete opposition due to the proposed commission's connection to the League of Nations.[100] Although no clear consensus existed among MAA members on the convention, the International Convention Committee under Bradley's leadership nonetheless endorsed ratification with Knapp and Patrick's six reservations—which included a reservation to the convention's patent provisions that Bradley himself had previously recommended—wholly "independent of" the Versailles Treaty. But in a portent of what was to come, members of the MAA-ICC also recognized that the United States could and should independently adopt the convention's provisions for "international, national and interstate traffic" to ensure "uniformity" of aerial "law, rules and regulations" even if the convention's connection to the League of Nations made official adherence politically difficult.[101] Crowell concurred, and he impressed upon Lansing the central role that "international regulation" would play in "the development of the existing industry."[102]

As the MAA's convention committee sought feedback from industry, interested executive departments were also analyzing the convention at the State Department's request.[103] To coordinate their responses, the NACA established a Special Subcommittee on International Air Navigation composed of representatives from the War and Navy Departments, the Post Office, the Geological Survey, the Patent Office, the Customs Service, and the Weather Service.[104] Like the MAA-ICC, this interdepartmental body agreed that the overriding importance of the convention and the proposed international air commission necessitated American participation "wholly independent of the peace treaty" and US membership in the League of Nations.[105] The NACA-approved report differed from industry's recommendations, however, in their preference that the United States ratify the convention *without any reservations whatsoever*. Without a voice from industry or a shared policy vision, the NACA concluded that official US membership in the new international air regime was more important than Knapp and Patrick's concerns over constitutionality and the desire of American manufacturers to secure their patents.[106] Shockingly, neither Director of Air Service Menoher or Rear Admiral David W. Taylor, the War and Navy Department representatives on the NACA, raised any concerns at this time about how the convention's requirement for universal applicability of prohibited zones could affect the United States.

The initial reactions of the MAA and the NACA to the convention are significant for several reasons. First, together they show that an overwhelming consensus existed in favor of US ratification of the convention among the two groups. Fully aware that international regulation would affect domestic aeronautical activities, policy entrepreneurs in both organizations recognized that the United States needed to have a direct say in the creation and modification of standard international rules through official membership in the new postwar regime. Second, the difference between the MAA's conditional ratification and the NACA's unconditional approach show a clear divergence in priorities that would need to be reconciled before an agreement could be reached on aviation policy. Finally, the fact that the first official interactions with the convention for both of these groups came in the form of an analysis over three months after its creation provides further evidence of the low status of postwar aeronautics within the Wilson administration. Rather than being active participants in the development of the nation's aerial priorities, the convention's creation, and the drafting of the US delegation's reservations, the two organizations with arguably the greatest interest in postwar aeronautics were forced to react after the fact. It was now up to the State Department to craft a suitable response to the convention, but they would not do so in isolation.

The Canadian Connection

On January 20, 1920, a rather inconspicuous meeting occurred in the Old Executive Office Building that would have a lasting effect on US aerial policy. Recognizing that Washington's interest in the convention existed "almost exclusively" in relation to their northern neighbor and desirous to avoid "anything out of harmony with Canada's attitude," the State Department had invited Canada's Judge Advocate General Lt. Col. Oliver M. Biggar to Washington in order to coordinate both nations' responses to the air convention now open for signature in Paris.[107] A respected Edmonton lawyer on the University of Alberta's Board of Governors, Biggar firmly believed that "Canada's immediate international interests in Air Navigation relates almost solely to the United States."[108] During this Tuesday meeting, he discussed how the recently promulgated Canadian Air Regulations of 1920 aligned with the convention's provisions as well as Canada's proposed reservations to the convention with W. Clayton Carpenter, special assistant to the State Department solicitor now spearheading the department's response to the convention.[109] While Carpenter was forced to recognize that the convention's ties to the League of Nations made US ratification increasingly unlikely, the conference nevertheless adjourned with "the suggestion that Canada and the United States would probably desire to follow the same line of action" in their response to the convention due to "their territorial proximity and comparative isolation from Europe."[110]

Although separated from Europe by the Atlantic Ocean, the development of an American regulatory ideology for aviation would occur within a decidedly international context thanks largely in part to the actions of the Canadian government. At the 1917 Imperial Conference, Canada's conservative prime minister Robert Borden had called for the "full recognition of the

Dominions as autonomous nations in an Imperial Commonwealth," and aviation policy would become an important expression of self-rule in the push to establish Canada as a fully sovereign state throughout his nine-year tenure.[111] Postwar questions about the airplane's place in the US–Canada relationship arose in an era of heightened focus on the international border between the two countries. Unhappiness with Great Britain remained strong in Ottawa over the disappointing results of the 1903 Alaska Boundary Commission, the demarcation of the border and territorial waters between the two nations had continued under the purview of the International Boundary Commission established five years later, and the 1909 Boundary Waters Treaty had created the International Joint Commission to adjudicate disputes involving transnational waterways.[112] Canada's wartime regulations discussed at the beginning of this chapter showed the inability to constrain the airplane on one side of the border, and the two North American neighbors now moved to coordinate their respective aerial policies. Spurred by a desire for increased political autonomy within the British Empire, the Borden government's proactive response to postwar aviation regulation provided both an impetus for federal action in the United States and a powerful reason for US pilots to operate in conformity with the international air convention even without Washington's official adherence. The dynamic between these two North American neighbors is generally portrayed as a one-way relationship wherein the military and economic dominance of the United States influences Canadian policy and culture, but in the case of early aviation policy it was the other way around.

Unlike the United States government, the Canadian government did not have any direct representation on the Inter-Allied Aeronautical Commission as it drafted the air convention. Borden's Privy Council—roughly equivalent to the US president's cabinet—found the British Civil Aerial Transport Committee's proposed draft convention "generally acceptable" but concurred with Minister of Justice Charles J. Doherty's assessment that "the Canadian Constitutional Acts" left aviation subject to "legislation either by the Parliament of the Dominion, or the legislatures of the provinces."[113] When the Canadian delegation to the Peace Conference finally received a copy of the nearly completed convention in early May 1919, Minister of Customs and Inland Revenue Arthur L. Sifton found it completely unacceptable. As a primary negotiator for the British Empire in the 1903 settlement of the Alaska–Canada border dispute, Sifton had an intimate appreciation for how Canada's unique situation differed from that of Europe, something that he believed was not properly accounted for in the convention's proposed international air commission.[114] While he recognized the benefits of uniform international regulations, he could "imagine . . . no circumstances [where] it would be of advantage to Canada to have her affairs in this important respect decided by an International body sitting in Europe and composed almost entirely of people representing countries with absolutely different conditions, many ignorant and practically all careless as to our particular circumstances."[115] In addition, the convention's requirement that members close their airspace to aircraft of nonmembers could create a barrier to North American aviation should Canada or the United States adhere while the other did not.[116] A British attempt to insert a clause authorizing a special agreement between Canada and the United States on account of their "particular geographic conditions" failed due to a French insistence on uniformity, and Borden was forced to inform the British delegation that the Canadian government's adherence would remain "wholly tentative and provisional."[117]

Despite concerns over the convention, there were clearly benefits to be had from conformity with its standardized operational criteria, and Borden's government moved to enact domestic regulations rooted in its provisions. Largely due to the prodding of John A. Wilson, assistant deputy minister of the Naval Service and future controller of civil aviation, the Canadian Parliament passed the Air Board Act on June 6, 1919.[118] This new interdepartmental board under Sifton's leadership tasked Biggar, its vice-chairman, with drafting domestic air regulations that came into effect prior to his meeting with Carpenter at the US State Department in late January 1920.[119] With the notable exception of aerial customs regulations that came from the Canadian Customs Act, these new Canadian Air Regulations of 1920 remained wholly "in accord" with the recently drafted air convention.[120] All aircraft flown in Canada had to be registered as per the convention. Foreign aircraft would be issued a prerequisite secondary Canadian registration only if a convention existed between Canada and the aircraft's country of registration. For US aircraft to legally fly in Canada, both nations would either have to ratify the new air convention or conclude a bilateral agreement *and* a comparable federal registration system had to exist within the United States. All pilots flying in Canada also needed to be licensed in accordance with the convention. The Air Board agreed to recognize foreign-issued licenses of pilots flying secondarily registered aircraft, but this would require the existence of

a complementary American licensing system. The near verbatim incorporation of the convention's rules of the air would mean that—once Canadian
registration and licensing criteria were met—US pilots would need to adhere
to international standard operating procedures while flying in Canada. As
detailed in subsequent chapters, the need to maintain a level of compatibility with Canada's new air law compelled the insurance industry, engineering
profession, and government policymakers to purposefully incorporate the
convention's provisions into their various regulatory systems even without
US ratification. As a result, the Canadian Air Regulations of 1920 transformed the US–Canada border into the membrane through which the convention's operational provisions permeated into the United States.

Operating under "former Secretary Lansing's instructions that it was
desirable for this Government to work in harmony with the Canadian Government," Carpenter began work on the US response to the convention after
his meeting with Biggar and a conference with Assistant Secretary of War
Crowell, Air Service representatives, and Bradley of the MAA.[121] Carpenter
narrowed US reservations down to three that aligned with those of Canada.
Dismissing Knapp and Patrick's concerns about patent protection and the
federal government's authority to ensure equal access to the nation's airfields, Carpenter carried over only their opposition to the convention's customs provisions. Second, Carpenter dismissed the convention's exclusionary clause and reserved the right for the United States to conduct bilateral
agreements with Canada and other American nations. With the Panama
Canal Zone in mind, his third reservation rejected the universal applicability
of prohibited zones and allowed US aircraft to fly in areas declared off-limits
to foreign aircraft. Although this final reservation would become a cornerstone of Washington's air policy in the years to come, it is important to note
that its inclusion in the US reservations originated not from the military
departments but rather from an analysis by William R. Manning, a career
bureaucrat in the State Department's Office of the Foreign Trade Advisor
with extensive experience in Latin American trade relations.[122] Under consultation with US aeronautical interests, Carpenter and other mezzo-level
advisors in the State Department devised a new set of reservations to the
convention specifically tailored to meet the perceived hemispheric needs
of the United States. President Wilson approved Carpenter's reservations
on April 7, 1920, and, as demonstrated in chapter 6, they would serve as the
three underlying assumptions that formed the foundation of the 1928 Ha-

Table 2.2. US and Canadian Reservations to the 1919 Air Convention

US Reservations*	Major Canadian Reservations**
The United States reserves the right to enter into special treaties, conventions, and agreements regarding aerial navigation with the Dominion of Canada and/or *any country in the Western Hemisphere if such Dominion or country be not a party to this Convention.*	That notwithstanding that the United States does not become a party to the Convention, *Canada may make reciprocal arrangements with the United States permitting the flight of aircraft which would under the Convention be properly registerable.*
The United States reserves complete freedom of action as to customs matters and *does not consider itself bound by the provisions of Annex H* or any articles of the Convention affecting the enforcement of its customs laws.	That the provisions of this *Annex [H] need not be followed.*
The United States expressly reserves, with regard to Article 3, the right to permit its private aircraft to fly over areas over which private aircraft of other contracting States may be forbidden to fly by the laws of the United States, any provision of said Article 3 to the contrary notwithstanding.	

* Carpenter to Colby, April 6, 1920, box 5614, Records of the State Department, RG 59, NARA II.

** Certified Copy of a Report of the Committee of the Privy Council, Approved by His Excellency the Governor General on the 7th February, 1920, enclosed in letter from Biggar to Carpenter, Mar. 30, 1920, box 5614, Records of the State Department, RG 59, NARA II.

vana Convention, an air agreement for the Western Hemisphere detached from the League of Nations.[123] Representatives for Canada and the United States signed the Convention Relating to the Regulation of Aerial Navigation with the complementary reservations above prior to its closing on May 31, 1920.[124]

World War I guaranteed that national security concerns and international elements would play a central role in the creation of US aviation policy. The rapid development of military aircraft and predictions of even greater destruction to come transformed the airplane's ability to fly over borders and barriers from a wonderous novelty into a primary threat to national security. In response, Allied delegates crafted an international air convention based on a state's absolute air sovereignty that established a set of principles and procedures to facilitate civilian flight in peacetime. Canada's assertion of its authority over aviation sparked a recognition of the need for closer cooperation between the two North American neighbors, and the inclusion of the new air convention's provisions within its regulations provided a powerful incentive for the United States to adopt them as well. Acutely aware of the connection between domestic and international aviation, members of the NACA and the MAA—two organizations with distinct constituencies and policy objectives—became united in their belief that only federal legislation could provide the national uniformity necessary to foster commercial aeronautics and ensure active US participation in the new international air regime. But the two groups remained split over the specifics of such legislation, and they developed separate administrative frameworks for placing aviation under federal control during the last months of Wilson's presidency. These proposals, based on a shared assumption that the United States would ratify the convention, would be reconciled in the Harding administration.

Debating the Administrative Framework
for Federal Control

Disaster stuck downtown Chicago as the business day came to a close on July 21, 1919. While employees of the Illinois Trust and Savings Bank completed their final tallies, they heard a thunderous crash as the Goodyear Tire and Rubber Company blimp *Wingfoot*, now a ball of flames, broke through the glass ceiling above them. The blimp's gas tanks exploded on impact with the floor, and "a wave of flaming gasoline" swept through the bank as employees struggled to reach the two exits of the "wire cage that surrounded the rotunda." An estimated crowd of twenty thousand people gathered to watch rescue workers pull out bodies "burned almost beyond recognition."[1] The press presented several possible reasons for the crash, but the cause of death of eleven people—two of them young bank messengers—ultimately rested with the Goodyear Company's decision to test the *Wingfoot* over a highly populated urban area.[2]

This horrific catastrophe in Chicago became front-page news throughout the country, where it joined similar headlines that summer such as "Airplane Kills Two Children," "Plane Crashes into Crowd on Take-Off; Kills a Girl," "Three Children Killed by Aero," and "Dinner Interrupted by Crash of Airplane in Yard."[3] Whether antebellum steam boiler explosions, the unprecedented concentration of wealth and power within railroad companies, or the deadly introduction of automobiles into crowded urban streets, Americans had turned to government to regulate transportation technologies in the interests of public safety and welfare. Aviation presented a clear threat to the lives and property of citizens, one that required similar regulatory action. But as discussed in the previous chapter, beginning in January 1919 the Joint Army and Navy Board on Aeronautical Cognizance began to simply issue pilot's licenses to all applicants, and even this illusion of safety ended when President Wilson rescinded his wartime licensing order less than two weeks after the *Wingfoot* incident.[4] Tragic headlines only increased as more unlicensed pilots took to the sky in unregistered and uninspected war surplus aircraft.

Just as with steamboats, railroads, and automobiles, municipal and state governments moved to meet this threat to public safety. City leaders in Chicago and New York City mulled over a possible municipal ordinance that would require anyone flying within city limits to possess either a license from the military or the Fédération Aéronautique Internationale—an "international record-keeping organization" based in Paris—and register their aircraft with the city police.[5] Other municipalities, such as Venice Beach, California, considered and in some cases passed more restrictive ordinances.[6] The potential for numerous conflicting local regulations portended a repeat of the same complications that had plagued automobile owners discussed in chapter 1, only this time the airplane's ability to move in three-dimensional space would further complicate the matter. Once again, state regulation seemed to offer a solution. Five months before US Ambassador Hugh Wallace signed the new air convention in Paris, New York governor Al Smith established the New York State Aviation Commission "to investigate . . . the regulation of flying."[7] Regulation by New York—the most prosperous and populous state in the union—would almost certainly begin a path-dependent process of state-based aviation regulation similar to the one that had undercut Terry's attempt at a federal automobile bill a decade earlier.[8]

This possibility ran directly counter to the federal control over aviation that both the American Aviation Mission and the National Advisory Committee for Aeronautics (NACA) viewed as vital to aeronautical development. As discussed in the previous chapter, events before and after World War I melded the domestic and international aspects of aviation within the minds of American Aviation Mission members such as Manufactures Aircraft Association (MAA) general manager Samuel S. Bradley, Hudson Motor Company vice president Howard Coffin, and director of the Air Service Maj. Gen. Charles T. Menoher and Smithsonian secretary Charles Walcott of the NACA. As a result, key individuals within these two primary interest groups became convinced of the need to ratify the new air convention and pass corresponding federal legislation. The establishment of a state-based regulatory system—one that they viewed as fundamentally incompatible with both the airplane's freedom of movement and the convention's declaration of national air sovereignty—had to be stopped, and Bradley and Menoher galvanized supporters of federal control within their respective networks to prevent Smith's commission from recommending anything "at variance with the principle of Federal control."[9] At Menoher's request, Air Service officials testified before the New York commission that no state "should initiate . . . rules and regulations for aerial navigation until after the Federal Government has outlined the fundamental policy . . . in line with any international rules and regulations which are now under consideration."[10] Coffin argued that aviation was "particularly international and interstate in character," and his compatriot on the American Aviation Mission George Houston, president of Wright Aeronautical Corporation, cautioned against "unfortunate" state regulations.[11] As reported by chief of the Air Service Information Group Col. Horace M. Hickam, this concerted lobbying effort succeeded "in persuading the Committee of the desirability of National rather than State control," and the New York State Aviation Commission recommended that legislators in Albany coordinate aerial policy with their representatives in Washington.[12] Victorious in New York, the MAA and the Air Service continued to actively discourage state regulation while they promoted federal control, and New York City and Chicago adopted minimum regulations in consultation with Bradley that would terminate "as soon as a Federal Act becomes a law."[13] Even with the highly publicized *Wingfoot* crash and the continued loss of life from unfettered flying, the MAA, the NACA, and their allies in the Air Service remained adamantly opposed to state regulation prior to federal legislation.

But what should federal control look like? As discussed in the previous chapter, by the end of September 1919 the MAA and the NACA had agreed that the United States should ratify the new air convention. Once ratified, the Constitution's Supremacy Clause—"all treaties made . . . under the authority of the United States shall be the supreme law of the land"—combined with Congress's authority to pass "all . . . necessary and proper" laws to carry out a ratified treaty would "eliminate constitutional objections" to federal regulation.[14] Working on the assumption that ratification would occur, industry and the interested executive departments independently began work on enabling legislation to bring the convention's provisions into effect within the United States. But the two constituencies adopted radically different approaches. Bradley, Coffin, and others linked to the manufacturing sector viewed a unified Department of Air as the best means to secure the appropriations necessary for industry's development and, therefore, national security. The NACA, on the other hand, had already gone on record in support of two different administrative approaches, one through a board composed of representatives from civilian and military executive departments and *another* through the secretary of commerce alone. As these two groups debated the best apparatus for federal control, recruited allies, and refined their regulatory visions, the relationship between civil aviation and the military became the primary point of contention. Other than Second Assistant Postmaster General Otto Praeger's pioneering (and constantly underfunded) airmail program, organized civil aviation was practically nonexistent in the years immediately after the armistice. The nation's aeronautical activity and the aircraft industry's survival revolved around the military services, and powered flight would continue to possess national security ramifications. The military services, therefore, naturally desired a say in the nation's aviation regulations. But the possibility that military considerations could stymie the development of civil aviation, as well as constitutional concerns about military control over the civil sphere, raised serious questions among the NACA's civilian members. United in their belief in the necessity of federal control, key individuals in industry, civilian executive agencies, and the military entered 1920 divided on the administrative structure that would best promulgate the convention's provisions throughout the United States, ensure the survival of the nascent aircraft industry, and meet the nation's security needs.

United in their shared belief in the need for uniform federal regulations compatible with the convention's provisions, relevant stakeholders within government and industry disagreed on the best administrative structure to achieve their goals. On one side stood Coffin, Bradley, Brig. Gen. William "Billy" Mitchell, and their allies in Congress, who viewed a unified department along the lines of the British Air Ministry as the most cost-effective means to both fully harness the airplane's military power and secure the procurement funds necessary to sustain an aircraft industry in peacetime. Opposed to them were members of the NACA concerned about the blurring of the civil and military spheres and those in the military who argued that land-based and water-based aviation possessed distinct needs and missions that necessitated their continued separation. Under the direction of Secretary of War Baker and guided by the NACA's Samuel Stratton, military and civilian departmental representatives developed an administrative framework based on cooperation between the Department of Commerce and a joint board that underwent further refinement within the NACA. The absence of widespread support for a unified Department of Air within the executive branch, industry's concern over a lack of representation on the proposed joint board, and anxiety from NACA Executive Committee chairman Joseph Ames over the NACA's continued existence prompted both sides to view regulation solely through a Commerce Department Bureau as a suitable compromise. Through an intricate process defined by intense debate, continuous legislative refinement, and a shrewd recognition of what was politically possible under the new Harding administration, key figures within executive branch agencies and industry settled on a shared administrative approach to civil aviation regulation.

Dueling Proposals in the Wilson Administration

Having supported US acceptance of the air convention by August 1919, the MAA and the NACA both began work on enabling legislation. At the request of the MAA's Board of Directors, the same International Convention Committee (MAA-ICC) tasked with analyzing the convention for industry turned its attention toward the "federal legislation necessary to make this Convention effectual within the United States, and . . . to control and stimulate production and use of aircraft and the licensing of aircraft."[15] Details were left to the MAA-ICC's new Legislative Subcommittee, composed of Bradley, MAA counsel W. Benton Crisp, and John M. Woolsey, an admiralty lawyer from the New York firm Kirlin, Woolsey, and Hickox retained at the expense of the Wright-Martin Aircraft Corporation. Concerns arose among the subcommittee's members as they discussed how to bring about the American Aviation Mission's recommended unified department. Bradley questioned whether a "constitutional limitation [against] military activities outside of [the] War and Navy department(s)" would limit a Department of Air to "civilian operation, production and technical development," but Woolsey—the subcommittee member most removed from the American Aviation Mission's prior discussions—went even further.[16] Arguing that "the military side" of a unified department would "over-shadow the commercial side," he drew upon the examples of the Steamboat Inspection Service and the Bureau of Navigation to assert that "the proper place for commercial aeronautics is in the Department of Commerce" and recommended

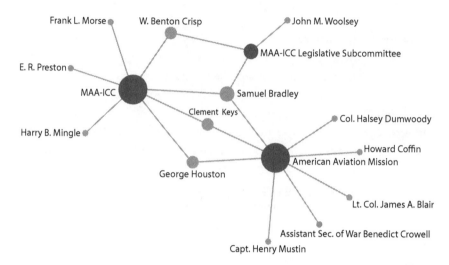

Network graphic of the American Aviation Mission, the Manufacturers Aircraft Association's International Convention Committee (MAA-ICC), and the committee's Legislative Subcommittee. Note the centrality of Bradley. Created with Palladio, http://hdlab.stanford.edu/palladio/.

the creation of a Bureau of Commercial Aeronautics to regulate civil aviation in conjunction with an advisory committee that included industry representation.[17] Although motivated by a desire to best promote commercial aeronautics, Woolsey simply could not dissuade the MAA's leadership of their shared belief "that a separate Air Service is the only solution."[18]

Bradley and his associates were not alone in this assessment, and they found support from like-minded individuals in both Congress and the Air Service. Viewed as an economical and efficient means of administration, bills for a Unified Air Service had been introduced well before the American Aviation Mission issued its recommendations in July 1919. Democrat Charles Lieb of Indiana first introduced such legislation in 1916, but it gained little traction.[19] Democrat George Murray Hulbert of New York submitted a similar bill hours before President Wilson asked a joint session of Congress to declare war on Germany on April 2, 1917.[20] Although this rendition won the unequivocal support of the Aero Club of America as a wartime measure, the proposal's sweeping nature and estimated $1,000,000 price tag ultimately undermined its passage.[21] But the idea of a unified department lived on among certain members of Congress, most notably Indiana's Republican Senator Harry S. New and Republican Representative Charles F. Curry of

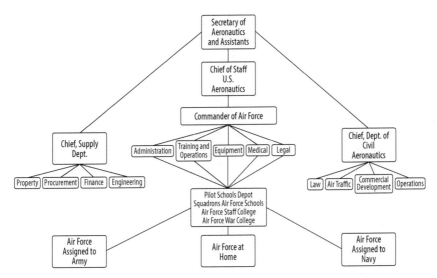

Mitchell's plan for a unified department would create a cabinet-level position to exercise unified command over all aspects of military and civil aviation. This graphic vividly illustrates Woolsey's fear that military concerns would overwhelm civil aviation within a unified department. Recreation of a graphic found in Proposed US Department of Aeronautics, Drafting Air Service, Washington, DC, July 28, 1919, folder 3, box 485, MAA Papers, AHC.

California. They found a fierce ally in Brig. Gen. William "Billy" Mitchell, the famed aerial leader of the Saint Mihiel and Meuse-Argonne offensives now in charge of the Air Service's Operations and Training Division. Convinced that he had successfully demonstrated the effectiveness of unified air command in the skies over Europe, Mitchell fervently believed in the need for aviation to operate as an independent force. This meant specific training, a distinct doctrine, and a tailored procurement plan developed in coordination with industry. Most importantly, it required the placement of military aviation under the control of aviators who, according to Mitchell, were the only ones who really understood the true potential of this revolutionary weapons platform.[22] For these reasons, Mitchell considered a unified cabinet-level department, one "under civil control," as absolutely necessary to ensure the continued development of US aviation "from an economical, industrial and international standpoint."[23]

The perceived economic benefits of a single department, Mitchell's doctrinal argument, and Assistant Secretary of War Benedict Crowell's signature on the American Aviation Mission's recommendations sparked renewed congressional interest in a unified department over the summer of 1919. Curry introduced legislation that roughly corresponded with Mitchell's centralized vision a week after the *Wingfoot* tragedy, and New submitted a similar bill in the Senate three days later.[24] As the MAA-ICC's Legislative Subcommittee, Bradley, Crisp, and Woolsey watched this activity with interest.[25] They concluded that either Curry's or New's bill could, with the "proper amendments," bring about a unified department that would "promulgate regulations for . . . aerial navigation" in accordance with the convention, provide the level of sustained procurement necessary for industry's survival and development, and offer manufacturers a voice on the convention's proposed International Commission for Air Navigation.[26] As had occurred with the American Aviation Mission's report, international concerns remained a primary reason that Bradley, Coffin, and other MAA leaders supported a unified department.

But significant resistance to this approach existed throughout the executive branch agencies potentially affected by such a major administrative change. In response to Hubbard's 1917 bill, the military and civilian members of the NACA concluded that "it does not appear to be necessary or wise to organize a Department of Aeronautics" at this time, and this remained the dominant view of the NACA for years to come.[27] Despite Mitchell's strong support for a unified department, feelings among Air Service officers remained mixed.[28] Based on his wartime experience as head of the US Air Service in Europe, Maj. Gen. Mason M. Patrick concluded that aviation at present "cannot act as a separate and independent force." Instead of a unified department, he believed that a bureau in the Department of Commerce analogous to the Steamboat Inspection Service could best implement the convention's provisions and avoid unnecessary "confusion caused by state laws and regulations."[29] Secretary of War Newton D. Baker agreed with the NACA and Patrick. The Progressive former mayor of Cleveland recognized the need to "lay down the necessary rules, national and international, for aircraft operation" and "prevent [a] discouraging lack of uniformity in State regulation," but he felt that the American Aviation Mission's recommendations for a unified department went "too far."[30] At his request, director of Air Service and NACA member Maj. Gen. Charles T. Menoher convened a board to analyze the Curry and New bills. Unsurprisingly, this "Menoher Board" declared that "the military air force must remain under the complete control of the Army and form an integral part thereof both in peace and war."[31] Before a Senate subcommittee in September, acting secretary of the Navy and future president Franklin D. Roosevelt testified that "not only the Navy Department officially, but the entire naval service, is absolutely opposed" to a unified military air service.[32] While the idea of a unified department along the lines of the British Air Ministry had its supporters within Congress, industry, and the Air Service, opposition ran deep among the civilian and military leadership of the War and Navy Departments.

The American Aviation Mission's recommendations and the New and Curry bills provoked a strong reaction against a unified department from the Wilson administration that prompted the creation of an alternative regulatory vision. Three days after the *Wingfoot* incident, and with the need to secure US participation in the recently drafted air convention in mind, the Joint Army and Navy Board on Aeronautic Cognizance under Menoher's chairmanship requested that Secretary of War Baker and Secretary of the Navy Daniels ask the president to appoint a board composed of representatives from the Army, Navy, Treasury, Commerce, Labor, Post Office, and Agriculture Departments as well as the Coast and Geodetic Survey to draft civil aviation legislation.[33] At President Wilson's insistence, W. Clayton Carpenter of the State Department—the same individual tasked with developing the administration's response to the convention—joined this new body "so that the international and domestic regulations can be harmonized and a single bill presented to the Congress."[34] In addition to Carpenter, this presidentially sanctioned "board to draft a proposed act covering Air Navigation and Civil Aviation in the United States and its island possessions" (hereafter referred to as the Interdepartmental Board) included chairman Col. John F. Curry of the Air Service; Commander John "Lanny" Callan of the Navy, who had helped draft the air convention that spring; head of the Post Office's airmail service Otto Praeger; Stanley Parker from the Treasury Department; John B. Lennon from the Department of Labor; and three NACA members: the Weather Bureau's Charles Marvin, director of the Bureau of Standards Samuel Stratton, and newly elected NACA Executive Committee chairman Joseph Ames.[35] With no industry representation, this Interdepartmental Board was stacked against the MAA's desired unified department from the beginning.[36]

Members of the Interdepartmental Board immediately recognized the link between their work and the new postwar air agreement recently opened

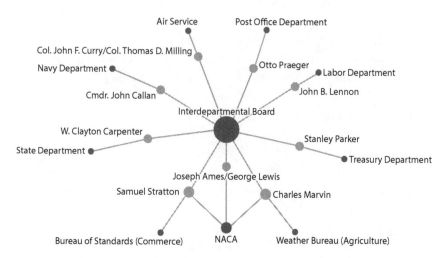

Network graphic of membership of the Interdepartmental Board to draft aviation legislation, Oct.–Dec. 1919. Col. Thomas D. Milling, General Mitchell's assistant in the Office of Training and Operations, replaced Curry as Interdepartmental Board chair at its November 26 meeting. Although appointed to the Interdepartmental Board, newly elected National Advisory Committee for Aeronautics Executive Committee chairman Joseph Ames never physically attended a single meeting, and NACA Executive Officer George Lewis sat in his stead on three separate occasions. Created with Palladio, http://hdlab.stanford.edu /palladio/.

for signature in Paris. After they concluded that their charge "had nothing to do with determining whether a Department of Aeronautics was desirable," both military and civilian delegates agreed that regulatory legislation "should be drafted . . . as soon as possible" and that "such laws [should] be in agreement with the provisions of the I.C.A.N. [International Convention for Air Navigation]."[37] At Stratton's urging, they decided that legislation should provide for three primary elements: (1) US representation to the convention's proposed international commission, (2) a "national commission to draft rules and regulations for aerial navigation," and (3) "a Bureau of Aerial Navigation in some Department of the Government . . . charged with . . . enforcement."[38] This framework, one that melded the NACA's two prior regulatory proposals, had already gained adherents in response to the push for a unified department. As discussed previously, individuals such as Major General Patrick and Woolsey looked to the precedent of steamboat regulation in their call for a civil aviation bureau in the Commerce Department, and NACA members William F. Durand of Stanford University and Major General Menoher also proposed some sort of joint board in coordination with a Commerce Department Bureau.[39] While members of the Interdepartmental Board agreed on these three principles, differences immediately arose over the proper home for the proposed bureau. Some, like the Labor Department's John B. Lennon, felt that the military's overwhelming dominance in postwar aviation necessitated placing it within the War Department. But the NACA's Stratton drew upon a deep-seated American belief in the separation of the civil and military spheres to vehemently assert that "Congress will not give to [the] military proper civil function. The civil should be free to make its own regulations."[40] His view carried the day, and a majority of members agreed that the regulatory agency belonged in a civilian department.

Stratton, however, believed that the Commerce Department alone should administer civil regulations, and he worked to secure this vision as a member of the Interdepartmental Board's drafting subcommittee alongside Curry, Callan, Praeger, and Carpenter. All agreed to use a new draft bill that merged the NACA's previous two proposals for regulation via a joint board and the secretary of commerce into a single framework. The subcommittee members readily agreed with Stratton that both the NACA and the nation's delegates to the international air commission should be represented on the joint board, the joint board should collectively develop regulations, and the secretary of commerce should promulgate them. But uncertainty remained over the scope of federal regulatory authority under the Constitution. The Interdepartmental Board began its work under the assumption that the United States would ratify the new air convention, but growing resistance in Congress to any connection with the League of Nations made official adherence increasingly unlikely. Carpenter and Stratton feared that the subcommittee's proposed placement of intrastate flights under federal control would be "undoubtedly unconstitutional . . . without the support of a treaty behind it."[41] Nevertheless, all five members of the subcommittee agreed that federal regulation of "intrastate operation would be desirable" and voted to retain language from the NACA's original draft bill that would place *all* flights under federal control. As detailed in chapter 5, it would take a sense of crisis before supporters of uniform control and a more conservative Congress would ultimately reconcile their divergent views on the extent of federal authority.

Members of the Interdepartmental Board accepted Stratton's proposal to place executorial power within the Commerce Department, but disagreement remained over the relationship between the secretary of commerce and the proposed joint board.[42] The issue revolved around what extent the military branches would influence civil aviation regulations through their membership on that body. Debate centered on three variations: one where a national board executed its own regulations, one where the secretary of commerce executed the board's regulations without question, and one where the secretary of commerce executed the board's regulations "subject to his approval."[43] Stratton and other civilian representatives wanted the secretary of commerce to have veto power over the joint board's proposals, while Army and Navy representatives preferred that he merely carry out the board's decisions.[44] At the Interdepartmental Board's November 26 meeting, a "general understanding" developed that the commission should draft regulations that would then be approved and enforced by the secretary of commerce.[45]

The Interdepartmental Board's draft bill—commonly known as the Milling bill after its new chairman Col. Thomas D. Milling—offered a suitable compromise between the military's national security interests in aviation and the desire to place civil aviation under civilian control. Although constitutional concerns still existed over the extent of federal authority, the joint board's regulations would apply to both interstate and intrastate flights, while provisions for "lights, signals, and rules of the air" would also apply to government aircraft (i.e., military and postal) to secure operational uniformity in the name of public safety.[46] Like the recently drafted Canadian regulations, foreign nationals could receive permission for temporary flights within US air space only if a treaty existed between their home country and the United States.

Although members of the Interdepartmental Board assumed that the joint board's provisions would be "the same regulations provided by the International Convention," shifting political realities required the bill to be sterilized of any connection to the postwar air agreement.[47] In light of the Senate's recent rejection of the Versailles Treaty, Carpenter removed any direct mention of the air convention, instead referring to more general "subsequent acts or treaties."[48] By the time Carpenter and Biggar met to coordinate their nations' responses to the convention in January 1920 (discussed in the previous chapter), representatives from the various executive agencies had approved a bill that offered both a means to secure uniform federal

Table 3.1. Administrative Structure Proposed in the Interdepartmental Board (Milling) Draft Bill

Board Composition	Commerce Component	Board/Commerce Relationship
One representative each from the Departments of War, Navy, Commerce, Post Office, State, Agriculture, Treasury, and the NACA	Secretary of Commerce	Board drafts regulations, commerce secretary approves and promulgates them

Commerce Department

Post Office Department

State Department

Navy Department

War Department

Air Navigation Board

Agriculture Department

Treasury Department

NACA

Secretary of Commerce

regulation without a unified department *and* connect the United States to the international air regime should political winds change.

But the Interdepartmental Board's compromise solution did not go far enough for the NACA's civilian leadership. As Carpenter worked to finalize the board's draft legislation, Interdepartmental Board members Stratton, Praeger, and Ames joined Charles Walcott as the NACA's Special Committee on Organization of Governmental Activities (SCOGA) to draft a bill that would ensure the primacy of civilian control.[49] Both Mustin's reservations to the American Aviation Mission's report and the Menoher Board's report "greatly struck" and "impressed" Ames, and he used their arguments in favor of separate military air services to support a distinct home for civil aviation.[50] Under Ames's leadership, the civilian-only SCOGA modified the proposed joint board into a body composed of the NACA and heads of four new separate aviation bureaus within the Army, Navy, Post Office, and Commerce Departments that would "settle questions" on "all . . . matters in which the several agencies may be jointly interested."[51] Creation and enforcement of civil aviation regulations would rest solely with the new Commerce Department Bureau, while the civilian-dominated joint board would serve in a strictly advisory capacity.

Unlike the rather benign Interdepartmental Board's joint board approach, the SCOGA's framework required a fiscally minded Republican Congress to approve four new bureaus during a postwar economic contraction. With the summer flying season fast approaching, Walcott met with fellow NACA members Capt. Thomas T. Craven, director of naval aviation, and chief of the Navy's Bureau of Construction and Repair Rear Adm. David W. Taylor to devise a more realistic proposal that could win congressional approval prior to the summer recess in early June. This Navy-dominated committee added one additional representative from both the War and Navy Departments to the Interdepartmental Board's proposed joint board and replaced representatives from the State and Agriculture Departments with one from the Interior. With a new commissioner of aeronautics as its chair, the joint board would draft regulations, the commerce secretary would approve them, and the commissioner would enforce them.[52] This seemed like a suitable compromise that would address congressional budgetary concerns, ensure a level of regulatory coordination among the relevant departments, and maintain the NACA's much-desired civilian control.

By the time US Ambassador Hugh Wallace signed the air convention with

Table 3.2. Administrative Structure Proposed in the NACA's Special Committee on Organization of Governmental Activities Draft Bill

Board Composition	Commerce Component	Board/Commerce Relationship
Heads of new aviation bureaus within the War, Navy, Commerce, and Post Office Departments and NACA members	Bureau of Aviation in the Commerce Department	Joint Board advisory only, power to create and enforce civilian regulations vested solely in a new Commerce Department Bureau

Bureau of Aviation (Commerce) — Bureau of Aviation (War) — NACA — Bureau of Aviation (Navy) — Bureau of Aviation (Post Office) → Joint Board

Table 3.3. Administrative Structure Proposed in the Bill Drafted by Walcott, Taylor, and Craven

	Board Composition	Commerce Component	Board/Commerce Relationship
	Two representatives each from the War and Navy Departments. One representative from Interior, Treasury, and Post Office Departments and the NACA	Bureau of Aeronautics in the Commerce Department under a commissioner of aeronautics	Commissioner of aeronautics as chair of board. Board drafts regulations that commissioner then promulgates

Carpenter's reservations on May 31, 1920, several different bills existed in Congress to provide the federal control over aviation needed to execute its provisions. New's MAA-supported unified department bill made it to the Senate floor, but War Department opposition and concern over an estimated budget of "something like $97,000,000" prompted its return to committee.[53] Although support for a unified department clearly existed among certain individuals within Congress and the Air Service, no similar legislation would ever get as far in the process as New's bill.[54] And two bills drafted within the executive branch to establish a joint board did not even make it out of committee. Despite Secretary of War Baker's personal appeal that Canada's new air regulations made rapid passage of the Interdepartmental Board's legislation "absolutely necessary," the House Interstate and Foreign Commerce Committee took no action before the 66th Congress adjourned on June 5, 1920.[55] Walcott, Taylor, and Craven's variation met the same fate.[56] While legislative passage proved elusive, the need to connect to the convention's new international aviation regime and the threat of a unified department had compelled key officials within the executive branch to agree that regulation of "international, interstate, and intrastate flying" should occur through a "civilian bureau of Air Service" in cooperation with a joint board to allow for consultation with the military.[57] These foundational principles, developed in the final months of Wilson's presidency, would crystallize into a regulatory ideology during the Harding administration.

With the executive agencies in general agreement, the focus now shifted to overcoming congressional concerns over costs and administrative bloat.[58] To increase the chance of success when the lame duck Congress reconvened in December, Ames replaced the joint board in the Walcott, Taylor, and Craven bill with the NACA, where it would serve as both the official advisory body on all aviation matters *and* the central coordinating agency for government aeronautics.[59] The NACA would then approve regulations drafted by the commissioner of air navigation, who—as head of the new Commerce Department Bureau—would also be an NACA member. Although historian Alex Roland labeled this "a sweeping grab for power," it constituted the next logical step in Ames's ideological development.[60] He had already reduced the joint board's membership with the SCOGA's proposal, the NACA could easily fulfill the role of the joint board with the addition of a few more departmental representatives, and the creation of a new administrative body that replicated much of the NACA's membership would run counter to the Republican desire to "prevent duplication of expenditures and effort in the military and naval air service."[61] But this cost-saving approach clearly clashed with the NACA's scientific mandate, and Ames's revision of the Walcott, Taylor, and Craven bill caused some of its members to question whether the body "should assume executive functions" or if it could even "function satisfactorily as an administrative body."[62] In response to these concerns, when Ames revised the Interdepartmental Board's bill to correspond with his modified version of the Walcott, Taylor, and Craven bill he eliminated the NACA as the coordinating body for government aeronautics and *did not propose any parallel joint board in its place*. As a result of Ames's various revisions, the NACA now supported two bills, one where the NACA served as the coordinating joint board and one where a Commerce Department Bureau regulated civil aviation completely independent of a joint board.

As the nation awaited president-elect Warren G. Harding's promised "return to normalcy," a shared ideology for aviation had already coalesced among key members of the executive branch in response to both the international air convention and attempts to create a centralized Department of Air. As the various proposals drafted during the final months of the Wilson administration show, members of the NACA and the interested executive departments agreed that the regulation of civil aviation should occur through the Commerce Department—preferably through a new bureau analogous to the Bureau of Navigation—working in tandem with a joint board. Ames's revisions, sparked by a desire to overcome budgetary concerns in Congress, opened the door to the complete elimination of the joint board as a coordinating body. Despite continued uncertainty as to the legislative particulars, discussions among the NACA and its affiliated executive branch departments increasingly cast Mitchell and his congressional allies as ideological outliers. Astute political players within the aviation industry, however, adjusted accordingly. As early as January 1920, Coffin, after "mature consideration," had come to accept "somewhat of a compromise" between a unified Department of Air and separate departmental control.[63] During the congressional recess, MAA board of directors member Frederick B. Rentschler of Wright Aeronautical Corporation suggested to Bradley that the MAA drop its support for a unified department to quickly secure the "rules and regulations" that its members viewed as "imperative."[64] By the time Americans went to the polls on November 2, 1920, the "tendency" among MAA members had shifted away from a completely unified department to one that would oversee "all aeronautical activities of the Government except" those of the Army, Navy, and Post Office.[65] Left out of the discussions within the executive branch and desirous of regulations that would encourage the public to fly, industry had moved away from the American Aviation Mission's recommendation of the past year. But as the Wilson administration came to a close, a large gulf still existed between the MAA's preferred approach and the regulatory vision coalescing within the executive branch.

Forging Consensus in the Harding Administration

In a special joint session five weeks after his inauguration, President Harding made plain the extent of the problems facing the Republican-dominated 67th Congress. The former newspaper publisher and senator had routed his fellow Ohioan James Cox in the previous November's election with a campaign centered on "normalcy, . . . restoration, . . . [and] triumphant nationality."[66] He had also called on Congress to pursue increased "administrative efficiency" and end the "unnecessary interference of Government with business."[67] Now in a speech that was anything but laissez-faire, Harding dismissed conservativism's traditional hands-off approach to the economy and instead sought the judicious exercise of government regulatory power to promote increased access to and efficient use of new technologies. He talked at length about the need for continued railroad regulation, federal guidance for state highway construction, "effective regulation of both domestic and international radio operation," and the establishment of "government-owned [radio] facilities . . . available for general uses." The airplane was no exception. Although Samuel S. Bradley and Benedict Crowell had tried to get Harding to publicly support "legislation authorizing and controlling aerial operations" on the "basis of . . . an international agreement" during the campaign, the Republican candidate had remained silent on the aviation issue.[68] Now, he forcefully declared that "it has become a pressing duty of the Federal government to provide for the regulation of air navigation" to prevent the passage of "independent and conflicting legislation . . . by the various States." To avert this possibility, he called for the creation of a civil aviation bureau in the Department of Commerce, the establishment of a Bureau of Aeronautics in the Navy, and the continued existence of a separate Army Air Service.[69] In contrast to the relative silence of his predecessor, Harding placed the full power of the presidency in favor of a Commerce Department Bureau and decidedly against a unified department.

Harding did not personally devise this unprecedented expansion of federal authority into the atmosphere; his advisors provided it to him. Unlike the highly determined and frequently stubborn Wilson, the laidback golf enthusiast Harding believed "the president's major function was to act as facilitator and compromiser, adjudicator and political counsel" rather than policy creator.[70] Throughout his administration—plagued by scandal and cut short by a heart attack in August 1923—Harding largely deferred to his cabinet on matters of policy, and he held the opinions of his secretary of state Charles Evans Hughes, secretary of the treasury Andrew Mellon, and secretary of commerce Herbert Hoover in particularly high esteem. These new cabinet secretaries, in turn, relied on the established administrative structures within their departments staffed by career civil servants, bureaucrats, and military officers responsible for the daily operations of government, an administrative class that historian Daniel P. Carpenter categorized as the "salaried and relatively durable . . . mezzo level."[71] In the last year of the Wilson administration, interested executive agencies came to support civil aviation regulation through a joint board working in tandem with the Department of Commerce in response to the new postwar air convention and calls for a unified Department of Air. This support did not simply end with the presidential transition. Rather, new secretaries, undersecretaries, and assistant secretaries entered agencies staffed by career civil servants who carried over previously established institutional priorities. This dynamic was perhaps most notable in the presidentially appointed NACA. Other than the replacement of Navy representative Capt. Thomas T. Craven with Rear Adm. William A. Moffett in early March, the committee's membership, and its commitment to civil control over civil aviation, remained unchanged under the new Republican administration. Harding and his "Ohio Gang" came into an executive branch that had, in many respects, already come to an agreement on a preferred solution for aviation.

It was now up to Walcott, Ames, Stratton, and the rest of the NACA to secure the new president's support for this shared regulatory vision, and they moved swiftly to convince recently confirmed Secretary of War John W. Weeks, Secretary of the Navy Edwin Denby, Secretary of Commerce Herbert Hoover, and Postmaster General William H. Hays of "the need for the establishment of a Bureau of Aeronautics in the Department of Commerce."[72] After forwarding Ames's recently revised bills to the new secretaries, NACA members Menoher, Stratton, and Moffett as well as Earl C. Zoll of the Post

Table 3.4. Membership of the National Advisory Committee for Aeronautics before and after the Inauguration of President Harding

NACA Membership as of December 1920	NACA Membership as of December 1921
Charles D. Walcott (Smithsonian, NACA chair)	Charles D. Walcott (Smithsonian, NACA chair)
Samuel W. Stratton (Bureau of Standards, Department of Commerce)	Samuel W. Stratton (Bureau of Standards, Department of Commerce)
Joseph S. Ames (Johns Hopkins University, NACA Executive Committee chair)	Joseph S. Ames (Johns Hopkins University, NACA Executive Committee chair)
Maj. Thurman H. Bane (Air Service, chief of Engineering Division)	Maj. Thurman H. Bane (Air Service, chief of Engineering Division)
Capt. Thomas T. Craven (director of Naval Aviation)	Rear Adm. William A. Moffett (chief, Navy Bureau of Aeronautics)
William F. Durand (Stanford University)	William F. Durand (Stanford University)
John F. Hayford (Northwestern University)	John F. Hayford (Northwestern University)
Charles F. Marvin (Weather Bureau, Department of Agriculture)	Charles F. Marvin (Weather Bureau, Department of Agriculture)
Maj. Gen. Charles T. Menoher (chief of Air Service)	Maj. Gen. Mason M. Patrick (chief of air service)
Michael I. Pupin (Columbia University)	Michael I. Pupin (Columbia University)
Rear Adm. David W. Taylor (navy chief constructor)	Rear Adm. David W. Taylor (navy chief constructor)
Orville Wright	Orville Wright

Sources: Aeronautics: Sixth Annual Report of the National Advisory Committee for Aeronautics, 1920 (Washington, DC: Government Publishing Office, 1921); *Aeronautics: Seventh Annual Report of the National Advisory Committee for Aeronautics,* 1921 (Washington, DC: Government Publishing Office, 1922); *Admiral William A. Moffett: Architect of Naval Aviation* (Annapolis, MD: Naval Institute Press, 1994), 80.

Note: This table illustrates the consistency in the NACA's membership across the transition from the Wilson administration to the Harding administration, one that ensured a level of ideological continuity both within the NACA itself and among its represented executive departments. The replacement of Navy representative Capt. Thomas T. Craven with Rear Adm. William A. Moffett in early March constituted the committee's only personnel change during the presidential transition. A strong proponent of a Bureau of Aeronautics in the Navy, Moffett had already accepted the general tenets of the NACA's National Aviation Policy prior to his appointment. The creation of a separate air bureau in the Navy in July 1921 elevated Moffett to the head of the first of the proposed bureaus that Ames's SCOGA had called for in late 1919. Mason Patrick succeeded Menoher in late 1921, but as noted previously he already viewed a Commerce Department Bureau as the best way to implement the provisions of the international convention.

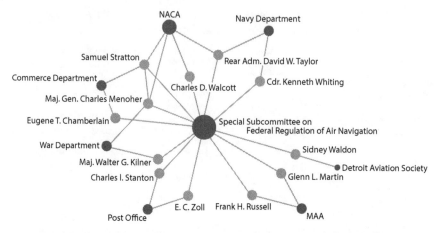

Network graphic of the membership of the National Advisory Committee for Aeronautics Special Subcommittee on Federal Regulation of Air Navigation. Representatives from executive agencies that had already affirmed support for the Commerce Department model, of which NACA members made up almost half, dominated the presidentially appointed subcommittee. Created with Palladio, http://hdlab.stanford.edu/palladio/.

Office's Air Mail Service successfully recruited them into the ideological fold at a March 31 meeting.[73] At Weeks's direction, Menoher, Stratton, Moffett, and Zoll drafted a letter for the new president's signature that directed the NACA to establish a special subcommittee to determine "what can and should be done without further legislative action" and "what legislative action and appropriations are necessary to carry into effect the recommendations of the subcommittee."[74] Walcott forwarded it to Harding immediately, and the president signed and returned it to the NACA chairman the same day. In typical Harding style, the new president left the details of aviation policy to "representatives from the War, Navy, Post Office and Commerce Departments, *and civil life*."[75] As a result, this presidentially sanctioned NACA subcommittee marked the first time that industry representatives had an official seat in high-level policy discussions since Lansing had requested the MAA's thoughts on the convention more than a year ago. Although the ideas of industry representatives such as Bradley and Coffin had softened since then, there was little chance that MAA members Frank Russell of Curtiss and aircraft designer Glenn Martin or Detroit Aviation Society president Sydney D. Waldon would convince this new presidentially appointed subcommittee to adopt any sort of unified department: four of its twelve members came from the NACA, and the new secretaries of the executive departments represented had already agreed to support the Commerce Department approach.

The four meetings of the Special Subcommittee on Federal Regulation of Air Navigation a week prior to Harding's address to Congress mark an important inflection point for US aviation policy. At the insistence of Chairman Walcott, subcommittee members used the NACA-drafted "General Considerations on a National Aviation Policy"—with its recommendations for a Bureau of Aeronautics in the Commerce Department, the continued separation of Army and Navy aviation, and the continuation of the NACA's limited research and advisory role—as the basis of discussion.[76] With the executive departments already in agreement, it was clearly assumed that the body would recommend a Commerce Bureau to Harding instead of a unified department. Of equal importance, the NACA's proposed policy did not include a joint board, most likely due to negative reactions to Ames's suggestion that the NACA assume that role. The elimination of this coordinating body, a central element in executive branch discussions to that point, did not elicit any objections. And, with a new president ardently convinced that "every commitment must be made in the exercise of our national sovereignty," it comes as no surprise that the NACA's proposed policy contained absolutely no mention of the international air convention.[77] Although the convention had prompted and propelled regulatory discussions for the past two years, domestic political realities now required aviation policy to be couched in purely domestic terms. As detailed in chapter 5, this conscious desire not to disturb the delicate sensibilities of politicians empowered to enact legislation could not change the fact that the International Convention continued to exert a profound influence on the United States.

With procurement foremost in their minds, attempts by Russell, Martin, and Waldon to shape the subcommittee's recommendations for Harding met with mixed results. Russell secured support for a committee with industry representation to survey the state of aircraft manufacturing and propose a policy of sustainment, but his request for industry representation on the NACA received a cool reception.[78] In consultation with subcommittee members Russel, Martin, and Mitchell's chief of operations Maj. Walter G. Kilner, Waldon removed any explicit support for separate Army and Navy air services within the subcommittee's draft final report and inserted language to allow for the "consideration of a Unified Air Service" and a separate independent air force "at some later date." Waldon's proposed changes prompted a sharp rebuke from executive department representatives, and of them only Kilner, a non-NACA member, voted to jettison the proposed policy's explicit support for separate air services. Waldon's modifications were tabled, and the subcommittee approved the NACA's "General Considerations on a National Aviation Policy" largely unchanged.[79] Based on the committee's membership, this comes as no surprise; the real shock lies in the half-hearted efforts of industry representatives to push for a unified approach. None pressed for the immediate creation of a unified department along the lines of the American Aviation Mission's report, and Russell and Martin even left Washington to "attend to important personal business" rather than support Waldon's revisions at the subcommittee's final meeting.[80] And their request that Harding charge the NACA—a body firmly committed to separate aviation bureaus—with investigating the feasibility of "a Department of Air, a unified service, and an independent air force" simply defies belief.[81] Eight months earlier, the MAA's Rentschler had pointed out that demands for a Department of Air were undermining prompt legislative action. The tepid response of Russell and Martin shows that political reality had begun to sink in; anything other than a Commerce Bureau was simply outside the realm of possibility.

Harding's call for legislative action before the new 67th Congress came practically verbatim from the special subcommittee's report. Gone was the joint board component, any mention of an international air agreement, and the possibility of a unified department or air force. With the president taking a clear position on his preferred form of legislation, bills providing for a Commerce Department Bureau gained increased support within Congress. But bills reintroduced from the last Congress would also establish a joint board, a coordinating mechanism that civilian members of the NACA had started to move away from and industry found wholly unsatisfactory in its current form.[82]

It was at this point that Coffin, Bradley, and other representatives from the aircraft industry and engineering profession turned to newly appointed commerce secretary Herbert Hoover. Largely sidelined from the regulatory discussions within the executive department for the past year, Bradley remained acutely aware of the need for business interests to have a say in any future regulatory apparatus. Harding's endorsement of a Commerce Bureau effectively ended any hope for a unified department, but industry did not have a single representative on the joint boards proposed in legislation currently before Congress. To make inroads into the regulatory debate dominated by the "more or less academic" NACA, Bradley, Coffin, Russell, Martin, and fifty-one other private sector individuals petitioned President Harding to instruct Hoover to convene an Aviation Consulting Board composed of representatives from "civilian aviation interests" that would develop a general policy for aviation, legislation, and a system of air routes.[83] Their call for Hoover to lead such a committee made perfect sense. A Stanford-educated mining engineer who had made his fortune overseeing projects around the globe, Hoover's efficient coordination of public and private wartime food aid secured him such a positive reputation that both Republicans and Democrats viewed him as a possible 1920 presidential candidate.[84] Using the close partnership between industry and government during the war as a guide, Hoover sought similar peacetime cooperation under the aegis of government to increase production, cultivate improved efficiency, and reduce the negatives of the business cycle.[85] Throughout his eight-year tenure as commerce secretary, Hoover personified the ideology of associationalism, one in which government actively supports and encourages voluntary attempts to rationalize, standardize, and codify industrial practices to save struggling industrial sectors and stimulate emerging ones.[86] Commercial aviation interests saw Hoover as a potentially powerful advocate within the new Republican administration. Although the proposed Aviation Consulting Board did not materialize, this petition did secure "the opportunity that the civilian interests in aviation had waited for"—a meeting with Hoover.[87]

With Harding on record in support of a Commerce Bureau and the meeting between industry representatives and Hoover fast approaching, the ideological gap that separated the MAA and the NACA for the past two years closed rapidly. First, the two constituencies agreed to drop the joint board component. After Senator William Borah of Idaho submitted a resolution that would divide the NACA's responsibilities among the Commerce, War, and Navy Departments in the name of "economy," Ames completely dropped any attempt to turn the NACA into the joint board. He informed Bradley that he was now "willing to accept almost any law which will put into effect a Bureau of Aeronautics with [the] power to license aircraft and pilots."[88] Bradley welcomed the Executive Committee chair's openness and—with the issue of industry representation on a joint board no longer an issue—he believed the two organizations could "get together on this question of legislation."[89] The recent creation of a Bureau of Aeronautics in the Navy had further undermined the possibility of a unified department, and Bradley, Coffin, and their industry associates recognized that civil regulation solely under the Commerce Department, particularly with Hoover at its head, would be much more amicable to their desires than a joint board with no industry representation.[90] The NACA's Stratton now found the MAA's leadership ready to support a Commerce Department Bureau empowered to create and enforce regulations aligned with those of the convention "to permit international flight."[91] The ideological gulf between the two groups had significantly narrowed.

On July 18, Bradley, Coffin, Crowell, A. R. Small of the Underwriters' Laboratory, and president of the National Aircraft Underwriters Association Edmund Ely secured a promise from Hoover that he would support a Commerce Bureau.[92] But this was not a new position for the secretary, and it is important to keep both this meeting and Hoover's future legislative contributions in proper perspective. Hoover had already agreed to this approach as a result of the NACA's lobbying nearly three months earlier. This gathering, therefore, did not "convince" Hoover of the need for federal regulation, nor did his reaction to industry mark the "awakening" of the Commerce Department to the subject. And Hoover's commitment to aviation remained shaky. According to one Commerce Department official close to the secretary, as legislative momentum slowed over the next two years, Hoover felt "reluctantly . . . drawn into the aviation proposition" and hoped to "proceed cautiously" in order to "not stir up too much activity and excitement among the various organizations interested in that field."[93] Often portrayed as the primary force and even the sole genius behind American aviation policy—an interpretation bolstered by his own savvy self-promotion—Hoover was instead a convert to an existing regulatory ideology who remained acutely aware of how aviation would affect his carefully cultivated public image.[94] With Hoover's backing secured, Coffin and Bradley turned their attention to the creation of legislation that would place civil regulation solely under a Commerce Department Bureau.[95]

The last year of the Wilson presidency and the first few months of the Harding administration represent a period of intense ideological activity. United in support of convention adherence and the need for uniform federal control, representatives from industry and the interested executive branch agencies developed two conflicting administrative frameworks for the regulation of civil aviation. Industry supported a unified department as the best way to secure the appropriations necessary for its survival, while executive branch agencies debated the proper relationship between the military branches and civil aviation. The NACA's Ames, Walcott, and Stratton worked to ensure that authority over civil aviation remained within civilian hands, and an intense lobbying effort converted key individuals within the new Harding administration to their position. With a unified department now highly unlikely, industry leaders came to believe that regulation through the Commerce Department alone would suit their interests better than a joint board without their representation. Although the executive branch and industry were now united in their support for a Commerce Bureau, aviation still remained largely unregulated two years after the *Wingfoot* tragedy. But to legislate, Congress needed to be convinced of both the need *and* its authority to do so. Faced with the reality that Harding would not submit the air convention to the Senate, advocates of uniform federal regulation needed a new justification for legislative action.

4

The Struggle for Legislation

By all outward appearances, the Noedings lived the American dream. Charles had immigrated to the United States from Germany in 1886 at the age of ten, and Ella, eleven years his junior, had arrived in 1907. Charles operated a New York City trucking business, and the couple raised three children in their home in Hoboken, New Jersey.[1] But this seemingly idyllic life all changed on August 24, 1924, when Charles, Ella, and their ten-year-old son Herbert Noeding traveled to Long Island for some summer fun with Herbert's uncle Theodore. There they came across pilot William Sharp selling fifteen-minute flights in his war surplus Standard J-1 biplane for five dollars. Theodore went up despite murmurs that "something seemed to be loose" on the airplane, and Charles elected to follow him. After some debate, young Herbert successfully convinced his parents that he should join his father. As the aircraft completed its circuit, those in the crowd below "saw one or two unusual puffs of smoke from the airplane engine, and heard a barking sound." Ella watched in horror as the plane plummeted "like a falling rocket," leveled off, and then plunged into a nearby field. There were no survivors.[2]

In the five years since the *Wingfoot* crash in downtown Chicago, the air above the United States became increasingly unsafe as more pilots took to the sky in war surplus aircraft. According to the Aeronautical Chamber of Commerce of America (ACCA)—a new, broad-based aviation industry organization under the management of the MAA's Bradley—lack of uniform regulation constituted the primary reason for aerial accidents.[3] In a report to Secretary Hoover, the ACCA flatly stated that "the uncontrolled itinerant pilot encounters and causes most of the danger in flying."[4] The report ap-

The remains of William Sharp's Standard J-1 biplane. *New York Daily News*, Aug. 26, 1924.

proximated that 1,200 nongovernmental aircraft flew 250,000 people a total of 6,500,000 miles in 1921. Of the 114 accidents that had resulted in 49 deaths and 89 injuries, the ACCA attributed nearly four-fifths to itinerant fliers (barnstormers, commonly referred to as "gypsy fliers") and pointed to stunt flying as the direct cause of 40 percent of casualties. The danger from itinerant fliers became a persistent refrain of the ACCA over the following years. Of the 419 "itinerant accidents" recorded from 1921 to 1923, the ACCA asserted that regulations for pilot qualifications, mandatory inspection, and rules of the air could have prevented 99.5 percent of the total.[5] The tragedy that forever changed the Noeding family constituted only one of eighty-nine such incidents in 1924.[6] The ACCA's message was clear: uniform, nationwide regulations would drastically reduce aeronautical accidents. This required Congress to take action.

In agreement on the need for federal regulation of all flights *and* a Commerce Department Bureau by the summer of 1921, representatives from industry, the NACA, and their allies struggled over the next five years to convince Congress to institutionalize their shared regulatory vision, one prompted by and refined in relation to the postwar international air convention. But the convention had become a political liability in the Harding administration due to its connection to the League of Nations, and the Senate refused to accept that the commerce clause allowed for federal control over intrastate flights. Advocates of uniform federal regulation then turned to the House, where the congressional calendar and a general lack of interest among representatives and the public left the desired legislation in limbo. Ultimately, it took a crisis that accentuated the national security implications of civil aviation to compel legislative action. Congress, however, refused to fully accept an expansive interpretation of the commerce clause necessary to extend federal authority to intrastate flight and airports. The subsequent Air Commerce Act of 1926 offered both a compromise to these difficult domestic sovereignty questions and, as we will see in the next chapter, a way to ensure US regulatory compatibility with the 1919 air convention, even without official adherence.

Chicago lawyer William P. MacCracken Jr. was central to this process. The son of a prominent doctor with a "booming voice and outgoing personality," MacCracken earned his law degree from the University of Chicago Law School before wartime service as a flight instructor solidified his interest in aviation.[7] Within a few years after the war, MacCracken emerged as a major voice in the nascent field of aviation law through his service as chair of the American Bar Association's Special Committee on the Law of Aviation, a position that initiated a mutually beneficial relationship with Coffin, Bradley, and leaders in the struggling aircraft industry. Although a 1934 contempt of Congress charge arising out of a Senate investigation into air mail contracts severely tarnished his reputation, MacCracken's work as Legislative Committee chair of the National Aeronautic Association (NAA), legal counsel for numerous early air carriers, and the nation's first assistant secretary of commerce for aeronautics secured him a prominent place in US aviation history.[8] In many ways, MacCracken played the same role in aviation that Charles Thaddeus Terry had a decade earlier for the automobile. Both believed that this new means of transportation required a federal solution based on an expansive interpretation of the commerce clause, both represented national special interest organizations (the NAA and the AAA, respectively), and both strove to convince the House Committee on Interstate and Foreign Commerce to accept their regulatory vision. But whereas Terry was unsuccessful in his endeavor, most of MacCracken's views came to be embodied within the Air Commerce Act of 1926. Two primary differences led to such radically divergent outcomes: the existence of an international air regime prior to national aviation legislation and the national defense implications inherent within the civilian aircraft industry.

Conflicting Interpretations of Federal Power

Harding's April 1921 call for Congress to regulate civil aviation through a Commerce Department Bureau discussed in the previous chapter did not include a legal rationale for such a major expansion of federal authority. Industry leaders and NACA members from the various executive departments and civilian life recognized the necessity for uniform regulations, but, as dean of Cornell University Law School George Bogert pointed out, "the desirability of uniformity does not confer on the national government power to legislate."[9] Under what constitutional authority could federal jurisdiction extend to the nation's atmosphere without ratification of the 1919 convention providing a constitutional justification for federal control? In the absence of a federal law, states would pass their own laws and aviation regulation could become entrenched at the state level as it had with the automobile, a doomsday scenario for Bradley, Coffin, Walcott, Patrick, and others who recognized the need for uniform national regulations compatible with those of Canada and the convention.

Advocates of federal regulation needed to find another constitutional mechanism to allow for federal control. Having received Hoover's personal assurance that he would support civil regulation solely through a Commerce Department Bureau, Coffin chaired a committee composed of Bradley, former assistant secretary of war Benedict Crowell, Wright Aeronautical Corporation president George Houston, and Curtiss vice president Clement Keys to revise the NACA-sponsored legislation establishing a joint board currently before Congress. Without the convention, they immediately confronted how to justify the uniform federal control that they viewed as an "absolute necessity."[10] In response to the *Wingfoot* tragedy, various lawyers proposed several possibilities: admiralty jurisdiction, a constitutional amendment, and the commerce clause.[11] Each, however, came with their own unique challenges.

The Constitution's extension of federal jurisdiction to "all cases of admiralty and maritime jurisdiction" offered one possible solution.[12] Over the previous decades, this provision had evolved into "a distinct category of federal power" that could, according to William Velpeau Rooker of the Indiana

State Bar Association, allow for uniform federal regulation of aerial naviga-tion.[13] Rooker was shocked to find that "no branch of jurisprudence" existed in the wake of "the recent balloon accident in Chicago," and as chairman of the Conference of State and Local Bar Associations he established an explor-atory committee on the matter under his direction.[14] In a widely distributed letter, he argued that similar complications in determining the "geometri-cal measurement" and "allocation" of both the sea and the air equated to a shared legal status between the two mediums.[15] Although his reliance on Aristotle, Francis Bacon, and Jeremy Bentham made Rooker's argument read more like a Western Civilization essay than a modern legal treatise, his idea that admiralty jurisdiction could provide the justification for uniform federal air regulation struck a chord with some. In the absence of conven-tion ratification, both the Interdepartmental Board's bill and its revision by Walcott, Craven, and Taylor discussed in chapter 3 explicitly relied on ad-miralty.[16] Desirous of any means to achieve rapid uniform federal control, Bradley eagerly distributed Rooker's letter among his network.[17]

But a clear difference existed between water and air navigation. The pull of gravity meant that airplanes threatened people below in a way ships did not, and an airplane's ability to penetrate a nation's interior raised unique national security concerns. Besides, precedent already existed against equat-ing waterborne and aerial navigation. In 1914, a federal court had ruled that by their very nature aircraft "did not lie within the federal admiralty power," a position the New York Court of Appeals upheld in 1921.[18] Travelers Insurance Company vice president Walter G. Cowles—a graduate of Yale University Law School, an aviation insurance pioneer, and a self-described "intensely practical" individual—advised Bradley that the courts would ultimately rule against the constitutionality of any bill based on admiralty.[19] MAA counsel W. Benton Crisp agreed, and warned that an analogy between the sea and the air would only lead to renewed calls for free navigation of the air, a pre-war concept now invalidated by the convention's declaration of national aerial sovereignty.[20] The ABA's Committee on Admiralty and Maritime Law cautioned that "admiralty rules . . . may serve as rough guides for the case where air planes are moving in exactly the same level" as ships but could not extend into the atmosphere.[21] Admiralty simply raised too many problems to form a sound constitutional basis for uniform federal regulation.

Some argued that, because the Constitution formed a government of enumerated powers, only a constitutional amendment could secure the federal authority necessary for "a uniform code of laws, rules, and regulations."[22] This became the preferred method among several legal advisors in the Air Service, such as Maj. Elza Johnson, who concluded that "the conflicting ideas of State legislatures" could not achieve the "universal and therefore international" regulations required for true safety in flight. For Johnson, a constitutional amendment constituted the only way to guarantee the requisite all-encompassing federal authority to align domestic regulations with those of Canada and the convention.[23]

A newly established ABA special committee supported this position as well. When the National Conference of Commissioners on Uniform State Laws (NCCUSL) established a committee to develop a uniform state law in August 1920, William P. MacCracken Jr.—convinced that "it ought to be the ABA leading the fight for adequate laws"—successfully lobbied the ABA to create a Special Committee on the Law of Aviation to analyze aviation's "interstate and international features."[24] Under the chairmanship of New York lawyer and future ABA president Charles A. Boston, and with MacCracken as a member, the five-person committee agreed on the absolute necessity of "a uniform law operative throughout the country" that would allow for regulations compatible with those of Canada to facilitate flights between the two countries.[25] Ratification of the air convention would allow for regulation under the treaty power, but as MacCracken himself pointed out its connection to the League of Nations would make the ABA's endorsement a "political action."[26] In their report, Aviation Committee members dismissed admiralty due to its conflict with the convention's declaration of aerial sovereignty and warned that "complete control over aeronautics" could not be based on the "unprecedented judicial recognition of an unprecedented claim." Nor did they believe federal authority could derive from a reinterpretation of any other existing clauses in the Constitution, although, as shown in chapter 1, federal steamboat and railroad regulation had grown out of a steady expansion of the commerce clause. Because "the Constitution neither expressly delegates to the United States powers over air flight as such nor prohibits them to the states," the committee asserted that only an amendment could allow for nationwide regulatory uniformity.[27] With four amendments ratified between 1913 and 1920, Americans were intimately familiar with the process, but it would take an organized nationwide effort and time to foster broad public support.[28] Both the MAA and the NACA preferred that Congress "enact the legislation deemed necessary . . . and let the constitutionality be tested in due course" rather than undertake a lengthy and uncertain amendment process.[29]

The commerce clause provided the third possible justification for federal aviation regulation. San Diego lawyer Warren Jefferson Davis personified this approach. A member of the Aero Club of America's Legal Advisory Committee who had served in the office of the Air Service's judge advocate general during the war, Davis—like his contemporary MacCracken—came from a new generation of lawyers more open to a broad interpretation of federal power as a means to address the ills of modern industrial society. Drawing upon Justice Marshall's expansive definition of commerce in *Gibbons v. Ogden* and the lessons of federal steamboat and railroad regulation discussed in chapter 1, Davis stipulated that the need for regulatory uniformity to protect interstate commerce provided "sufficient basis" for Congress to regulate *all* flights under the commerce clause in the absence of convention ratification.[30] But federal steamboat and railroad regulation had developed through a lengthy process of court rulings and public demands for action, and both Congress and the US Supreme Court had rejected the commerce clause as justification for federal automobile regulation before the war. Convincing dually elected congressional representatives to surrender the authority of the very sovereign states they represented without broad public support would be a struggle, particularly in the Jim Crow South, where states and local governments zealously guarded a system of institutionalized racial segregation. Davis's expansive interpretation also faced challenges from his fellow ABA members. Boston's committee reported that the explicit power to regulate foreign and interstate commerce under the commerce clause must leave intrastate flights to the states, and Rooker believed that the inability to regulate intrastate flight under the commerce clause would prevent the creation of "a proper [federal] law."[31]

Tensions over the best way to secure a uniform federal law boiled over at the ABA's 1921 annual meeting, where the Aviation Committee's report stirred up a "hornet's nest."[32] Boston argued for the necessity of an amendment, Rooker defended admiralty, Davis pointed to convention ratification as the best solution, and MacCracken distanced himself from the committee's conclusions. The discussion created such an acrimonious atmosphere that Boston "announced that under no circumstances would he serve again on the committee."[33] Although united in the desire for federal action, the legal profession remained divided over the constitutional means to achieve it. And as ABA members squabbled over the particularities of constitutional law, George Bogert, a member of Boston's Aviation Committee, continued to work on the NCCUSL's uniform state law, one that threatened to undermine the passage of federal legislation.[34]

So when Coffin, Bradley, Keys, Houston, and Crowell began revising legislation to allow for federal regulation solely through a Commerce Bureau after meeting with Hoover on July 18, 1921, deep uncertainty remained over how best to provide for the uniform federal control that industry, the NACA, and now the ABA's Special Committee on the Law of Aviation viewed as necessary. For guidance, they turned once again to the law firm of Kirlin, Woolsey, and Hickox, the New York–based attorneys for the Wright Aeronautical Corporation. While working on a response to the convention as a member of the MAA's International Convention Committee in late 1919, John M. Woolsey had compiled a draft aeronautical law—one easily revised to allow for regulation through either a Department of Air or a Commerce Department Bureau—that focused solely on the ability to promulgate and readily revise rules of the air in conformity with the convention.[35] Although initially set aside in the hope that a unified department bill would make its way through Congress in the last year of the Wilson administration, Coffin's Drafting Committee now adopted Woolsey's broad, flexible approach to legislation as the best means to establish the safety requirements necessary to promote civil aviation *and* avoid nonessential elements that could imperil rapid passage.

With Woolsey's earlier draft and Kirlin's counsel, Coffin's Drafting Committee revised the NACA-sponsored legislation currently in Congress into a Commerce Bureau–only bill that met industry's desire for nationally uniform regulations.[36] Introduced in the Senate by Republican James W. Wadsworth Jr. of New York in August 1922, this industry-drafted bill charged the head of the new Commerce Department Bureau with promulgating and enforcing national air regulations drafted "in conformity with and carrying into effect international aeronautical agreements and treaties."[37] Rather than directly assert a federal power to regulate intrastate flight that would no doubt generate constitutional objections, the Drafting Committee ingeniously relied on the airplane's speed and freedom of movement to ensure uniform regulations under the commerce clause. Rules and regulations established under this act would apply to any aircraft engaged in interstate or foreign commerce as well as those flying "upon or along or through navigable streams, rivers, or waters, or post roads or routes of the United States, or the air or atmosphere above the same or in, over, or through the District of Columbia, the Territories, dependencies, reservations, national parks, or over any place or building over which the United States has jurisdiction."[38] With the high probability that flights would either cross state lines or fly over these areas already under federal jurisdiction, this careful wording would allow federal regulations to "cover practically all aviation" while staying "entirely within the limits of constitutional authority for Federal control."[39] By indirectly providing federal control over intrastate flight, Coffin's committee hoped to secure a uniform federal law to avoid "the unsatisfactory legislative experiences of the motor car industry."[40]

While NACA members generally supported this industry bill, some of its lesser provisions prompted concern. Both Chief of Air Service Mason M. Patrick and Chief of the Navy's Bureau of Aeronautics William A. Moffett worried about the complete absence of any mechanism for coordination between military and civil aeronautics, Charles Marvin of the Weather Service took issue with the possible creation of a separate weather service within the Commerce Department, and NACA Executive Committee Chairman Ames saw a new aeronautical research agency within the proposed Commerce Bureau as a threat to the NACA's very existence.[41] In meetings over the first week of December, Coffin and Bradley—anxious to secure support for legislation that would "head off [the] indiscriminate enactment of local legislation throughout the country"—jettisoned the draft bill's aeronautical research and weather provisions and incorporated a clause that required the head of the new Commerce Bureau to cooperate to "the fullest degree" with "established governmental agencies."[42] After two years of debate prompted by the postwar air convention, industry and the NACA were now in agreement on the need for uniform federal regulation, the desirability of regulation through a Commerce Bureau, *and* the legislative particulars. Now came the hard part—convincing Congress to accept their shared regulatory vision.[43]

Hearings before the Senate Committee on Commerce showed the work still needed to convince Congress that the commerce clause would allow for complete federal control. Although NACA members Ames, Patrick, and Moffett; Second Assistant Postmaster General E. H. Shaughnessy; and Curtiss vice president Frank Russell all successfully argued the *need* for a federal law— most notably as it related to the "highly undesirable" Canadian situation— members of the Senate subcommittee took serious issue with the *extent* of federal authority within the bill. Republican chairman Wesley L. Jones of Washington posited that the inclusion of flights above navigable waters and post roads effectively extended federal control to all flights, but staunch states' rights advocate Democrat Duncan Upshaw Fletcher of Florida countered that the direct mention of "foreign commerce and interstate commerce" in the bill specifically left intrastate flights to the states. Kirlin and Woolsey's attempt to provide an unassailable means to secure uniform federal regulation had instead left the extent of federal authority open to interpretation and uncertainty.[44] Not even the combined weight of subcommittee testimony, a brief by Kirlin and Woolsey on the bill's constitutionality, and the endorsements of numerous aeronautical clubs and organizations could overcome concerns of federal overreach among certain senators.[45] Although committee members agreed on the benefits of uniform federal regulations, they nonetheless found the bill "most sweeping in its terms and scope" and revised it to eliminate any "doubtful constitutional provisions." The bill that emerged expressly limited federal regulation to flights engaged in interstate and foreign commerce. After an unsuccessful attempt during debate to reinstate the original bill's more expansive definition of federal jurisdiction, this restricted version passed the Senate on February 12, 1922.[46]

After striving for legislative action for more than two years, this first successful passage of a federal aviation bill represented both a victory and a defeat for industry, the NACA, and their allies in the executive branch. On the one hand, a plurality of senators had voted in favor of a bill to establish a Civil Aviation Bureau in the Commerce Department without the need for a prerequisite constitutional amendment. On the other, the strict interpretation of interstate commerce within the bill showed that senators remained unconvinced that the airplane's speed and inherent freedom necessitated an all-encompassing interpretation of federal authority under the commerce clause at the expense of state sovereignty. For those convinced of the need for uniform national regulations compatible with the convention, the committee's revised bill amounted to "nothing but a scrap of paper," and they redirected their focus to getting "a proper bill put through the House."[47] With the advice of Chief of Air Service Patrick, Curtiss vice president Clement Keys, and "other representatives of the aircraft industry," Commerce Department solicitor William E. Lamb began revising the disappointing Senate bill to allow for "full and complete control" under the commerce clause, including flights made by "privately owned aircraft used for pleasure."[48] An endorsement from the ABA would go a long way toward convincing members of the House Committee on Interstate and Foreign Commerce of the constitutionality of this approach.

A February 25 meeting at the Willard Hotel in Washington, DC, to coordinate the work of the ABA and NCCUSL aviation committees offered Coffin, Patrick, and their compatriots an unprecedented opportunity to secure that endorsement. Under the leadership of the ABA Aviation Committee's new chair William P. MacCracken Jr., representatives of the interested executive departments, industry, and aero clubs gathered together to advise the two legal committees on legislation.[49] Opinions ran the gamut as representatives from the various constituencies discussed the legislative situation collectively for the first time. During the course of the meeting, Patrick and Coffin argued for rapid passage of federal legislation compatible with the air convention, former ABA aviation chairman Charles Boston and Air Service Judge Advocate Maj. Elza Johnson insisted on a constitutional amendment as a prerequisite to federal legislation, and George Bogert—now chairman of the NCCUSL's Aviation Committee—believed that any federal law should be limited to interstate and foreign commerce and promoted his uniform state law as a necessary complement.

This debate over the legality of a uniform federal law no doubt created a sense of déjà vu for Aero Club of America trustee Charles Thaddeus Terry. As discussed in chapter 1, the Columbia University law professor had been unable to convince Congress that it possessed the power to regulate automobiles under the commerce clause before the war, and a divergent system of automobile regulation had developed under the control of state governments rather than a central authority. Now over a decade later, Terry drew upon his experience fighting for national automobile regulation to stress the urgent need for action in an impassioned speech to his fellow attendees. He warned that pilot licensing, aircraft inspection, and rules of the air "must be regulated by the Federal Government" as soon as possible to ensure safety in operation; failure to do so risked the creation of conflicting state laws as had occurred with the automobile. For Terry, federal aviation regulation could occur just as it had for steamboats—under the commerce clause through the broad interpretation of commerce espoused in *Gibbons*. Only after the passage of a uniform federal law should states be permitted to legislate on "anything they think is left for the protection of their little territory *within their imaginary geographical limits.*"[50]

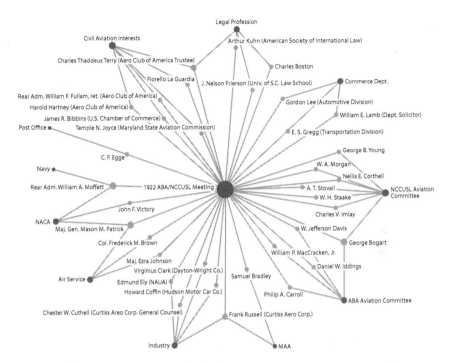

The joint meeting of the American Bar Association and the National Conference of Commissioners on Uniform State Laws on February 25, 1922, constituted the largest and most comprehensive gathering of organizations to discuss aviation legislation to date. Network graphic based on Minutes of the Joint Meeting of Committee of the Conference of Commissioners on Uniform State Laws and Committee of the American Bar Association on the Law of Aeronautics, Feb. 25, 1922, box 36, MacCracken Papers, HHPL. While not listed in the official minutes, Commerce Department Solicitor William E. Lamb attended on Hoover's behalf. Hoover to MacCracken, Feb. 15, 1922, Commerce Department, Aeronautics, Bureau of Legislation, 1922, box 121, Herbert Hoover Papers, HHPL; Michael Osborn and Joseph Riggs, *Mr. Mac: William P. MacCracken, Jr. on Aviation, Law, and Optometry* (Memphis, TN: Southern College of Optometry, 1970), 39. Created with Palladio, http://hdlab.stanford.edu/palladio/.

The arguments of Patrick, Coffin, Terry, and others had the desired effect. Members of both aviation committees recognized the overwhelming need for nationwide regulatory uniformity, and Bogert agreed to remove any mention of regulations from the NCCUSL's uniform state law. Although unconvinced that the commerce clause provided the federal government with authority over intrastate flights, members also decided against the need for a lengthy and uncertain amendment process. Whereas the ABA's committee under Boston had recommended the passage of a constitutional amendment only five months earlier, MacCracken and his colleagues now preferred, like the NACA and the MAA, to let the courts determine the constitutionality of legislation after the fact. In agreement on the desirability of legislation, the ABA Aviation Committee directed MacCracken to assist Commerce Department Solicitor Lamb in his revision of the recently passed Senate bill.

MacCracken left the ABA-NCCUSL meeting uncertain about whether the commerce clause would allow Congress to regulate intrastate flight, but such doubts did not remain for long.[51] The Supreme Court's decision in *Railroad Commission of Wisconsin v. Chicago, Burlington and Quincy Railroad Co.* issued two days after the joint meeting of the aviation committees settled the issue for the Chicago lawyer. In an opinion upholding the newfound power of the Interstate Commerce Commission under the Transportation Act of 1920, Chief Justice William Howard Taft built on the prewar Shreveport Case to declare that when "interstate and intrastate commerce . . . are so mingled together that the supreme authority, the Nation, cannot exercise complete effective control over interstate commerce without incidental regulation of intrastate commerce, such incidental regulation is not an invasion of state authority."[52] NCCUSL Aviation Committee chair George Bogert now felt certain that the Taft court "would sustain the constitutionality of a Bill giving the federal government exclusive control over Aeronautics" under the commerce clause.[53] By March 1922, ABA Aviation Committee chair Mac-Cracken, Bogert, industry representatives, and members of the NACA all agreed that the commerce clause provided sufficient justification for uniform federal regulation through legislation. It had taken three years to achieve this consensus. It would take another four before a federal aviation bill became law.

The Winslow Bill

On May 25, senior Republican representative Samuel Winslow of Massachusetts took time out of his busy schedule to personally visit the Commerce Department to "again [seek answers] as to the status of the Aviation Bill."[54] The day after the joint meeting of the ABA and NCCUSL aviation committees, MacCracken had visited Winslow, the powerful chairman of the House Committee on Interstate and Foreign Commerce, to offer his services in drafting legislation. Although Winslow "wasn't slightly interested" in aviation, the two bonded over horse racing. Winslow introduced MacCracken to assistant legislative counsel Frederick P. Lee and promised that "whatever you agree upon, I'll introduce and get through the House."[55] Now, three months had passed as MacCracken, Lamb, and Lee worked on a House bill with Hoover's blessing that would provide federal control over aviation under the commerce clause. As winter turned to spring and the temperature rose, so too did the frequency of crash reports in the press, and pressure mounted on the Commerce Department to deliver. Secretary of War John Weeks, Assistant Secretary of the Navy Theodore Roosevelt Jr., and Bradley expressed serious concerns over the legislative delay to Hoover, while Coffin fretted that "the counsels of our timid friends who fear States Rights and other bugaboos" would undermine the uniform control that both industry and the NACA viewed as essential.[56] But no one at the Commerce Department knew the status of the bill, and Winslow could only share "a few emphatic sentiments on the 'efficiency' of [the Commerce Department's] Solicitor's office" before he stormed off empty-handed.[57]

The Commerce Department bill was postponed for nearly a year as MacCracken played legislative ringmaster. The first delay, and the one that prompted Winslow's unceremonious visit to the Commerce Department, arose as MacCracken vigorously worked to convince a majority within the ABA of the legal soundness of federal control over intrastate flight. By the end of May, Lamb delivered a bill to Hoover with the assurance that their meticulous work would "practically remove opposition."[58] Hoover forwarded this new bill to a frustrated Winslow on June 12, only to disavow it the next day. Left out of prior discussions within the executive branch, MacCracken and Lamb had, at Winslow's insistence, spent three months crafting a bill to establish a Department of Air, an administrative approach that went "beyond anything that has ever been discussed in the Government."[59] Winslow received a revised version that authorized the creation of a Commerce Bureau in late June, but Ames and his fellows on the NACA refused to support a bill that also called for their committee's absorption into Hoover's Commerce Department.[60] Convinced that the bill as currently worded would only "stir up meritorious criticism in large amounts," Winslow turned it over to Lee "to put in shape for the Committee."[61] The Legislative Counsel's vacation from Washington in August caused further delay. In November, MacCracken traveled to Washington at Lee's request, where they "spent two days . . . going over the preliminary draft of the bill," and MacCracken elicited feedback from his fellow ABA Aviation Committee members Davis and Bogert—two vocal proponents of US adherence to the convention—after the first of the year.[62] Winslow finally introduced this Commerce Department–sponsored legislation in the House on January 8, 1923, roughly seven weeks before the end of the 67th Congress and almost a year after Wadsworth's bill had passed the Senate.

The Civil Aeronautics Act of 1923—popularly known as the Winslow bill—differed considerably from the Senate's bill. As Winslow himself stated, MacCracken's labors resulted in a comprehensive bill that addressed the "constitutional questions involved; the situation presented by the International Air Navigation Convention; certain departmental differences; the adaptation of the existing customs, immigration, public health and other regulatory legislation to air travel; some necessary administrative details, as well as certain questions in respect of torts, crimes and court jurisdiction of matters relating to air navigation."[63] In regards to the constitutional questions, MacCracken's bill provided for the federal regulation of *all* flight. With language ripped from the recent *Wisconsin* decision, it stipulated that because those "elements" of air navigation "ordinarily subject to regulation by the States are so mingled with those elements subject to regulation by the Federal Government" the "Federal Government cannot effectively regulate, prevent interference with, and safeguard interstate and foreign commerce by air navigation without incidental regulation of intrastate commerce." As head of a new Commerce Department Bureau, a commissioner of civil aeronautics would designate air routes, establish and maintain airports, and promulgate rules of the air as well as criteria for inspection, licensing, and registration. Like previous bills, this one would allow for the creation of domestic regulations aligned with those of both Canada and the convention, but under MacCracken's legalistic eye the Civil Aeronautics Act of 1923 became a fully formed foreign policy document. Although the current political climate precluded any direct mention of the convention within the bill, its creators readily admitted that it "conforms to the principles of the International Air Navigation Convention, and prevents conflict between our air navigation laws and the terms of the Convention."[64] The federal government asserted "complete sovereignty in the airspace" over US territory and territorial waters (including the Panama Canal Zone), the president received the authority to designate restricted areas applicable only to foreign aircraft as stipulated in the US reservations to the 1919 air convention, and foreign access to US airspace became contingent on reciprocity.[65] In addition to the establishment of a planned national air system, federal authority to create and operate airports would enable the United States to guarantee foreign access to airports as called for the 1919 convention should it be ratified at a later date.[66] Convinced that "uniform regulation of aeronautics is . . . absolutely indispensable to the effective development of aerial transportation," MacCracken had produced a bill that sought to both secure federal control and facilitate international flight.[67]

Coffin and MacCracken worked to galvanize support for the Winslow bill through the newly established NAA. Having become acquainted at the joint meeting of the ABA and the NCCUSL in February, the two men worked together over the following months to create a new organization that went beyond the sports focus of the Aero Club of America. In October 1922, their efforts culminated with the merger of the National Air Association and the Aero Club of America into the NAA at the Second National Aero Congress in Detroit.[68] With Coffin as president and MacCracken as both one of five governors-at-large and Legislative Committee chairman, the NAA committed itself to "voicing a vigorous public opinion upon beneficial and essential legislation."[69] This included "the establishment and maintenance of a uniform . . . system of laws," uniformity in "customs, ground rules, flying rules, plans and routes for aviation," and—with the convention still politically troublesome—the ratification of aerial treaties with Canada and Mexico.[70] Within two days of the Winslow bill's introduction in the House, Coffin placed the association "firmly behind the measure."[71] Through its "First in the Air" public relations campaign, the NAA portrayed itinerant fliers as the primary threat to public safety and promoted federal regulation as the best solution alongside the MAA and the ACCA.[72] MacCracken also secured endorsement of the Winslow bill from the ABA, and additional support came from the United States Chamber of Commerce, the American Society of Mechanical Engineers, the American Legion, and numerous editorial pages across the country.[73]

The Winslow bill faced significant challenges despite this concerted public relations effort. First, the NAA's public assault on itinerant fliers caused many pilots who had joined the association during its initial membership drive to view the expansive regulations within the Winslow bill as an attack on their personal freedom. Second, MacCracken's sixty-four-page bill was a far cry from the broad, flexible legislation that had emerged from the Senate.[74] While Ames supported the bill's proposed expansion of the NACA, he believed that several provisions not included within previous legislation would only "open serious objectives" and "provoke a storm of criticism in the House."[75] The most significant revolved around the federal government's power to create and operate airports.[76] Both the NACA and the War Department agreed that this provision's potential to generate "tremendous unknown expenditures" and accusations of regional favoritism could undermine the passage of "constructive legislation" in an aerial reenactment of the internal improvements debate of the antebellum era.[77] To avoid a legislative delay, they strongly advised applying the precedent of waterway navigation to aviation: the federal government would establish airways, erect navigational aids, and provide weather information while state and local governments constructed airfields, or "airharbours" in the waterways analogy. The extent of federal authority to establish airfields would remain a point of contention between the NACA and MacCracken up until the passage of the 1926 Air Commerce Act. Despite these misgivings, Bradley successfully worked behind the scenes to secure guarantees from the Air Service, the Navy's Bureau of Aeronautics, and the NACA that they would not actively oppose the airport provision if it would derail the bill's passage.[78]

As the number of aeronautical deaths continued to climb and calls for legislation grew over the next two years, McCracken's bill remained in limbo in Winslow's Interstate and Foreign Commerce Committee.[79] The congressional calendar bears much of the responsibility for the first year of delay. After the 67th Congress adjourned in March 1923 with no forward movement on the Winslow bill, the summer recess and presidential election pushed any chance for passage back to December. Failure to act in the 68th Congress, however, rests ultimately with a lack of leadership. Unlike his predecessor, President Coolidge did not place the full weight of his office behind aviation legislation. In his first State of the Union address four months after Harding's fatal heart attack, President Calvin Coolidge—nicknamed "Silent Cal"—simply stated that "laws should be passed regulating aviation," and when Winslow requested comment on his bill four days later the president merely replied, "Seems ok to me."[80] Winslow himself proved a poor legislative standard-bearer in the absence of strong executive support. Other than adding some amendments, the House committee took little action until three days of hearings in December 1924.[81] There, testimony from Hoover, representatives from various executive departments, and MacCracken himself could not fully alleviate concerns over federal control of intrastate flight.[82] A particularly frank exchange between MacCracken and staunch states' rights advocate George Huddleston—one where the ABA Aviation Committee chair unabashedly asserted an unlimited federal authority over aviation—left the Alabama Democrat in shock and foreshadowed a potentially rocky House debate.[83] The inclusion of a section within the House committee's report affirming the aerial rights of states helped to assuage misgivings over rampant federal power among some of its members, and Winslow finally moved the bill forward without a potentially devastating minority report.[84]

Advocates of uniform federal legislation, left crestfallen after the Senate limited its bill to interstate and foreign commerce, had achieved a victory. But their enthusiasm remained short-lived, as the House failed to take up the bill before the 68th Congress adjourned on March 4, 1925. With Winslow's subsequent retirement, the detailed legislation that MacCracken had created and shepherded over the previous two years quickly fell out of favor. Overly complicated and guaranteed to provoke opposition from states' rights supporters in the House and the Senate, the Winslow bill never enjoyed overwhelming and enthusiastic support. By June 1925, the Curtiss Company's Frank Russell could confidently proclaim that "the Winslow Bill is dead."[85] Much to their disappointment, it seemed that advocates of uniform federal regulation would need to start all over again in the House. But by the time the 69th Congress convened in December, an aura of crisis would compel Congress to act on the matter of aviation regulation.

Although the previous two years must have seemed like a waste of time to Coffin, Ames, and others, it would be a mistake to simply dismiss the Winslow bill as another legislative dead end. While MacCracken's "penchant for comprehensiveness" reduced the chances of the bill's passage, the Chicago lawyer's work notably expanded the parameters of the regulatory dialogue.[86] By incorporating extensive particulars, the law starkly illustrated the full scope of aviation's impact on American law and national security. In addition, both the ABA and a majority in the House Committee on Interstate and Foreign Commerce had come to support uniform federal regulation through the commerce clause. Equally important, MacCracken ensured that the international air convention became a central element of the House committee's discussions.[87] Members knew that numerous provisions within the Winslow bill—most notably the statement of national air sovereignty—

came directly from or purposefully aligned with the international air agreement, and they "presumed" that subsequent US regulations would "conform to the similar provisions of the convention."[88] Like Coffin, Bradley, Walcott, and their compatriots, future assistant secretary of commerce for aeronautics William P. MacCracken Jr. recognized that domestic regulation could not exist within a vacuum. Thanks to his efforts, so did the House Committee on Interstate and Foreign Commerce.

The Legislative Breakthrough

Howard Coffin couldn't have said it better himself. For the past four weeks, he and eight other presidentially appointed members of the President's Aircraft Board had heard testimony on the state of aviation in the United States. Now, on the second to last day of testimony, Godfrey Lowell Cabot neatly encapsulated the regulatory issues facing civil aviation. A scion of two distinguished Boston families, cousin to powerful Republican Senator Henry Cabot Lodge, and a natural gas magnate, Godfrey Cabot learned to fly in his fifties, received a commission in the Naval Reserve Flying Corps during World War I, and now held Coffin's prior position as president of the National Aeronautic Association.[89] Before the nine-member panel, Cabot called for an "all-embracing Federal law" to prevent the creation of conflicting state requirements, regulation through a Commerce Department Bureau, and the need to either ratify the international air convention or "promulgate and enforce air traffic rules" in accordance with its provisions to ensure safety in flight between the United States and Canada.[90] For Cabot, any attempt to address civil aviation policy needed to take into account the international context, the very argument that Coffin and Walcott had made to President Wilson more than six years earlier.

The President's Aircraft Board constituted only the most recently established investigation to address a perceived crisis in American aviation underway in the fall of 1925. The first, the House Select Committee on Inquiry into Operations of the United States Air Services, began its investigation in December 1924 in response to calls of corrupt practices in aircraft procurement. Concern over the lack of regulations as the Post Office transitioned to contract air mail service under the 1925 Air Mail Act prompted the creation of the second, the Joint Committee of the Department of Commerce and the American Engineering Council, in the summer of 1925. Only after two highly publicized military accidents created a potential political disaster did President Coolidge establish the President's Aircraft Board that September. These three investigative committees—one an official congressional investigation, one cloaked in the expertise of the engineering profession, and one with presidential sanction—focused public attention on aviation as a national security issue and demonstrated the need for legislative action. Together, they served as vessels through which Congress and the public came to accept the administrative framework that executive branch agencies and industry had agreed on three years earlier. Faced with a need to act upon these three largely unanimous reports, Congress finally passed legislation to regulate aviation in the United States. As a statement of domestic policy, the subsequent Air Commerce Act represented a compromise between the complete federal control that Bradley, McCracken, and others viewed as vital and perceived limitations of federal authority under the commerce clause. As a statement of foreign policy, it institutionalized the international air convention's principles and norms while providing a mechanism to promulgate domestic regulations compatible with the convention's provisions.

The House Select Committee on Inquiry into Operations of the United States Air Services—popularly known as the Lampert Committee after its Republican chairman Florian Lampert of Wisconsin—began in response to concerns over the effectiveness of military aviation. On one side stood those anxious about the level of preparedness. The precipitous decline in Air Service appropriations from $33 million in fiscal year 1921 to $12.9 million by 1923 under the extreme fiscal conservatism of the Harding administration prompted Chief of Air Service Patrick to warn that "the Air Service is now entirely incapable of meeting its war requirements and, if immediate and adequate remedial measures are neglected, will become practically demobilized within a few years."[91] At the same time that anxiety arose over the effectiveness of military aviation, others renewed charges that the MAA constituted a nefarious "aircraft trust," a specter that had haunted the organization since its creation in 1917.[92] With Wisconsin representative John M. Nelson leading an attack against waste, inefficiency, and corruption in aircraft procurement, the House established the Lampert Committee in March 1924.[93]

Lampert and his fellow committee members deserve credit for their extensive work.[94] Empowered to investigate the activities of the Air Service, the Navy's Bureau of Aeronautics, the Post Office's air mail service, and "any corporations, firms, or individuals or agencies having any transactions with or being in any manner associated with" those activities, members visited various hubs of government aeronautics before public hearings began in October 1924.[95] Over the course of the next five months, more than a hundred witnesses from the Departments of War, Navy, Post Office, and Commerce as well as industry and the private sector testified on the aeronautical situation. While the MAA's cross-licensing agreement, the question of a unified Department of Air, aircraft appropriations, and the technical differences between naval and land-based aircraft remained dominant themes, the need to foster commercial aviation to support an industrial base for military production meant that civil aviation and its regulation remained a subject throughout the hearings. Some feared that Brig. Gen. William "Billy" Mitchell and his allies might convince the House committee to support a unified Department of Air, but the preponderance of testimony, unsurprisingly, supported the continued separation of Army and Navy aviation and the creation of a bureau within the Commerce Department.[96] The announcement of Coolidge's personal opposition to a separate Air Service further undercut Mitchell's cause.[97] Witnesses also reminded committee members that US aviation existed within a larger international context. Lt. Leigh Wade informed the committee that the recent Army World Flight—in which he served as pilot of the ill-fated *Boston* and its replacement, the *Boston II*—proved "there is no part of the globe" that Americans could not reach, and Howard Coffin advised that US aerial expansion would greatly benefit from "at least an equal voice" in the development of international regulations that would, by their "very nature," affect American "air activities in the future."[98]

As the Lampert Committee toured aeronautical sites and collected documents, the second investigative committee arose out of Coffin's continuing efforts to secure a federal law that would both promote commercial aviation and ensure compatibility with the convention's provisions. At his behest, the American Engineering Council "voted to further an economic survey of commercial aviation" in May.[99] Hoover promised the assistance of the Commerce Department in this endeavor, and both Coffin and Curtiss vice president Clement Keys funneled $1,000 of their own money through the American Engineering Council to support this investigation.[100] With Assistant Secretary of Commerce J. Walter Drake as its chairman, the Joint Committee of the Department of Commerce and the American Engineering

Council embarked on a systematic study of commercial aeronautics both in the United States and abroad to draft recommendations for Congress.

The Joint Committee had an agenda from the very beginning. While Coffin could honestly claim in 1922 that he possessed "no commercial interest of any kind in connection with aviation," this had changed by May 1925.[101] The passage of the Air Mail Act that February provided the revenue stream necessary for sustained private air operations, and these initial air mail contractors would morph into storied commercial carriers such as United, American, and Eastern.[102] This foundational legislation not only authorized the postmaster general to establish new air mail routes and "encourage commercial aviation" through private air mail contracts but also to "make such rules, regulations, and orders as may be necessary to carry out" the Air Mail Act's provisions.[103] This would, by necessity, include safety regulations.[104] A discussed previously, leaders within the Post Office had supported the regulation of civil aviation first under an Interdepartmental Board in the Wilson administration and then through a Commerce Bureau. Under the Air Mail Act, the Post Office now became the sole congressionally mandated regulatory authority over commercial aviation in the United States. Although some hoped that the Post Office would simply apply the convention's provisions to all aircraft, it did not possess the budget or expertise to properly regulate civil aviation.[105] Assurance of public safety through universal regulations would ultimately benefit a carrier's bottom line, and Coffin and Keys bankrolled the Joint Committee to protect their most recent enterprise. Only days prior to the American Engineering Council's vote to support a commercial aviation survey, the two had, with the aid of MacCracken, incorporated National Air Transport to secure an air mail contract with "an astonishing $10 million" capitalization.[106] For the previous seven years, Coffin and Keys recognized that aviation would not become a profitable enterprise capable of global expansion until the United States adopted nationwide regulations aligned with the convention. With a member of NAT's board of directors and NAT's traffic manager on the Joint Committee and MacCracken as an unofficial legal advisor, there was little doubt as to its final recommendations.[107]

Established under presidential aegis, the President's Aircraft Board—popularly known as the Morrow Board after its chairman—held the greatest esteem among the public. Its creation arose in response to a potential and an actual political crisis. The potential crisis revolved around whether the

Lampert Committee would recommend the creation of a Department of Air, an approach diametrically opposed to long-standing executive branch policy. Coolidge remained largely silent while the committee heard testimony, but as the public hearings wound down he informed his old Amherst College friend and current J. P. Morgan partner Dwight W. Morrow "that I may like to have you look into the subject of airplanes for me."[108] As the summer dragged on, concern grew over the House committee's report, but Coolidge, never one to rush to a politically tenuous position, remained above the fray until an actual crisis compelled the taciturn president into action. On September 1, news arrived that a Navy PN-9 flying boat had crashed into the Pacific Ocean on a pioneering nonstop flight from San Francisco to Honolulu. Testimony before the Lampert Committee had already raised concerns about the state of military aeronautics, and press reports that the crash resulted from a lack of fuel further undermined the Navy's image.[109] Less than forty-eight hours into this blossoming public relations nightmare, fourteen servicemen died when the Navy airship USS *Shenandoah* succumbed to a storm over Ohio.[110] Things now seemed fundamentally amiss in military aviation, and questions swirled in the press as to the competence of its leadership.[111] But the greatest challenge to the administration came from Mitchell.[112] In a seventeen-page statement "filled with sarcasm and criticism," Mitchell forcefully declared that "these accidents are the direct result of the incompetency, criminal negligence and almost treasonable administration of the national defense by the Navy and War Departments. Were it not for the patriotism of our air officers and their absolute confidence in the institutions of the United States, knowing that sooner or later existing conditions would be changed, I doubt if one of them would remain with the colors, *certainly not if he were a real man.*"[113]

Coolidge's response to this public relations fiasco took two forms. To address Mitchell's blatant insubordination, he signed off on a War Department investigation that led to the colonel's court martial.[114] Next, to ensure a measured response to the outcry for institutional change Coolidge turned to his friend Morrow to chair a presidentially appointed investigative committee. Morrow Board members William Durand of the NACA, Howard Coffin, ranking Republican on the House Committee on Interstate and Foreign Commerce James S. Parker of New York, and Connecticut's freshman senator Hiram Bingham—a wartime aviator and former Yale professor of Latin American history famous for his "discovery" of the Incan city of Machu Picchu—

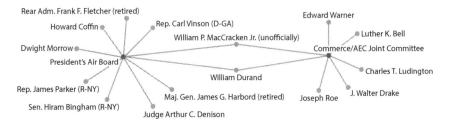

Network graphic of the personnel on the President's Aircraft Board and the Joint Committee of the Department of Commerce and the American Engineering Council. In addition to his membership on the President's Aircraft Board, Coffin also bankrolled the Joint Committee. Created with Palladio, http://hdlab .stanford.edu/palladio/.

had all supported a Commerce Department Bureau for some time.[115] Of the remaining members, none held any radical ideas about aviation.[116] With Durand appointed to both the Morrow Board and the Joint Committee and MacCracken an unofficial legal advisor to them both as well, the Coffin-financed Joint Committee quickly morphed into an information clearing-house for the Morrow Board.[117] With so many adherents to the Commerce Bureau approach among its members and affiliates, the Morrow Board's "investigation" constituted a mere formality.[118]

By the end of January 1926, Congress and the American public could read the three investigative committee reports.[119] Unsurprisingly, all agreed that a Commerce Department Bureau could best promote civil aviation to ensure the existence of a robust industrial base vital to national defense. Differences existed, however, among the reports. In addition to a Commerce Bureau, the Morrow Board's recommendations called for the creation of three new assistant secretary positions in the Departments of War, Navy, and Commerce to coordinate air policy; the elevation of the Air Service to the status of corps; and the creation of a farsighted procurement plan to support the aircraft industry.[120] Instead of coordination through assistant secretaries, the Lampert Committee proposed the creation of a cabinet-level Department of Defense and a single aeronautical procurement agency, but both of these farsighted ideas ran counter to the current regulatory consensus.[121] Whereas the Morrow Board remained silent on the air convention for obvious political reasons and the Lampert Committee simply noted "that the United States has failed to ratify" it, the Joint Committee—dominated by industry interests and engineers operating through the largess of Coffin and Keys—directly placed the domestic situation within a broader international context. Under consultation with MacCracken, the Joint Committee explicitly called for the ratification of the 1919 convention to not only prevent discrimination again US aircraft in international flight but also to ensure US "participation in the formulation of amendments and regulations *to which American aircraft must eventually conform.*"[122] Members further asserted that because the convention constituted "the general basis for Air Regulation," it "should be followed in its main lines by the United States," even in the absence of formal ratification.[123] Even though the presidential and congressional committees were not as overt in their discussion of the convention as the Joint Committee, the recommended Commerce Bureau that they all agreed on would allow for the promulgation of regulations compatible with the convention, something Coffin, Bradley, Walcott, Patrick, and others had consistently called for since the Wilson administration.

The three investigative committees recommended an administrative framework for civil aviation regulation but remained silent on the scope of federal control. Since 1922, the Senate had continued to pass legislation that limited federal regulation to interstate and foreign flights, and MacCracken had fought in the House for federal authority to establish uniform regulation and airports to ensure safety and foster civil aviation. The Senate's consistent position coupled with the three-year struggle to move the Winslow bill out of committee portended a prolonged battle over the extent of federal power, one that could undercut the current legislative momentum. Having waited for Congress to rouse itself to action for years, advocates of a federal law now faced two possibilities. On the one hand, they could continue to fight for complete federal control and risk missing their best opportunity to prevent the entrenchment of state-based regulation similar to the automobile. On the other hand, a bill based on a strict interpretation of the commerce clause would surely make it to the president's desk in the current political climate. Although such a limited bill would not secure universal federal regulation, experience would likely prompt Congress to expand on it in the future, just as they had in the case of federal steamboat and railroad regulation. As legislation to establish a Commerce Bureau made its way through Congress in early 1926, key supporters of a federal law backed away from MacCracken's all-encompassing vision of federal power.

Although members of Congress submitted a variety of aviation bills when they convened in December, only one had any real hope of passage in the wake of the Morrow Board's recommendations.[124] Two days after the Morrow Board began public hearings, Hoover wrote Bingham to offer the assistance of Commerce Department solicitor Judge Stephen Davis to address "the legislative problem."[125] When their duty on the Morrow Board came to a close three weeks later, Bingham and Representative James Parker took Hoover up on his offer with the assurance that whatever bill Davis drafted would meet with the secretary's approval.[126] Bingham introduced this new Commerce Department–backed bill in the Senate on the same day that Coolidge declared his support for the Morrow Board's recommendations in his 1925 State of the Union address, and Parker introduced the House version two days later.[127] Rather than a version of the Winslow bill that Hoover had supported for the past two years, this new bill resembled the more limited Senate version passed in the last two Congresses. It strictly confined federal regulations to interstate and foreign flights, left the creation and operation of airfields to state and local governments, and—in an abrupt departure from the recommendations of the three investigative committees—called for the creation of a new assistant secretary of commerce position rather than an entirely new bureau.[128] Not less than three months earlier, Hoover had testified before the Morrow Board that a new bureau constituted "the essential provision" of any federal legislation, but he now believed that a sole assistant secretary working in conjunction with the current resources of the Commerce Department could provide the same level of efficiency without arousing concerns over administrative bloat and fiscal waste from "an economy-minded Congress."[129] Hoping to avoid "contentious questions" that could delay passage of much-needed legislation, Hoover now supported this safer Senate bill with its limited interpretation of federal power as a matter of expediency.[130] Proponents of uniform federal regulation had lost a powerful ally, and Bingham's bill received rapid approval in the Senate.[131]

For the third time in three years, a more restricted Senate bill made its way to the House Committee on Interstate and Foreign Commerce, now under Parker's chairmanship. By this time, a majority of the committee's members agreed with McCracken on the need for more thorough legislation than the Bingham-Parker bill. Under intense lobbying from both Cabot and Coffin, the House drafting subcommittee accepted the Senate bill's shift to a sole assistant secretary but once again refused to limit federal power to strictly interstate and foreign commerce.[132] With the aid of MacCracken, the subcommittee thoroughly revised the Bingham-Parker bill to retain key features from the Winslow bill, such as federal power over intrastate flight, Congress's ability to establish and operate airports, a declaration of national air sovereignty, and presidential authority to establish airspace reservations.[133] In response to a request for comment on this revised bill, Ames cautioned the House committee to "avoid the introduction of controversial elements which are not essential, such as the extension of Federal control to intrastate air commerce or the Federal establishment of airports."[134] The NACA's Stratton remained uncomfortable with the idea of complete federal control in the absence of convention ratification, and Ames shared his concerns over constitutionality. Convinced that practical experience would persuade Congress of the need to extend federal authority to intrastate flights in the future, Ames—another longtime proponent of a federal law—now decided that limited legislation was better than no legislation at all.[135] Despite Ames's concerns, the House committee approved its more expansive bill.

By a vote of 229–80 on April 12, 1926, the whole House finally passed a bill that provided the uniform federal control that Bradley, Coffin, and MacCracken desired. Despite this large margin, an intense debate about federal control over intrastate flights on the House floor showed the difficulty such a bill would face in the equally apportioned Senate. With a clear recognition that any erosion of state sovereignty could affect their states' ability to sustain institutionalized racial segregation under Jim Crow, several southern Democrats vehemently opposed the bill's unprecedented extension of federal power.[136] Members of the House Commerce Committee rallied to the defense of the bill, most notably Democrat Clarence Lea of California, a former Lampert Committee member who had fought for more comprehensive legislation. Declaring that "whenever an airplane . . . rises from the ground and enters the air . . . it is in interstate commerce," Lea espoused the long-held position of Bradley, Coffin, and MacCracken that safety required uniformity in regulation. He argued that "practically everything that is in the bill is in the international air navigation convention," and as such the bill provided a connection with the international air regime and allowed the secretary of commerce to promulgate complementary regulations to guarantee the greatest "economy and convenience" in intrastate, interstate, and foreign flight. Lea's arguments convinced a majority of representatives to reject amendments that would limit federal authority to foreign and interstate commerce.[137] But the tenacity of resistance among the House minority, the need to convince a majority of senators clearly opposed to complete federal authority over aviation, and the reservations of Ames and Stratton convinced the NACA to formally back away from the House bill's more controversial aspects. In a hastily conveyed Special Subcommittee, NACA members who had supported federal control for years, such as Patrick, Moffett, and Durand, now joined Ames and Stratton in their refusal to risk legislation on an all-or-nothing gamble.[138] Having come this far, Hoover and the NACA were unwilling to fight for uniform federal control or federal appropriations for airports.[139]

The bill that emerged from the reconciliation committee offered a nuanced solution to the thorny issue of domestic sovereignty. With Lea and the rest of the House contingent convinced of the need for regulatory uniformity and Hoover and the NACA now in support of the Senate's more limited interpretation of federal power, the "conferees . . . agreed upon a compromise."[140] To ensure safety in flight, federal air regulations (rules of the air) would apply to *all* aircraft—civil, military, and foreign alike—but federal registration and licensing requirements would only apply to those aircraft and pilots engaged in "foreign or interstate commerce," a term strictly defined as "operation [of an aircraft] in the conduct of business." MacCracken's years of seemingly fruitless work in the House to secure federal regulatory uniformity had paid off, while the ability of states to register aircraft and license pilots engaged in solely intrastate flight assuaged concerns of states' rights advocates. Although the bill did not require universal federal licensing and registration, advocates of uniform federal control took heart in the belief that the vast majority of aircraft and pilots would opt for federal documentation due to the airplane's speed and ability to readily disregard state borders. The secretary of commerce obtained the power to establish airways and provide necessary navigational aids, but, with little support for the creation of federal airports that could spark runaway appropriations, the reconciliation committee explicitly excluded airports from the list of federally funded navigational facilities. As we will see in chapter 6, this decoupling of airports from the federal government in response to domestic sovereignty concerns would profoundly influence US foreign aerial policy.

The reconciliation committee also incorporated the international elements of the House bill that MacCracken had fought for over the past three years. The bill's declaration of sovereignty perfectly matched that of the 1919 convention, while the declared right to establish reciprocal air agreements with other nations, the ability to create restricted zones applicable solely to foreign aircraft, and the modification and application of existing US customs law to aviation aligned with Carpenter's reservations to the convention discussed in chapter 2. The State Department found no conflict between this compromise bill and the 1919 convention. Future ratification of the convention would automatically meet the Air Commerce Act's required reciprocity provision, and if the United States remained officially outside of the international air regime the secretary of commerce could still enact regulations fully compatible with the convention's provisions.[141] Thanks largely to the early vision of individuals such as Bradley, Coffin, and Walcott as well as MacCracken's tireless legislative efforts, the Air Commerce Act—signed by Coolidge on May 20, 1926—provided the United States a means to facilitate international aerial exchanges through the promulgation of nationwide regulations in conformity with those of Canada and other convention members.

The Air Commerce Act marked the end of an eight-year ideological process in which numerous relevant interest groups grappled with the extent of federal power to regulate aviation under the commerce clause and the need to ensure complementarity with the principles and rules of an unratified air convention. Viewing uniform federal regulation as a vital prerequisite to a civilian market, Bradley, Coffin, and other industry leaders placed their argument for a Department of Air within an international context. In agreement on the need for a federal law, civilian members of the NACA led the charge to place civil aviation within the Commerce Department. Concerned over their lack of voice on an Interdepartmental Board, industry secured Commerce Department support for a bureau alone and turned to the legal profession for aid. When the Senate failed to accept universal federal control, ABA Aviation Committee chair MacCracken—working in close consultation with Coffin—labored on an all-encompassing House bill expressly tailored to complement the international regime. After an atmosphere of crisis prompted a congressional response, MacCracken continued to fight against watered-down legislation. To placate those in Congress concerned about federal overreach and burgeoning appropriations, a congressional reconciliation committee agreed on a bill that replaced the proposed bureau with an assistant secretary, rejected federal funding for airports, and provided for uniform rules of the air but left intrastate licensing and registration under state control. Throughout this process, precedents in the application of the commerce clause to steamboats and railroads, the "failure" of the federal government to exert authority over the automobile, and the need to ensure compatibility with Canada and other convention adherents guided advocates of universal federal control. The Air Commerce Act allowed the uniform safety regulations necessary to stimulate commercial aviation, provided a framework for international air relations, and established a legislative foundation open to future modification based on experience. Compelled by a desire to strengthen national defense, stimulate business, and protect the public safety, a Republican-controlled Congress and Republican president had secured an unprecedented legislative expansion of federal power.[142]

After his appointment as the first assistant secretary of commerce for aeronautics in August, MacCracken recognized that "the first thing we needed was a system of Air Commerce Regulations."[143] He would not begin this massive undertaking from scratch. As an increasing number of Americans took to the skies in the years following World War I, the convention's provisions offered a ready-made set of criteria for them to follow in the absence of government regulations. By the time MacCracken began work on the nation's first Air Commerce Regulations, various aeronautical interests had already adopted and diffused the convention's provisions throughout the United States, most notably through voluntary regulatory attempts on the part of the insurance industry and the engineering profession. These efforts to apply the convention's operational (rules of the air) and registrational (licensing, registration, etc.) provisions ensured that when federal regulations finally materialized they would remain compatible with those of Canada and other convention adherents.

5

The Need for Regulatory Compatibility

On March 3, 1919, William Boeing and test pilot Eddie Hubbard landed their Boeing CL-4S seaplane on the water of Lake Union, Seattle, with sixty letters from Vancouver, British Columbia. With this 120-mile flight, "authorized by the Canadian Post Office," these two aeronautical pioneers—one the president of Boeing Airplane Company and the other a future vice president of Boeing Air Transport—successfully demonstrated the viability of international air mail service in the Pacific Northwest.[1] Using the only Boeing B-1 flying boat ever produced, Hubbard initiated regular international air mail service between Seattle and Victoria, British Columbia, in October 1920 under US Post Office Foreign Airmail Contract (FAM) #2.[2] After a year, Hubbard had "carried over 1,110,000 first-class letters in a total of 14,976 miles of aerial travel without a single delay or accident" and established an unparalleled reputation for safety and efficiency.[3] His success and operational expertise proved central to William Boeing's decision to bid on the San Francisco–Chicago air mail route in 1927, a choice that eventually led to the creation of United Airlines four years later.[4]

Like all pilots in the immediate postwar period, Hubbard did not have to adhere to any federal regulations when flying in the United States. But this was not the case when he entered Canada. With the Air Regulations of 1920, the Borden government adopted the provisions for pilot licensing, aircraft inspection and registration, and rules of the air found in the 1919 Convention Relating to the Regulation of Aerial Navigation. According to these new regulations, foreign aircraft needed to be registered in their home country *and* an air agreement had to exist between their home country and Canada to legally fly in the dominion's airspace. But the United States did not pos-

When Hubbard applied for a Canadian registration for his Boeing B-1, he initially received registration number G-CADS ("G" corresponding to the nationality letter for Great Britain as per the 1919 convention and "CA" denoting a Canadian-issued number), which was changed to N-CADS to reflect the convention's "N" designation for US aircraft. This revised registration number was issued on November 26, 1920. Peter M. Bowers, *Boeing Aircraft since 1916* (Annapolis, MD: Naval Institute Press, 1966, 1968). Image courtesy of the Boeing Company.

sess a national system of licensing and registration in 1920, nor did an aerial agreement exist between the two North American nations. Without a complementary regulatory system south of the border and an air agreement, the new Canadian regulations effectively banned Americans such as Hubbard from flying into or through Canada.

Hubbard was able to begin his international air mail service thanks to the Canadian government's desire to facilitate cross-border flight while its southern neighbor got its regulatory house in order. With overly optimistic assurances from Washington that the United States would soon adhere to the 1919 convention and pass domestic legislation, the Canadian Air Board issued a temporary courtesy to American pilots in May 1920 that exempted "qualified American military pilots" from its new licensing requirements and allowed aircraft that "would under the Convention relating to International Air Navigation be registerable in the United States of America" to enter Canadian airspace. To take advantage of this courtesy US pilots needed to (1) provide "full particulars of the aircraft" in advance of their flight, (2) obtain a temporary certificate of airworthiness after inspection at a designated Canadian customs station, (3) display a Canadian-issued nationality and registration mark on their aircraft that conformed with the 1919 convention, and (4) pay all applicable fees.[5] Under these temporary provisions, "properly qualified [US] military pilots and machines certified as airworthy" could now cross the border "under the same conditions [as] if that Government had passed regulation similar" to Canada, and nonmilitary US pilots like Hubbard, who could provide "proof of qualifications," could receive a Canadian-issued license.[6] As a result of this courtesy, American citizen Eddie Hubbard flew an American aircraft with a Canadian-issued pilot's license and registration number to fulfill a US airmail contract.[7] Just as Coffin had warned President Wilson in his February 1919 letter discussed in chapter 2, Canada's adoption of the convention's provisions had compelled US pilots to adhere to the stipulations of the new postwar international air agreement.[8]

Hubbard's case shows that the 1919 air convention created an international regime—a set of "principles, norms, rules, and decision-making procedures"—that Americans could not ignore.[9] As detailed previously, Bradley, Coffin, McCracken, and other like-minded individuals sought uniform federal regulation to increase public perceptions of safety in order to stimulate commercial aviation. This desire did not simply end at the nation's border, nor could governments erect aerial barriers to separate domestic aviation from the larger world. After the creation of the 1919 convention, the United States existed within a global aeronautical system, and advocates of regulatory uniformity readily understood that "the rules and regulations in this country should conform to the standard practice abroad."[10] In the immediate postwar years, the City and County of Los Angeles, the Army Air Service, and the NACA all looked to the convention's provisions for operational procedures (lights, rules of the air, logbook requirements, etc.) and registrational requirements (pilot and crew licenses, aircraft registration number, etc.) as the best way to secure both aeronautical safety and conformity with standard international practices.[11] Although the convention's con-

This image of an attempt to close off airspace to German aircraft during World War I shows the impracticality and near impossibility of completely closing political borders to the airplane. *Flying* 8, no. 1 (Feb. 1919): 42.

nection to the League of Nations precluded official adherence, those seeking to foster US aviation recognized the benefits to be had from aligning domestic practices with those of the International Air Convention.[12]

This chapter details how Americans used the convention as a guide in their efforts to create a regulatory system for aviation both before and immediately after the passage of the 1926 Air Commerce Act. As Coffin, Mac-Cracken, the NACA, and their allies struggled to convince Congress of the need to take legislative action, the insurance industry and engineering profession adopted an associationalist approach to the issue of aeronautical safety.[13] For National Aircraft Underwriters Association president Edmund

Ely and A. R. Small of the Underwriters Laboratories, a common set of standard practices would assist underwriters in their attempts to assess risk in a relatively unknown field. For Henry M. Crane of the Society of Automotive Engineers, the Bureau of Standards' Arthur Halstead, and others working on the American Aeronautical Safety Code, the creation of a shared set of aeronautical practices complemented the engineering profession's broader fascination with standardization as a means of social improvement. Viewed largely as failures, these two voluntary efforts nonetheless successfully diffused the convention's provisions among key constituencies in the United States.[14] By the time MacCracken began work on the Air Commerce Regulations in the fall of 1926, the need for US regulations to align with the provisions of Canada and the 1919 convention had already become firmly entrenched in the aeronautical community.

For Insurance's Sake: The National Aircraft Underwriters Association and the Underwriters Laboratories

Things had not gone as planned for America's famed "Ace of Aces" Eddie Rickenbacker. On June 6, 1922, he had taken off from Mitchel Field, Long Island, with flight endurance recordholder Edward Stinson and two others in a surplus Air Mail Service JL-6—an imported German all-metal Junkers F-13 monoplane—on a planned aerial tour of sixty-six cities to inspect his recently established automobile dealerships.[15] After an uneventful takeoff, things quickly took a turn for the worse. Lack of fuel forced the plane down on its way to Cleveland, a lightning strike on the hanger in Detroit "fus[ed] parts of the machine," and mechanical difficulties brought the plane down in Michigan and Iowa.[16] After an overhaul of the airplane in Omaha, Nebraska, "the propeller and landing gear were demolished" on takeoff. Having shot down twenty-six confirmed enemy aircraft in the skies over Europe, Rickenbacker blamed the debacle on the poor state of American aviation. The famed military pilot and future head of Eastern Airlines declared that "there is not a plane in the United States fitted to make such a long tour" and that he "was through with commercial aviation for the time being." His aircraft unusable, Rickenbacker "was forced to abandon" his highly publicized aerial tour.[17]

This outcome was no surprise to former Air Service test pilot Rudolph William "Shorty" Schroeder. A week before the ill-fated flight, he had in-

Eddie Rickenbacker (*second from right*) and Edward (Eddie) Stinson (*far right*) pose with two other individuals, presumably Lloyd Bertaud and Steve Hannigan, next to their JL-6 after they arrived in Chicago. DN-0074995, Chicago Sun-Times/Chicago Daily News collection, Chicago History Museum.

spected the machine in his capacity as chief aviation engineer for the Chicago-based Underwriters Laboratory (UL). Upon joining the UL the previous September, Schroeder had developed a system of airworthiness inspections derived from "the Engineering Division of the Army Air Service, the Air Mail Service, and the Air Board of Canada" and now oversaw a network of thirty-two resident aircraft engineers in twenty-two states.[18] With a UL airworthiness certificate required to obtain aviation insurance, Rickenbacker had approached his friend Schroeder to inspect the JL-6 before his planned aerial tour. Schroeder refused to certify the aircraft based on "substantial evidence of lack of care."[19] Confident in his abilities as a pilot, Rickenbacker nevertheless chose to use the airplane without insurance coverage. Although the lightning strike certainly didn't help, Rickenbacker's particular JL-6 had begun the voyage in poor condition.[20] Had America's "ace of aces" listened to Schroeder, he could have avoided a wave of bad press just as he launched his automobile business.

Schroeder inspected Rickenbacker's JL-6 as part of a voluntary postwar system to enable the insurance industry to more accurately access risk and rates. But prior to the armistice, insurance companies actively shied away from this new field. Two weeks before the United States entered World War I, the Aircraft Manufacturers Association, the precursor to the MAA, brought to the NACA's attention the fact that "no insurance company will issue a life policy or an accident policy to any man engaged in aviation."[21] Over the next several months, efforts on the part of these two organizations to convince insurance companies to underwrite aviators encountered a "strong tendency" among them "not to solicit or obtain more of this class of insurance than is actually forced upon them."[22] Wartime patriotic duty spurred a few insurance companies to cover military pilots under a special war hazard premium, and some even extended coverage to mechanics and individuals engaged in aircraft manufacture. But even those willing to offer coverage to active service aviators bluntly refused to "insure men who are to make a business of aviation and who are not in the military service."[23] When the NACA surveyed the nation's preeminent fire and accident insurance companies about possible coverage for "airplanes engaged in civil service" in late 1917, they received an overwhelmingly negative response due to "a lack of statistics to establish the relative risk."[24]

For Bradley, Coffin, and others seeking to foster commercial aviation, insurance took on renewed importance after the armistice. In addition to the aircraft themselves, commercial aviation needed two things to become a successful enterprise: investors willing to invest and Americans willing to fly. Although the airplane inspired near-messianic visions of a new aerial age, its recent use in war and continued press coverage of aerial crashes meant that "the airplane appear[ed] to the average person thrilling and fast, but dangerous."[25] As Curtiss Company sales representatives struggled to persuade prospective buyers to buy a new aircraft over a war surplus one, they also had to "convince prospective buyers that flying is safe."[26] The American aviation industry—desperate to sell aircraft to a civilian market after the abrupt end of wartime appropriations—found itself in a Catch-22: the public refused to fly in large numbers due to the negative perceptions of aviation while investors remained leery to commit to an enterprise that lacked a secure source of revenue. Individuals within industry and the NACA recognized that uniform federal regulations would assuage public concerns and worked to convince Congress to act. In the meantime, the development of aviation insurance would lend the field much-needed legitimacy, for "without insurance facilities, men will not risk their capital, pilots will not risk their limbs, passengers will not risk their lives and merchants will not risk their goods in schemes of air transportation."[27]

Although they had resisted the NACA's appeal to enter the civil aviation field in 1917, the fear of British competition after the armistice prompted US insurers to reconsider. Already the center of the global maritime insurance industry, Lloyd's of London underwrote the world's first aviation insurance policies before the war. In early 1919, some of its members came together as the Aviation Insurance Association "to establish a worldwide organization" for aviation underwriting under its manager Horatio Barber.[28] With the prestigious Lloyd's name attached to the endeavor, the British now seemed poised to dominate the postwar aviation insurance market. Unwilling to simply surrender this new field to foreign competition, US insurance firms nevertheless remained legally restrained from underwriting aviation under current state regulations. With a large number of the nation's major firms centered in New York City, US insurers turned their attention to Albany. On May 5, 1919, Governor Al Smith signed three bills into law that allowed marine, fire, and casualty companies registered in New York to cover aviation hazards, and other states soon followed.[29] With the legal ability to underwrite aviation secured, American insurance firms began to tentatively venture into this new field.

But the airplane provided unique challenges for the underwriter. As a device constantly fighting against the pull of gravity, it represented a near constant threat to those aboard and below while in flight, and "every landing of an aircraft, whether forced or intended," constituted "a collision."[30] In addition to collision coverage, a fully insured aircraft needed fire, public liability, property damage, personal injury, workmen's compensation, and death and dismemberment coverage.[31] Properly determining risk and therefore premium costs for each line required reliable statistical data, but—with civil aviation effectively banned until July 1919 due to a presidential order— no central authority existed to collect such material. Without accurate data on civilian aviation, providing affordable coverage seemed a daunting task.

Convinced that insurance companies had a "duty" to provide coverage,

Travelers Insurance Company vice president Walter G. Cowles, "a pioneer in the development of liability and compensation insurance," turned his attention to aviation in the spring of 1919.[32] To make sense of a completely unknown field, Travelers' actuaries sought to obtain the Aero Club of America's "considerable amount of [prewar] data," but they were unceremoniously "dismissed by [ACA governor] Mr. Woodhouse" when they attempted to do so.[33] Denied whatever information the ACA possessed, Cowles was forced to rely on Air Service data that did not accurately reflect commercial conditions.[34] Cowles nonetheless persisted, and Travelers began "a comprehensive insurance program" in May 1919 based on the assumption that rates of death and "permanent total disability" would be much higher "than similar cases in other lines."[35] As the year came to a close, Cowles was forced to admit that the field remained "largely in the realm of conjecture" and wholly detached from "fundamental issuance principles."[36]

This assumption of high loss rates forced Travelers and those insurance companies that followed them into the field to set exorbitantly high premiums. In a witty quip not too far from the truth, one insurance industry periodical lamented that "by the time the owner has taken out . . . coverage he discovers that the cost of the insurance rivals the cost of the plane."[37] And the situation only grew worse as the years passed without a federal law. Legally mandated workmen's compensation insurance in the state of New York alone cost the Fairchild Aerial Camera Corporation $584 a year for *each* employee that *might* take to the air, regardless of that individual's actual flight time.[38] Such high rates not only threatened the solvency of legitimate firms but also meant that only "the most dangerous risks"—in other words, those most likely to result in a total loss—sought coverage.[39] The inevitable costly claims from these high-risk clients forced premiums to remain prohibitively high and further disincentivized others from purchasing what was, in the absence of any legal requirement, an optional policy. The optimal solution would be for a government agency to make reliable data available to underwriters much like the Department of Commerce's Steamboat Inspection Service and Bureau of Navigation did for water transportation. Without such an authority, the insurance industry needed to collect its own data.

Edmund Ely, manager of Aetna Insurance Company's recently created Aircraft Department, worked to address the situation as president of the National Aircraft Underwriters Association (NAUA). Established in March

1920 and composed of the nation's major insurance companies with the notable exception of Travelers, the NAUA served as "a medium of exchange of information" among its members and a forum through which they could standardize "suitable policy forms and endorsements."[40] Ely hoped that the collection and exchange of uniform statistics would allow NAUA members to transform aviation insurance from a haphazard collection of "judgement rates" to a field based "squarely on experience."[41] But he also understood that statistical data alone would not address the problems in aviation insurance. His work over the past seven years as head of Aetna's automotive department had occurred within a system of state-based automobile regulations that had provided insurance companies with a basis for understanding risk and assessing fault. Unlike British underwriters who benefited from government regulations that established a set of operational standards for aviation aligned with the 1919 convention, Ely and his associates were on their own.[42] With no federal legislation and both the MAA and the Air Service actively discouraging the passage of state aviation laws, anyone anywhere could ascend in any machine in any condition at any time. For aviation insurance to provide the benefits that Bradley and others hoped, underwriters would need a set of standardized procedures to determine a perspective policyholder's level of risk and fault after a claim. To fully meet the NAUA's commitment to "devise, advocate and promote all reasonable and proper means of eliminating or reducing . . . hazards" in the absence of governmental regulations, the insurance industry would need to establish its own operational requirements.[43]

To do so, Ely turned to the UL. Initially established in 1893 by William H. Merrill to research fire prevention for the National Board of Fire Underwriters, the nonprofit had since morphed into a materials testing and standardization clearinghouse.[44] Merrill assigned future UL president A. R. Small to work with Ely on standards for aircraft construction, safe operation, and safety devices. With limited experience in the aeronautics field, Small contacted the MAA's Samuel Bradley for assistance in drafting "a practical outline," but persistent scheduling conflicts prevented an official meeting of the MAA's new Special Insurance Committee.[45] Drawing on abstracts of the 1919 convention printed in the MAA's 1920 *Yearbook*, Small developed a plan in consultation with Bradley and Frank Russell of the MAA's Special Insurance Committee and aeronautical engineer Alexander Klemin.[46] Under the guid-

After an inspection by Hubbard's friend and Boeing test pilot Hunter A. Munter, the Boeing B-1 used on the Seattle-to-Victoria air mail route received this UL-issued certificate of airworthiness and new N-dash number N-ABNA, which no longer included the "C" designating it as Canadian-issued. Note the "copy sent to Canada" scrawled on the top left. Because Alaska Airways had won FAM-2 for fiscal year 1923–1924 and contracted operations to Hubbard, the certificate was not issued to Hubbard directly. Hubbard successfully regained the contract the next year. Unprocessed Material, UL Archives.

ance of Bradley and Russell, Small created a private regulatory system for the insurance industry based on the postwar air convention that was fully compatible with the Canadian Air Regulations of 1920.[47]

The UL system consisted of four primary elements. The aircraft and pilots registers, "based upon the provisions of the Convention" and "the Air Regulations of the Canadian Air Board," opened on July 1, 1921.[48] After submission of the required information and fee, an aircraft received an N-dash registration number as stipulated in the convention, and the Canadian Air Board readily recognized these UL registrations.[49] (Small initially toyed with the idea of replacing the "N" with an "A" for America, but Bradley cautioned him on the need to conform with the established international system.)[50] To secure a place on the pilot's register, one needed to provide the necessary information and the results of a medical examination based on the requirements of the convention.[51] For the third element of the UL's system, Small compiled thirty-two rules of the air drawn directly from the convention and the Canadian Air Regulations of 1920. These operational guidelines not only allowed insurance companies to assess fault but also ensured that UL-registered pilots knew and adhered to standard international procedures. But registration and pilot's licensing did not guarantee the structural integrity of an aircraft, and Schroeder's airworthiness inspection program constituted the final element of the UL system.

Although its architects had high hopes, the voluntary nature of the UL system proved its undoing. By the beginning of 1923, only thirty-four aircraft (including Hubbard's Boeing B-1) and roughly the same number of pilots had registered, while a mere six aircraft had received airworthiness certificates.[52] Without a requirement for compliance, little incentive existed to participate in the UL system beyond the ability to purchase expensive and completely optional insurance from NAUA member companies. And because most insurance policies explicitly forbade trick flying, barnstormers and those who made their living through aerial entertainment—a large percentage of aviators in post–World War I America—were completely excluded from the system.[53] Even legitimate commercial operations with impressive safety records, such as Aeromarine Airways, the first regularly scheduled postwar airline under the leadership of MAA member Inglis M. Uppercu, struggled under exorbitant premiums.[54] Although the company "carried 30,000 passengers and flew more than 1,000,000 passenger-miles with only one serious accident in three and a half years," insurance coverage accounted for roughly 17.25 percent of total expenses.[55] Such extensive outlays for insurance limited Aeromarine's ability to accumulate the fiscal reserves necessary to replace two flying boats in 1923, a loss that forced the company to cease commercial operations the next year.[56] Even as they worked to create and implement the UL's system, both Ely and Small recognized that their efforts were only a stopgap measure.[57] As Ely himself admitted, aviation insurance required a federal law "setting forth the rights, duties, and limitation of owners and operators" as well as regulations that "shall apply uniformly within federal jurisdiction" to thrive.[58] The limitations of the UL's effort to regulate aviation reinforced the belief among Small, Ely, Bradley, and others that only federal regulations compatible with the convention could fully address the needs of civil aviation.

Table 5.1. Comparison of Selected Rules of the Air among the 1919 Convention, the 1920 Canadian Air Regulations, and the UL System

Rules of the Air (Selected)	1919 Convention, Annex D	Canadian Air Regulations of 1920, Part 7	Underwriters Laboratories, August 1920
Right of way	21. Flying machines shall always give way to balloons, fixed or free, and to airships. Airships shall always give way to balloons, whether fixed or free.	63. Flying machines shall always give way to balloons, fixed or free, and to airships. Airships shall always give way to balloons, whether fixed or free.	7. Airplanes shall always give way to balloons and airships, whether fixed or free. Airships shall always give way to balloons.
Risk of collision	23. Risk of collision can, when circumstances permit, be ascertained by carefully watching the compass bearing and angle of elevation of an approaching aircraft. If neither the bearing nor the angle of elevation appreciably change, such risk shall be deemed to exist.	65. Risk of collision can, when circumstances permit, be ascertained by carefully watching the compass bearing and angle of elevation of an approaching aircraft. If neither the bearing nor the angle of elevation appreciably change, such risk shall be deemed to exist.	6. Risk of collision can, when circumstances permit, be ascertained by carefully watching the compass bearing and angle of elevation of an approaching aircraft. If neither the compass bearing nor the angle of elevation appreciably change, such risk shall be deemed to exist.
Safe distance	25. A motor-driven aircraft must always maneuver according to the rules contained in the following paragraphs as soon as it is apparent that, if it pursued its course, it would pass at a distance of less than 200 metres from any part of another aircraft.	67. A motor driven aircraft must always manoeuvre according to the rules contained in the following paragraphs as soon as it is apparent that, if it pursued its course, it would pass at a distance of less than 200 yards from any part of another aircraft.	9. A motor-driven aircraft must always maneuver, according to these rules, as soon as it is apparent that, if it adhered to its course, it would pass at a distance of less than 600 ft. from any part of any other aircraft.

Aircraft whose courses cross	27. When two motor-driven aircraft are on courses which cross, the aircraft which has the other on its own right side shall keep out of the way of the other.	69. When two motor driven aircraft are on courses which cross, the aircraft which has the other on its own right side shall keep out of the way of the other.	12. When two motor-driven aircraft are on courses which cross, the aircraft which has the other on its own right side shall keep out of the way of the other.
Flight over cities	No parallel	92. No aircraft shall fly over any city or town except at such an altitude that will enable the aircraft to alight outside the city or town should the means of propulsion fail through mechanical breakdown or other cause.	29. No aircraft shall fly over any city or town below such heights as make it impossible or difficult to alight by gliding outside the city or town.
Disclaimer	50. Nothing in these rules shall exonerate any aircraft, or the owner, pilot or crew thereof, from the consequences of any neglect to carry lights or signals, or of any neglect to keep a proper lookout, or of the neglect of any precaution which may be required by the ordinary practice of the air, or by the special circumstances of the case.	No parallel	2. Nothing in these rules shall interfere with the operation of any special rule or rules duly made and published relative to navigation of aircraft in the immediate vicinity of any aerodrome or other place, and it shall be obligatory on all owners, pilots, or crews of aircraft to obey such rules.

As months turned into years without a federal law, the UL's voluntary system collapsed and the bottom fell out of the nascent aircraft insurance market. According to the Aeronautical Chamber of Commerce of America, Aetna, the National Liberty Insurance Company, the Home Insurance Company, the Automobile Insurance Company, Fireman's Fund, Globe and Rutgers, and Travelers underwrote "95% of the total number of insured planes in the United States" in 1921.[59] Revenue figures for 1921 in Massachusetts offer a glimpse of the difficulty in underwriting aviation without government regulation. When it came to automobiles, these companies paid out 76.95 percent of premium income in claims, leaving a profit of 23.05 percent. But total aviation claim payouts came to 299.40 percent of premium income, making aviation underwriting an expensive line of business rather than a profitable enterprise.[60] During the first half of 1921 in Massachusetts, the National Liberty Insurance Company paid out $12,747 on $1,015 in aircraft premiums alone, a loss of 1,255 percent.[61] Publicly citing the low volume of business, National Liberty abandoned aviation underwriting that July, and Aetna followed the next year.[62] Hartford Fire Insurance Company and Hartford Accident and Indemnity Company brought over Horatio Barber of the London-based Aviation Insurance Association to salvage the situation in 1922, but even his expertise could not prevent the company's steady disengagement from the field.[63] The remaining NAUA members ceased underwriting aircraft in 1924, leaving Barber to reinsure select Hartford policies through Lloyd's.[64] The UL issued its fifty-third and final Bulletin of Information in December 1923, and Schroeder left for Ford's Stout Metal Airplane Company in early 1925.[65] By the time Coolidge signed the Air Commerce Act in May 1926, the US aircraft insurance market had practically ceased to exist.

Throughout the life of the UL's roughly three-year program, forty-seven pilots and forty-five aircraft registered under the system and resident aircraft engineers issued ten airworthiness certificates.[66] When compared to the 1,000 to 1,200 estimated aircraft in operation in 1923, the UL's effort had little direct impact on the increasingly dangerous situation unfolding in the nation's skies.[67] But the influence of the UL's work goes beyond its immediate success or failure. As *Aviation* editor Lester Gardner recognized, the UL's pioneering efforts constituted the "first serious work . . . of subjecting American civil aircraft to the provisions of the International Air Convention, and thus bringing them within the purpose of the world's law code of the air."[68] By adopting the convention's provisions for aircraft registration, pilot's licensing, and rules of the air with the express purpose of facilitating US–Canadian flight, the UL system promoted the diffusion of international standard practices into the United States. But its influence paled in comparison to another private attempt to establish uniform operational and registration standards for aviation that also engaged the efforts of Small and Ely—the American Aeronautical Safety Code.

The American Aeronautical Safety Code

Henry Middlebrook Crane must have felt a sense of accomplishment as he addressed his fellow members of the Society of Automotive Engineers gathered for their annual aeronautic banquet. One of a new generation of professionally trained engineers, Crane had received his electric engineering degree from MIT in 1896 before pursuing experimental telephone work for both American Telephone and Telegraph and Western Electric. His interests soon shifted to the new technology of the automobile, and he became president and chief engineer of what became known as the Crane Motor Company. After a series of mergers, Crane—armed with the rights to domestically manufacture the Hispano-Suiza motor—applied his talents to aviation as vice president and chief engineer for George Houston's Wright-Martin Aircraft Corporation and became the postwar director of the Loening Aeronautical Engineering Corporation as well as president of the SAE in 1924.[69] As toastmaster for the 1925 annual banquet, Crane now presided over a head table that included former SAE president Howard Coffin, chief of Air Service Gen. Mason Patrick, NAA president Godfrey Cabot, Curtiss president Clement M. Keys, Cmdr. Holden C. Richardson of the Navy's Bureau of Aeronautics, and Orville Wright alongside representatives from France, Italy, and Canada.[70] For Crane, this year's festivities held special significance. For the past four years, he had overseen the creation of the American Aeronautical Safety Code, a document that those present believed would "serve as a basis" for federal regulations once Congress "placed" aviation "under the jurisdiction of the Department of Commerce."[71] With work on the code recently completed, Crane undoubtedly felt a sense of pride as he lifted his glass in celebration. Thanks to his guidance, American aviation could now develop under a standardized system specifically drafted to align with the Canadian Air Regulations and the International Air Convention.

A relatively new society, the SAE's history mirrored Crane's career. The establishment of the Society of Automobile Engineers in 1905 fit into a larger trend of engineering professionalization that stretched back decades. As the first professional societies emerged in the second half of the nineteenth century, newly organized engineers looked to the standardization of parts and processes as a means to reduce industrial waste and delineate an area of expertise for their developing fields.[72] Momentum for and public acceptance of professionally established standards only increased in the wake of Fredrick Winslow Taylor's *The Principles of Scientific Management*, as the technocratic promise of efficiency embedded within standardization transformed it from an obtuse industrial practice into a potent tool of Progressivism. Henry Ford, Howard Coffin, and like-minded members viewed standardization of the exciting and chaotic automotive field as a central element of the SAE's mission. After the development of the airplane, several members became interested in the new field of aviation, and the SAE fully welcomed its "aeronautical brethren into the fold" and rechristened itself the Society of Automotive Engineers in 1916 due in no small part to Coffin's efforts.[73] Industrial preparedness intensified the desire to apply Fordist mass production techniques to aircraft manufacture, and the SAE quickly established an Aeronautic Division within its Standards Committee. Within the first year of the division's existence, its members established standards for propeller-shaft ends and "formulated recommendations for thirty-two other standards and recommended practices."[74] This standardization work took on a decidedly international dimension once the United States official entered the war in April 1917. Under the chairmanship of the Curtiss Aeroplane and Motor Corporation's Charles M. Manly, Crane and other members of the Aeronautic Division worked closely with the Bureau of Standards to devise thirty-four standards for Howard Coffin's Aircraft Production Board that were "adopted by the International Aircraft Standards Board" for use by Allied governments as well as "the Aviation Section of the Signal Corp."[75] By war's end, the SAE's newly created Aeronautic Division had become an important body for the development of both domestic and international aircraft standards.

A desire to continue such beneficial standardization work after the war prompted the creation of a permanent international body. Established three weeks prior to the armistice with Charles le Maistre, the "father of international standardization," as secretary, this new International Aircraft Standards Commission (IASC) would "study and recommend . . . industrial and technical standards," revise them as appropriate, and "secure approval" of recommended standards by their respective governments through officially recognized "National Aircraft Committees."[76] To participate in the IASC, the United States needed an officially recognized national standards body. Fortunately, two days after the IASC's creation five major professional societies, in conjunction with the Bureau of Standards and the War and Navy Departments, formed the American Engineering Standards Committee (AESC) after nearly two years of organizational activity.[77] According to its operating rules, the AESC did not directly establish standards but rather designated organizations as a standard's sponsor. This sponsor would then form a Sectional Committee composed of representatives from all interested parties to formulate a standard that, upon AESC approval, would become the accepted standard.[78] Over the next decade, the AESC established more than a hundred national safety standards in a variety of fields under the management of former Bureau of Standards researcher Paul G. Agnew, who, like Coffin, firmly believed in the socially transformative power of standardization.[79] But the AESC could not represent the United States on the IASC. Despite the lobbying efforts of the MAA's Bradley, Agnew, and the SAE, the AESC never received official government sanction and therefore could not constitute the National Aircraft Committee required for IASC participation.[80] As with the Versailles Treaty and the International Air Convention, the United States remained formally detached from this new international standards body. While the engineering profession's postwar aeronautical standardization efforts would not occur under the aegis of the IASC, wartime experiences nonetheless instilled an internationalist worldview among standardization advocates in the AESC, the NACA, the SAE, the Bureau of Standards, and industry.[81]

Despite a lack of official connection to the IASC, the Bureau of Standards under the direction of the NACA's Samuel W. Stratton initiated a formal process of aeronautical standardization under the AESC in early 1920.[82] First Lyman J. Briggs—a Johns Hopkins University–trained physicist who would lead the National Bureau of Standards from 1933 to 1945—obtained the support of other government agencies and industry as a member of the recently established Sub-Committee on Commercial Aviation of the Economic Liaison Committee.[83] Created to "formulate a program for promoting commercial aviation through public and private initiative" in January 1920, the subcommittee's members immediately agreed that a lack of uniform national regulation constituted the greatest barrier to the development of commercial aviation, supported the passage of the Interdepartmental Board's recently drafted bill (discussed in chapter 3), and declared that the Bureau of Standards should "proceed with the codification of flying rules, aircraft construction rules, and aircraft equipment."[84] Armed with the subcommittee's blessing, overlap in personnel between the Bureau of Standards and the AESC paved the way for the code's creation. Chief of the Bureau of Standards' Electrical Division and head of the AESC's National Safety Code Committee Edward B. Rosa proposed bringing the SAE in as a cosponsor.[85] Morton G. Lloyd, chief of the Bureau of Standards' Safety Section and its representative on the AESC's National Safety Code Committee, tasked his associate engineer Arthur Halstead with creating an initial "draft of a safety code."[86] Months before Harding sailed to victory in the election of 1920, the Bureau of Standards had committed itself to the creation of a system of aeronautical standards that included both technical elements and operational procedures. But would the SAE, the Bureau of Standards' wartime partner, go along with such an expanded vision of standardization?

Fortunately, leadership of the Aeronautic Division of the SAE's Standards Committee had already come to believe that a broad approach to standardization offered the best way to prevent the loss of life and property from unregulated aircraft and pilots. Like Coffin and Agnew, Aeronautic Division chair Charles Manly and his successor Henry Crane believed standardization not only provided a service to industry but also offered a means to improve society. In the wake of the *Wingfoot* incident in Chicago, it was beholden on engineers to use their knowledge and expertise to rationalize and contain the growing aerial chaos in the name of the public good. Together, the two worked to shift the postwar Aeronautic Division away from its myopic wartime focus on "the details of construction" to "important matters which really affect the international operation of aircraft and airships," such as "the legal regulation of the manufacture and use of airplanes."[87] Although some of the SAE's members no doubt found the nontechnical aspects of the proposed code "a little out of the line of the regular standards work," the Aeronautic Division under Crane's leadership supported the SAE's active participation in the development of "a system of national registration" and called on the society to "back Federal legislation with all its strength as against State legislation."[88] In the meantime, Crane and his fellows in the Aeronautic Division argued that the SAE had a duty to apply its wealth of expertise to the airplane problem.[89] The SAE's Standards Committee concurred, and

Table 5.2. Member Bodies and Designated Representatives of the American Aeronautical Safety Code Sectional Committee as of August 18, 1921

Organizations Represented on the Sectional Committee of the American Aeronautical Safety Code	Sectional Committee Representatives
Bureau of Standards	Lyman Briggs, H. C. Dickinson, Morton Lloyd, Arthur Halstead
Society of Automotive Engineers	Virginius Clark, Henry Crane, Grover Loening, Inglis Uppercu, Ralph Upson
American Society of Mechanical Engineers	Edward Warner (MIT)
Navy Department	Patrick N. L. Bellinger, Bruce Leighton
Aero Club of America	Maurice Cleary, Caleb Bragg
Weather Bureau	W. R. Gregg
Manufacturers Aircraft Association	Albert Flint, Samuel Bradley
American Society of Safety Engineers	H. M. Hine (Travelers Insurance Co.)
Post Office Department	C. F. Egge
Underwriters Laboratories	A. R. Small
National Aircraft Underwriters Association	Edmund Ely
Forest Service	T. R. C. Wilson
NACA	Joseph Ames, George Lewis
Coast Guard	C. E. Sugden
National Safety Council	G. E. Patterson (Travelers Insurance Co.)
War Department	Walter R. Weaver
Rubber Association of America	A. L. Viles
American Society for Testing Materials	—

Source: Personnel, AESC Sectional Committee on Safety Code for Airplanes, folder 10, box 222, MAA Papers, AHC.

Note: The American Society for Testing Materials had yet to name a representative. Bernard Mulvihill and Harold E. Hartney of the National Aeronautic Association, John V. L. Hogan of the American Institute of Electrical Engineers, George Mead of the SAE, and N. M. Du Chemin of the American Society of Safety Engineers were added to the Sectional Committee in 1922.

the SAE agreed to officially cosponsor the AESC's proposed aeronautical code alongside the Bureau of Standards.

In keeping with the AESC's practice of consulting relevant stakeholders, the Bureau of Standards and the SAE called a preliminary conference on May 13, 1921. Gathered in the NACA's conference room, representatives from the

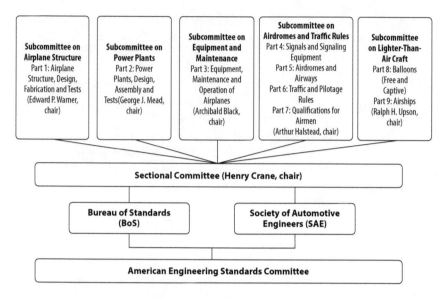

Organizational chart of the American Aeronautical Safety Code. *American Aeronautical Safety Code Approved by the American Engineering Standards Committee Oct., 1925* (NY: Society of Automotive Engineers, 1925).

interested civilian and military organizations agreed to create a permanent Sectional Committee, divide the code among subcommittees, and use Halstead's preliminary draft as their starting point.[90] Ever cognizant of the international aspects of aviation, Coffin stressed that the code should adopt "best known international current practices" to ensure "conformity with international engineering specifications and standards."[91]

At the first official meeting of the Sectional Committee in early September, Agnew impressed upon its members that their work "will without doubt have considerable influence on [the] Federal . . . regulation of aircraft," and all present agreed to send "tentative drafts" of their work to the IASC and the Canadian Engineering Standards Committee "for criticism" to ensure compatibility with international practices.[92] With the unratified convention as a guide and federal legislation the goal, Sectional Committee members clearly viewed the American Aeronautical Safety Code as the connection between the two. As head of the SAE's Aeronautic Division, Crane was the logical choice to chair the Sectional Committee, and the NACA's Joseph Ames joined him as vice-chair while Morton Lloyd and Arthur Halstead of the Bureau of Standards served as secretary and assistant secretary, respec-

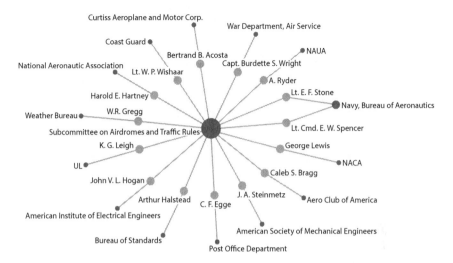

Members of the Subcommittee on Airdromes and Traffic Rules, ca. 1922. Note the wide range of organizations represented, a hallmark of the AESC's associationalist process of standard-setting. Created with Palladio, http://hdlab .stanford.edu/palladio/.

tively. With its leadership selected, the Sectional Committee now turned its attention to the monumental task of drafting technical, operational, and registrational standards for aviation.

The Sectional Committee met a total of seven times between 1921 and 1925 while the yeoman's work occurred among five subcommittees. Operational provisions such as lights, airways, licensing, and rules of the air fell under the purview of the Sub-committee on Airdromes and Traffic Rules under Halstead's chairmanship. A former Air Service pilot who would go on to assist MacCracken in the creation of air routes after the passage of the Air Commerce Act, Halstead recognized the need to align the code's operational and registrational provisions with those of the convention and the Canadian Air Regulations to ensure safety in international flight.[93] When drafting the Bureau of Standards' preliminary code at Morton's request, Halstead had used "what he thought best from the rules and regulations of the Canadian Air Board and of the British Air Ministry and from the French regulations."[94] Now, he stressed to his fellow subcommittee members the need to "keep very close to international convention."[95] Under Halstead's guidance, 94.4 percent of the 161 provisions decided on by the end of the subcommittee's first year either directly corresponded with or did not contradict those

Table 5.3. Comparison of Selected Rules of the Air among the 1919 Convention, the 1920 Canadian Air Regulations, and the AASC

Rules of the Air (Selected)	1919 Convention, Annex D	Canadian Air Regulations of 1920, Part 7	American Aeronautical Safety Code, Part 6
Right of way	21. Flying machines shall always give way to balloons, fixed or free, and to airships. Airships shall always give way to balloons, whether fixed or free.	63. Flying machines shall always give way to balloons, fixed or free, and to airships. Airships shall always give way to balloons, whether fixed or free.	601. Power-driven aircraft shall always yield the right of way to balloons, captive or free, and airplanes shall yield to airships.
Risk of collision	23. Risk of collision can, when circumstances permit, be ascertained by carefully watching the compass bearing and angle of elevation of an approaching aircraft. If neither the bearing nor the angle of elevation appreciably change, such risk shall be deemed to exist.	65. Risk of collision can, when circumstances permit, be ascertained by carefully watching the compass bearing and angle of elevation of an approaching aircraft. If neither the bearing nor the angle of elevation appreciably change, such risk shall be deemed to exist.	600f. Risk of collision can, when circumstances permit, be ascertained by watching carefully the relative bearing, the compass bearing and angle of elevation of an approaching aircraft. If the bearing does not appreciably change, a risk of collision should be deemed to exist.
Safe distance	25. A motor-driven aircraft must always maneuver according to the rules contained in the following paragraphs as soon as it is apparent that, if it pursued its course, it would pass at a distance of less than 200 metres from any part of another aircraft.	67. A motor driven aircraft must always manoeuvre according to the rules contained in the following paragraphs as soon as it is apparent that, if it pursued its course, it would pass at a distance of less than 200 yards from any part of another aircraft.	605a. Except when operating in a predetermined formation, an aircraft is considered to be in dangerous proximity to another aircraft when it approaches in flight within 1500 ft.

Aircraft whose courses cross	27. When two motor-driven aircraft are on courses which cross, the aircraft which has the other on its own right side shall keep out of the way of the other.	69. When two motor driven aircraft are on courses which cross, the aircraft which has the other on its own right side shall keep out of the way of the other.	605e. When two airships or two airplanes are crossing so as to involve risk of collision, the aircraft on the right has right of way. The aircraft that has the other on her own right side within a horizontal angle of 110 deg. from dead ahead shall give way to and keep clear of the other.
Flight over cities	No parallel	92. No aircraft shall fly over any city or town except at such an altitude that will enable the aircraft to alight outside the city or town should the means of propulsion fail through mechanical breakdown or other cause.	612. An aircraft shall not fly over closely populated areas except at such altitude or following such a course as will enable the aircraft to land outside the area or upon areas that do not involve danger to the public below, in the event that the means of propulsion or sustentation fail through mechanical breakdown or other cause; provided, that this prohibition shall not apply to any area comprised within a distance of 1 mile from the boundary of an airdrome.
Disclaimer	50. Nothing in these rules shall exonerate any aircraft, or the owner, pilot or crew thereof, from the consequences of any neglect to carry lights or signals, or of any neglect to keep a proper lookout, or of the neglect of any precaution which may be required by the ordinary practice of the air, or by the special circumstances of the case.	No parallel	600a. Nothing in these rules shall relieve any aircraft, or the owner or crew thereof, from the responsibility and consequence of any neglect to carry lights or signals, or of any neglect to keep a proper lookout, or of the neglect of any precaution that may be required by the ordinary practice of the air or by the special circumstances of the case.

of the convention.[96] Halstead's efforts to ensure compatibility with international practice within his own subcommittee accentuated the need to do so throughout the entire AASC, and Crane secured a resolution from the Sectional Committee that required *all* subcommittees to justify *any* deviations from the convention's principles and provisions.[97] According to Canadian officials, these concerted efforts to ensure operational compatibility produced prodigious results. After attending a meeting of Halstead's subcommittee in January 1923, British air attaché Wing Commander Malcolm G. Christie believed that the "projected regulations for aero navigation for [the] U.S.A. are becoming almost identical in both form and principle with the International Convention and its annexes."[98] By the Sectional Committee's final meeting on April 23, 1925, Controller of Civil Aviation for Canada John Wilson could confidently assert that—while it incorporated additional provisions—the AASC contained "few, if any, divergences in practice from the International Convention and the Canadian Air Regulations."[99]

The sections of the AASC that dealt with aircraft inspection, pilot qualifications, and rules of the air closely aligned with, and in many cases outright copied, the 1919 convention and the Canadian Air Regulations.[100] A few examples clearly demonstrate the heavy reliance on these two earlier documents. For instance, the code incorporated the convention's stipulations for lights—white at the front and rear with green on the right side and red on the left—and explicitly mentioned that a ten candlepower front light would "comply with the requirements of the International Convention." The AASC also included the convention's requirements for technical and practical pilot tests nearly verbatim, and its rules of the air complemented both the 1919 convention and the Canadian regulations. Archibald Black and the UL's Rudolph Schroeder of the Subcommittee on Equipment and Maintenance ensured that the AASC recognized "N" as the US nationality mark and stipulated that "identification and records for airplanes shall be in accordance with the current revision of the International Air Convention."[101] With a suitable final draft in hand, Crane shepherded the AASC through the approval process. Twenty-nine of the Sectional Committee's thirty-three members voted in favor with the other four abstaining, and by the end of July 1925, the Bureau of Standards and the SAE had both approved the code.[102] The AESC declared the AASC the Tentative American Standard, and Crane celebrated the successful conclusion of this monumental effort at the Aeronautic Division's banquet in the ballroom of New York City's resplendent Hotel Astor that October.

Approved as a tentative standard just as the Morrow Board heard public testimony on the nation's aviation situation, the AASC never entered into independent operation. But the code's importance lies in what it tells us about the regulatory values of those who created it rather than its ultimate utility. With the AASC, representatives from the engineering profession, civil and military government agencies, and private interests publicly affirmed their belief that US procedures should conform as much as possible to the 1919 convention's provisions. Due to the leadership of Crane and Halstead, the code became a primary means by which key constituencies within the United States came to accept international aeronautical practices. But as voluntary programs, neither the UL system nor the AASC could compel the compliance necessary to guarantee aerial safety in the United States. That required the coercive power of the state.

The Air Commerce Regulations of 1926

Upon becoming the nation's first assistant secretary of commerce for aeronautics in August 1926, MacCracken immediately turned his attention to the creation of official regulations, and the AASC provided a logical starting point for his work. By mid-September, MacCracken, Ira Grimshaw of the Commerce Department solicitor's office, and Alexander Klemin—someone intimately familiar with the convention from both his work on the UL system and his efforts on the Coffin-financed Joint Committee of the Department of Commerce and the American Engineering Council alongside Mac-Cracken—had produced an initial draft of government regulations based on "the Aeronautical Safety code, the rules and regulations in the International Convention for Air Navigation, and the rules and regulations of the Canadian Air Board."[103] Despite additions, omissions, and language variances, such as an additional Mail (M) classification for aircraft, these draft regulations offered sufficient complementarity with the convention's operational and registrational practices.[104] Aircraft engaged in international flight needed to display the US nationality letter "N" as stipulated in the convention in addition to their Commerce Department–awarded registration number. Requirements for color, location, and size of aircraft markings fully adhered to the convention and Canadian regulations. Other than some differences in distance and altitude requirements, practical tests for a commercial pilot's license and rules of the air effectively mirrored those of the convention and the AASC.[105] Already acutely aware of the international situation through his work on the ill-fated Winslow bill, MacCracken ensured that his draft regulations would allow US aircraft to safety engage in cross-border flight.[106]

The various constituencies that had created the American Aeronautical Safety Code no doubt supported MacCracken's efforts to align US operational and registrational provisions with those of the convention, but his reliance on the code's technical aspects caused concern among aircraft manufacturers. Although Winslow's House Commerce Committee had "presumed" that the recently completed code, "together with the corresponding regulations in the annexes of the International Air Convention, would serve as a basis for the formulation of . . . regulations," representatives from industry had advised the Commerce Department against incorporating the AASC's "long, complicated, and wholly theoretical" technical provisions into either the Air Commerce Act or subsequent regulations.[107] Warnings from MAA members that the AASC's "high sounding semi-official rules" for aircraft, engines, and safety apparatuses went too far had forced general manager Samuel Bradley to abstain from the Sectional Committee's final approval vote.[108] Manufacturers took no issue with *suggested* standards, but they remained opposed to the adoption of rigid and legally enforceable technical requirements they believed would stifle innovation and increase costs.[109] Lester Gardner, editor of *Aviation*, implored the aircraft industry to push back against MacCracken's "hastily prepared . . . wholly unnecessary and restrictive" regulations "based on the preliminary work of the Safety Code."[110] At a special committee meeting of the Aeronautical Chamber of Commerce of America, industry representatives readily accepted the inclusion of rules of the air and registrational provisions compatible with the International Convention in MacCracken's draft regulations but felt that the criteria for commercial pilot testing, airfield certificates, and certificate renewal remained "too severe"; complained that requirements for safety equipment and instruments were "too detailed"; and called for the replacement of complicated airworthiness tests with a more general requirement for observed "flight and landing tests."[111] Long desirous of federal regulation to assuage the public's concern over aviation's safety, members of the ACCA remained determined to push back against perceived overregulation.

In keeping with Hoover's preferred method of regulation through indus-

try consultation, MacCracken held a series of conferences with aircraft manufacturers, engine manufacturers, air transport operators, pilots, the insurance industry, and aviation media throughout October to elicit feedback on
his draft regulations.[112] There, aircraft manufactures successfully convinced
the new assistant secretary to accept most of their recommended revisions,
but he refused to accept any major changes to the list of required safety devices due to Hoover's widely publicized emphasis on safety.[113] The revised
regulations arising out of these conferences would become the nation's first
legally enforceable regulations. Through its stipulations for aircraft airworthiness, pilot and mechanic licenses, aircraft markings, inspections, and rules
of the air, the "Air Commerce Regulations, effective December 31, 1926," provided not only a domestic framework for commercial aviation but also a level
of operational complementarity that facilitated international flight with
Canada and other air convention members.[114] As the following table shows,
the Air Commerce Regulations retained the core requirements necessary to
ensure safety in international operation, although in some instances they
deviated considerably (safe distance) or incorporated additional stipulations
beyond those of even the AASC (flights over congested areas). A Department of Commerce analysis pointed out that the International Commission
on Air Navigation's airworthiness requirements proved "more stringent"
and those proposed by the Commerce Department provided more "definite
instruction," but neither document directly conflicted with the other, even
after industry's desired modifications.[115]

Curiously, the Air Commerce Regulations of 1926 did not include the
convention's required "N" designation present in MacCracken's initial draft
regulations. Considering its inclusion in both the UL system and the AASC,
the requirements of the Canadian courtesy, and industry's long-held interest in facilitating international operation, this exclusion remains puzzling.
Operational necessity would nevertheless prompt its reinsertion in short
order. Less than four months after the Air Commerce Regulations took effect, the Commerce Department required "the letter 'N' on all aircraft engaged in international air commerce."[116] When Charles Lindbergh departed
on his historic transatlantic solo flight on May 20, 1927, his aircraft, *Spirit of
St. Louis*, carried a Commerce Department–issued, convention-compatible
registration number: N-X-211.

The UL system, the American Aeronautical Safety Code, and the first
Air Commerce Regulations all show that key aeronautical constituencies

Table 5.4. Comparison of Selected Rules of the Air among the 1919 Convention, the American Aeronautical Safety Code, and the Air Commerce Regulations Effective December 31, 1926

Rules of the Air (Selected)	1919 Convention, Annex D	American Aeronautical Safety Code, Part 6	Air Commerce Regulations effective Dec. 31, 1926, Chapter 5
Right of way	21. Flying machines shall always give way to balloons, fixed or free, and to airships. Airships shall always give way to balloons, whether fixed or free.	601. Power-driven aircraft shall always yield the right of way to balloons, captive or free, and airplanes shall yield to airships.	(B) Giving-way order.—Craft shall give way to each other in the following order: 1. Airplanes. 2. Airships. 3. Balloons, fixed or free.
Safe distance	25. A motor-driven aircraft must always maneuver according to the rules contained in the following paragraphs as soon as it is apparent that, if it pursued its course, it would pass at a distance of less than 200 metres from any part of another aircraft.	605a. Except when operating in a pre-determined formation, an aircraft is considered to be in dangerous proximity to another aircraft when it approaches in flight within 1500 ft.	(C) No engine-driven craft may pursue its course if it would come within 300 feet of another craft . . . other than military aircraft of the United States engaged in military maneuvers and commercial aircraft engaged in local industrial operations.
Meeting head-on	26. When two motor-driven aircraft are meeting end on, or nearly end on, each shall alter its course to the right.	605c. When two airplanes or two airships are meeting head on or approximately head on, so as to involve risk of collision, each shall alter its course to the right to pass.	(E) When two engine-driven aircraft are approaching head-on, or approximately so, and there is risk of collision, each shall alter its course to the right, so that each may pass on the left side of the other.

Aircraft whose courses cross	27. When two motor-driven aircraft are on courses which cross, the aircraft on its own right side shall keep out of the way of the other.	605e. When two airships or two airplanes are crossing so as to involve risk of collision, the aircraft on the right has right of way. The aircraft that has the other on her own right side within a horizontal angle of 110 deg. from dead ahead shall give way to and keep clear of the other.	(D) Crossing.—When two engine-driven aircraft are on crossing courses the aircraft which has the other on its right side shall keep out of the way.
Flight over cities	No parallel	612. An aircraft shall not fly over closely populated areas except at such altitude or following such a course as will enable the aircraft to land outside the area or upon areas that do not involve danger to the public below, in the event that the means of propulsion or sustentation fail through mechanical breakdown or other cause; provided, that this prohibition shall not apply to any area comprised within a distance of 1 mile from the boundary of an airdrome.	(G) Height over congested and other areas.—Exclusive of taking-off from and landing, and except as otherwise permitted by section 86, aircraft shall not be flown— (1) Over the congested parts of cities, towns, or settlements except at a height sufficient to permit of a reasonably safe emergency landing, which in no case shall be less than 1,000 feet. (2) Elsewhere at height less than 500 feet, except where indispensable to an industrial flying operation. (H) Height over assembly of persons.—No flight under 1,000 feet in height shall be made over any open-air assembly of persons, except with the consent of the Secretary of Commerce. Such consent will be granted only for limited operations.

In this photo from late 1927, Assistant Secretary of Commerce for Aeronautics William P. MacCracken Jr. hands Santa Claus a pilot's license while Director of Aeronautics Clarence Young looks on. It is unclear whether Santa's sleigh also displayed an "N" designation number. "Santa Claus Receives Aeroplane Pilot's License from Assistant Secretary of Commerce," Library of Congress, https://lccn.loc.gov/2016888549, accessed 28 Sept. 28, 2018.

recognized the need to align any domestic regulatory system with the convention's internationally recognized operational and registrational provisions, even in the absence of official ratification. And the United States was not the only nation compelled to do so. By 1929, International Commission for Air Navigation Secretary General Albert Roper could safely claim that the convention's practices "were applied in the whole world by all the States, signatories and non-signatories."[117] With the passage of the Air Commerce Act and subsequent Air Commerce Regulations, the United States now possessed an enforceable means to foster commercial aviation compatible with the convention and the Canadian Air Regulations of 1920. But US aircraft also needed permission from foreign governments to enter their sovereign airspace. The 1919 convention offered a convenient way to obtain overflight rights from its various member states, but discrepancies between it and US hemispheric policy would prompt policymakers in Washington to pursue a parallel international agreement more aligned with the nation's security concerns and economic interests.

Shattered Expectations

An Air Convention for the Western Hemisphere

By the time he landed in Havana on February 8, 1928, Charles A. Lindbergh, the "Lone Eagle," was arguably the most famous American alive. Nine months earlier, he had completed the first successful solo transatlantic flight in his specially designed aircraft, *Spirit of St. Louis*, and the public response to this feat of rugged individualism had been overwhelming. Mobbed by tens of thousands of people upon his arrival outside Paris, Lindbergh went on to meet with leaders of France, Belgium, and Great Britain before returning home to a hero's welcome. In an era of newsreels, vivid newspaper photographs, and sensational tabloids, the soft-spoken and reserved Lindbergh—portrayed in the press as the personification of "middlebrow WASP values" and the embodiment of western technological proficiency—became the first true international celebrity.[1] With his newfound fame, Lindbergh instantly morphed into the nation's most prominent aeronautical spokesperson and aerial ambassador. After a Commerce Department–sponsored tour of the United States to promote commercial aviation from July to October, he began a two-month goodwill tour through Central America and the Caribbean in coordination with the State Department.[2] Lindbergh's arrival in Havana, his final destination before returning to the United States, played out in much the same way as his previous stops.[3] Cavalry reinforcements worked to contain the estimated crowd of 100,000 people awaiting him at Columbia Field, and well-wishers lined the streets as a procession of what seemed like "all the automobiles in Havana" followed him to the presidential palace.[4] Upon his arrival, President Gerardo Machado officially greeted Lindbergh as "an ambassador of the air, an envoy of democracy," and offered his

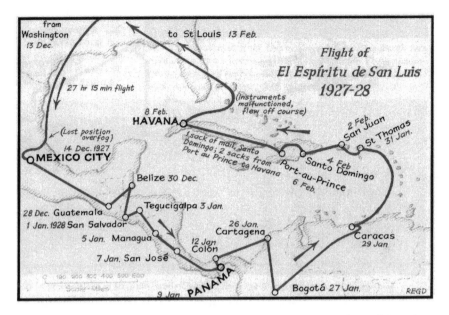

from Washington 13 Dec.

to St Louis 13 Feb.

Flight of
El Espíritu de San Luis
1927-28

27 hr 15 min flight

(Lost position over fog)

14 Dec. 1927
MEXICO CITY

(Instruments malfunctioned, flew off course)

8 Feb.
HAVANA

1 sack of mail, Santa Domingo; 2 sacks from Port au Prince to Havana

2 Feb. San Juan

St Thomas 31 Jan.

Santo Domingo 4 Feb.

Port-au-Prince 6 Feb.

Belize 30 Dec.

28 Dec. Guatemala

1 Jan. 1928 San Salvador

5 Jan. Managua

7 Jan. San José

Tegucigalpa 3 Jan.

12 Jan. Colón

9 Jan. PANAMA

26 Jan. Cartagena

Bogotá 27 Jan.

Caracas 29 Jan.

100 200 300 400 500 600

Scale – Miles

REGD

A map detailing Lindbergh's circuitous Goodwill Tour through Central America and the Caribbean in the *Spirit of St. Louis*. Map created by R. E. G. Davies and used with permission from the Smithsonian National Air and Space Museum (NASM 9A06773).

steadfast assurance that Cuba would remain "always on the side of the United States."[5]

Although undoubtedly the most famous individual in the Cuban capital, Lindberg arrived in a city inundated with American dignitaries. Three weeks earlier, eighty-three delegates and a small army of technical aides from all twenty-one member states of the Pan American Union (PAU) had converged on Havana for the Sixth International Conference of American States. President Calvin Coolidge himself had opened the conference on January 16. In a speech that emphasized the importance of "consideration, cooperation, friendship, and charity" among equally sovereign PAU members, Coolidge called on the assembled delegates to work toward agreements that would tighten the "eternal bond of [hemispheric] unity." Most notably, he called for the creation of international "aviation routes, primarily for the transportation of mails" that would promote "closer cultural and commercial ties" among the American republics.[6] But the creation and operation of these air routes would not be a multilateral affair. As Coolidge had pointed out to Congress five weeks prior, the United States was determined to "take a leading part in this development."[7]

Having recently initiated airmail operations between Key West, Florida, and Havana, Cuba, under a US airmail contract the previous October, Pan American Airways stood ready to serve as the "chosen instrument" of this US aerial expansion into Latin America under the leadership of its charismatic and ambitious president Juan Trippe. For all practical purposes, Lindbergh's goodwill tour constituted "a survey flight" for Pan Am, the first of many he would perform for Trippe's company over the coming years.[8] While in Havana, Lindbergh personally took President Machado and delegates to the Sixth Pan American Conference up on flights to help secure their support for US civil air operations.[9] Lindbergh's aerial diplomacy during the Sixth Conference helped Trippe obtain additional operating concessions from the Cuban government that—when combined with generous US airmail subsidies under the 1928 Foreign Air Mail Act—transformed the island into a springboard for Pan Am's expansion into Latin America.[10]

Personal flights with the world's most famous aviator were not the only activities at Havana designed to promote the southern expansion of US aviation. One of eleven conventions approved by the time the Sixth Conference adjourned on February 20, the Pan American Convention on Commercial Aviation—popularly known as the Havana Convention—marked the culmination of a five-year State Department–led effort to craft a multilateral hemispheric air agreement. As shown in previous chapters, although the United States remained officially apart from the 1919 convention the 1926 Air Commerce Act and subsequent Air Commerce Regulations offered a level of compatibility with the dominant global civil air regime. But compatibility was not the same as adherence. The airplane provided the United States with a new means to express its economic and military might, but the 1919 convention's universally accepted declaration of aerial sovereignty also granted Latin American states complete authority over their own airspace. As Pan Am representatives worked to obtain lucrative air mail contracts from Central American and Caribbean states along the "Lindbergh Circle" after the Havana Conference, they also needed to negotiate for overflight rights with sovereign governments fully aware that the airplane had transformed their nation's airspace into a thing of value.[11] As an outsider to the 1919 convention, the United States could not take advantage of its innocent passage clause. For Trippe, the inclusion of a similar provision within the Havana Convention provided a powerful tool in US "efforts to establish international commercial air lines" throughout Latin America.[12] This grant of innocent passage would not only benefit Pan Am. To facilitate Lindbergh's goodwill tour, the United States government had to obtain special permission from each nation along his proposed route, an onerous process that placed "a considerable burden . . . upon the Department of Commerce, the Department of State and the American diplomatic missions abroad."[13] The Havana Convention not only promised to help Pan Am in its quest for overflight rights but allow all US civil aircraft similar access to Latin American skies.

But why create another air convention? With its right of innocent passage among member states and several American republics already signatories, the 1919 convention could easily meet the US desire for access.[14] The 1919 convention proved unsuitable to US interests for several reasons. First, the 1919 convention's connection to the League of Nations made its approval in the Senate highly unlikely. Second, as discussed in chapter 2, Ambassador Hugh Wallace signed the 1919 convention on behalf of the United States with reservations to its requirements that members apply restricted zones to *all* aircraft, close their airspace to nonmembers, and adopt its detailed aerial customs provisions. Perhaps most important, simply securing overflight access was not enough for policymakers in Washington. A multilateral air agreement needed to remain as unobtrusive as possible to US interests, prevent European aerial competition in the Western Hemisphere, and allow the United States to restrict foreign access to the all-important Panama Canal Zone. This would require an entirely new convention.

The air agreement that came out of Havana did not live up to these expectations. Contemporaries generally considered the Havana Convention an unnecessary and ineffective agreement, and subsequent studies echo this refrain.[15] But the historical significance of the United States' first multilateral air convention rests not in its effectiveness but in what can be learned from its creation, modification, and interpretation. Even more than successes, policy failures can provide valuable insight into the constraints that policymakers faced, how they responded to those constraints, and the limits of state power.[16] As Assistant Secretary of Commerce for Aeronautics William P. MacCracken Jr., Ambassador Henry Prather Fletcher of the State Department, and other career bureaucrats struggled to craft a hemispheric air regime more amicable to US economic and national security priorities,

Latin American representatives such as Colombia's future president Enrique Olaya Herrera and Felipe A. Espil of Argentina harnessed established aerial principles and norms to force the United States to temper its hegemonic aerial ambitions. Hoping to use a hemispheric air agreement as a centerpiece of US foreign policy, policymakers in Washington instead encountered a set of entrenched principles and norms that proved extraordinarily difficult to overcome. Just as the 1919 convention influenced domestic regulatory thought, it also delineated the contours of acceptable practices in international aviation, much to the chagrin of US policymakers. Ratified by the United States in 1931, the Havana Convention faded quickly into irrelevancy as members disagreed over its interpretation and application, administrations changed in Washington, and Pan Am continued to negotiate with each individual nation as it steadily advanced into Latin America and then the Pacific. Rather than a tale of triumphant US diplomacy, the history of the Havana Convention tells a story of unfulfilled expectations and frustrated imperial aspirations that warns of the pitfalls states face when they pursue a policy of "independent internationalism" in the face of established global principles.[17]

Wrestling with the 1919 Convention

Assistant Secretary of State Leland Harrison had just about had it with Godfrey Cabot, the Boston brahmin and president of the National Aeronautics Association (NAA). Continuing the NAA's support for the 1919 convention initiated under his predecessor, Howard Coffin, Cabot had petitioned the War, Navy, and State Departments in early March 1925 to dispatch representatives to the upcoming session of the International Commission for Air Navigation (ICAN)—the permanent committee empowered to create and revise the new air regime's technical and operational provisions—so that future US efforts toward "the development of international commercial airlines" maintained operational uniformity with international practices.[18] His mechanizations had prompted Admiral William Moffat, head of the Navy's Bureau of Aeronautics, to order the Navy's air attaché in London John H. Towers to attend the ICAN's upcoming eighth session "as the representative of the United States Government."[19] But the United States Senate had not yet received the 1919 convention from the president let alone ratified it, and Congress's recent adjournment without passing the MacCracken-drafted Winslow bill had left the United States without a stated domestic

Assistant Secretary of State Leland Harrison in 1922. Library of Congress Prints and Photographs Division, Washington, DC, https://lccn.loc.gov/2016832091, accessed Oct. 16, 2018.

aviation policy. Harrison firmly believed that, in the absence of ratification of the 1919 convention, the United States should "not take part in the meeting either officially or unofficially" or even appoint an observer without "an appropriate invitation, which we should not seek to obtain."[20] The Navy quickly dropped its plans for Towers to attend, but Cabot was not so easily deterred. He suggested that NAA representatives attend the meeting and serve as intermediaries between the ICAN and US officials.[21] Taken aback by this blatant affront to diplomatic protocol, Harrison coordinated a joint response from the Departments of State, War, and Navy that if NAA representatives did secure an invitation to attend, they should in no way intimate that they spoke for or conveyed the positions of the United States.[22] When Cabot further asserted an "urgent" need for the United States to join the 1919 convention to avoid "disastrous consequences," Harrison simply replied that the issue "has been receiving and will continue to receive the Department's careful consideration."[23] Even a delegation from the NAA to the State Department led by Arthur Halstead—someone fully aware of the international aspects of aviation from his important work on the American Aeronautical Safety Code discussed in chapter 5—could not sway Harrison's concerns over the convention's connections to the League of Nations nor his firm belief in the need for domestic legislation prior to its ratification.[24]

This was not the first time Harrison had worked to forestall the 1919 convention's submission to the Senate. A Harvard-educated civil servant who joined the State Department in 1907, Harrison's career spanned three administrations and Foreign Service assignments on three continents. Appointed to a State Department post in Washington in 1915, Harrison served as secretary for the US delegation to the Peace Conference and remained in Paris until he was recalled to Washington in the first days of the Harding administration. After a stint as an "expert assistant" on communications matters at the 1921 Washington Conference for the Limitation of Armaments, Harrison rose to the position of assistant secretary of state in charge of "all matters pertaining to foreign commercial policy, commercial treaties, transportation and communications, etc."[25] Until his retirement in 1927, Harrison pursued these objectives within a political environment that differed from that of the more overtly internationalist former president Wilson. In light of Harding's overwhelming electoral victory grounded on a "return to normalcy" and the Senate's rejection of the Versailles Treaty, the State Department of the three Republican administrations of the long 1920s—Harding, Coolidge, and Hoover—pursued a policy that historian Joan Hoff Wilson labeled "independent internationalism," "a pragmatic method for conducting foreign affairs" based on the "implicit assumption . . . that the United States should cooperate on an international scale when it cannot, or does not want to, solve a particular diplomatic problem through unilateral action."[26] As the Conference for the Limitation of Armaments, the 1924 Dawes Plan, the 1928 Kellogg-Briand Pact, and the Young Plan of 1929 and 1930 show, Harding and his successors did not fundamentally reject international cooperation and multilateral action. They did, however, firmly oppose any restrictions on the ability of the United States to act independently and refused to sign onto any agreement that would restrict the nation's freedom of action.

Even with reservations, Harrison believed the 1919 convention clearly fell into the latter category. As discussed in chapter 2, the NACA and the MAA both immediately supported convention ratification, and the Interdepartmental Board of the Wilson administration, detailed in chapter 3, began working on domestic legislation under that very assumption. After consultation with Canadian authorities, US Ambassador Hugh Wallace signed the 1919 convention on May 31, 1920, with reservations, and the Departments of War, Navy, Post Office, Treasury, and Commerce supported ratification. But Harrison noted that the placement of the International Commission for Air Navigation (ICAN) under the League of Nations meant US adherence would "constitute a recognition of the League."[27] Although British director of civil aviation Sefton Brancker offered assurance that membership in the ICAN would in no way "implicate" the United States in the League of Nations, the potential limitations on US freedom of action and the Senate's continued opposition to the League of Nations made ratification politically difficult, if not impossible.[28] With President Harding expressing no interest in an air convention negotiated under his Democratic predecessor, Assistant Secretary Harrison orchestrated a bureaucratic veto: the State Department simply refused to recommend the submission of the 1919 convention to the Senate for its advice and consent.

Despite concerns within the State Department over the 1919 convention, the NACA continued to support ratification well into the Coolidge administration for several reasons. First, ratification would address the suboptimal aerial relationship with Canada. The Canadian courtesy discussed in the previous chapter allowed US aircraft to cross the border provided they adhered to Canada's regulations based on the 1919 convention, but no such understanding existed for Canadian aircraft to enter the United States. Second, ratification would provide a justification for federal control under the Constitution's Supremacy Clause, rendering irrelevant the contentious and lengthy domestic sovereignty debate that delayed the passage of a national law. Third, ratification meant that—with the official adoption of the 1919 convention's operational and registration provisions—US aircraft would follow the exact same regulations as those of other member states. The 1919 convention's right of innocent passage would grant US aircraft overflight rights to the airspace of fellow adherents, while official membership on the ICAN would allow the United States a voice in the creation and revision of the convention and its technical provisions. Its questionable connection to the League of Nations notwithstanding, ratification of the 1919 convention promised numerous benefits for the United States.

In many ways, the convention's connection to the League of Nations provided a convenient reason for the United States to reject a multilateral air agreement that conflicted with its interests. Although it would provide clear benefits, the convention's grant of innocent passage among member states constituted a double-edged sword. Not only would the United States need to allow European member states access to its airspace, but other American nations that ratified the 1919 convention would need to as well. With the British, French, and Italians eager to sell war surplus aircraft and expand their aerial influence, the 1919 convention would facilitate the creation of foreign airlines in Latin America that could serve as a stepping-stone to even more direct control in hemispheric affairs. This possibility directly conflicted with the Monroe Doctrine, a dominant principle in US foreign policy since President James Monroe had warned against further European imperialism in the hemisphere and asserted the existence of a special relationship between the United States and newly independent Latin American republics in 1823.[29] Although the meaning and practical application of the Monroe Doctrine remained "a contested process . . . throughout the nineteenth century," it proved a valuable ideological tool "to keep out the Europeans, to safeguard order and stability in areas of special concern, and to ensure open access to markets and resources."[30] From the 1880s through the early 1930s, the Monroe Doctrine manifested itself in two related yet often contradictory approaches to Latin America: a cooperative yet paternalistic approach known as Pan-Americanism, which emphasized collaboration via hemispheric institutions, and an imperialistic approach centered on direct military intervention to secure and maintain US hegemony.[31] The 1919 air convention ran counter to both these expressions.

Pan-Americanism presupposed the existence of a deep and unalterable bond that warranted increased hemispheric cooperation, coordination, and goodwill. For idealistic US adherents, it denoted "common or concerted action" on the part of the American republics to promote their "welfare . . . without infringement of their sovereignty or integrity."[32] But in reality, Pan-Americanism constituted a "discourse of U.S. hegemony" that aimed to supplant British economic dominance in Latin America through the elimination of trade barriers and the standardization of policies and procedures.[33] Although the extension of US hegemony manifested in various ways, a series of Inter-American Conferences from 1890 to 1920 and the creation of the Bureau of the Pan American Union best illustrate how US policymakers viewed Pan-Americanism as a means to establish a suitable periphery for the nation's burgeoning industrial economy.[34] At these conferences, delegates discussed and signed multilateral conventions and resolutions on transportation, customs, copyright and patent protection, pecuniary claims, consular rights, and other subjects designed to facilitate the penetration of US interests into Latin America.[35] Between conferences, the Bureau of the Pan American Union—evolved from a resolution at the First Conference of American States—became the primary institutional mechanism for the realization of Washington's vision of Pan-Americanism.[36] Headquartered three blocks from the offices of the State, War, and Navy Departments in Washington, DC, financially dependent on appropriations from the US Congress, and with the US secretary of state as the ex officio chair of its Executive Committee, the PAU became a de facto extension of the US State Department responsible for drafting conference agendas and distributing commercial information on the American republics to US industry.[37] The creation of hemispheric "transportation utopias" to facilitate US penetration of southern markets remained a primary objective of Washington throughout this process, but such desires were often thwarted by Latin American governments that did not share the same policy priorities.[38] With its ability to rise above thick jungles and impassable mountains, the airplane promised, in the words of famed Brazilian aviator Alberto Santos-Dumont, to "knit the various [American] States . . . into an integrally united, cooperating and friendly combination."[39] For Harrison and others in the State Department, the economic underpinnings of Pan-Americanism required that these be US airplanes and airlines, an assumption undermined by the 1919 convention's right of innocent passage among *all* members.

An international air agreement that opened the skies of the American republics to European aircraft also conflicted with Washington's national security objectives in the Western Hemisphere, most notably the protection of the Panama Canal. Concerns that European powers would seek to reassert control over their former Latin American colonies had originally prompted the Monroe Doctrine, and anxieties over European intervention only increased when the United States transformed into a full-fledged colonial power after it obtained control over Cuba, Puerto Rico, Guam, and the Philippines following the Spanish-American War of 1898.[40] Now faced with the need to defend far-flung imperial possessions, President Theodore Roosevelt—an ardent adherent of expansionism and a fierce advocate of naval power—acted to bring the long-dreamed-of Central American canal linking the Atlantic and Pacific Oceans into reality. Within a two-week period in November 1903, he fomented revolution in the Colombian province of Panama, dispatched US troops to actively support it, and recognized Panama as the newest American Republic.[41] The Isthmian Canal Convention, commonly known as the Hay-Bunau-Varilla Treaty, granted the United States "in perpetuity the use, occupation and control of a zone of land and land under water for the construction, maintenance, operation, sanitation and protection of said Canal of the width of ten miles."[42] Controlling and protecting the Panama Canal Zone became a primary aim of US hemispheric policy, and it weighed heavily in Roosevelt's decision to intervene when European creditors threatened action to collect on the Dominican Republic's debt. The Roosevelt Corollary to the Monroe Doctrine made plain the United States' intention to "exercise . . . an international police power" to address "chronic wrongdoing" on the part of Latin American states to prevent European intrusion into hemispheric affairs.[43] As Roosevelt's secretary of state Elihu Root bluntly stated in 1905, "the inevitable effect of our building the Canal must be to require us to police the surrounding premises."[44] Subsequent US interventions in Central America and the Caribbean over the next thirty years, from Roosevelt's creation of a customs receivership in the Dominican Republic to Coolidge's deployment of US Marines against Augusto Sandino's forces in Nicaragua, must all be placed within the context of Washington's chronic anxiety over Canal Zone security.[45]

The airplane added further complications to Canal Zone security, ones that membership in the 1919 convention would only exacerbate. Signed two weeks before the Wright brothers achieved powered flight on December 17, 1903, the Isthmian Canal Convention did not explicitly grant the United States control over the Canal Zone's airspace, and the 1904 Taft Agreement's delineation of the jurisdictional rights of Panama and the United States also did not address aviation in the Canal Zone.[46] Panama nevertheless recognized the United States' ability to secure and defend the Canal Zone, and the airspace above it came under US military control with President Wilson's wartime executive order of February 1918. But, just as occurred within the United States, uncertainties over peacetime aerial control returned after the order's cancellation in July 1919. Postwar European aerial activities in Latin America further heightened anxieties over Canal Zone security. As the Curtiss Aeroplane and Motor Corporation sought export markets for its aircraft in the wake of canceled wartime appropriations, their representatives in Latin America met fierce competition from European companies aggressively seeking to sell aircraft and establish airlines with the full support of their home governments.[47] A postwar naval inspection of the Canal Zone's defenses found that "a bombing attack on the Canal" would be "the most promising enterprise that an enemy could undertake," and the US embassy in Guatemala warned of a rising threat to Canal Zone security from increased foreign aerial activity.[48] The primary focus of US concerns centered on the Sociedad Colombo-Alemana de Transportes Aéreos (SCADTA), a German-backed Colombian airline established in December 1919 under the guidance of Austrian Peter Paul von Bauer.[49] Supported by government air mail contracts, SCADTA's imported F13 Junkers under the control of German pilots rapidly established a reputation for efficiency that became a source of pride for Colombia.[50] And von Bauer had no intension of limiting SCADTA's operations to Colombia. Any expansion of SCADTA northward into Central America and the Caribbean required access to Panamanian airspace.

The need to secure the Canal Zone, combined with the economic priorities at the heart of Pan-Americanism, prompted the creation of a two-tiered aerial policy for Latin America within the Coolidge administration. The first element addressed the immediate situation with Panama. At the request of Secretary of War Weeks, Canal Zone Governor Jay Morrow initiated discussions with the Government of Panama to "devise rules to govern flights" over the Canal Zone for codification into a future agreement.[51] Meeting to formulate foundational principles prior to negotiations, representatives of the Army, Navy, and Canal Governor's office concluded that restricting flight above the Canal Zone alone would prove insufficient. Because the airplane provided an unprecedented aerial gaze that could be used to obtain information "of military value" even at some distance from the Canal, the United States should "view as a distinct menace . . . any aerial activities not under its control" throughout Panama.[52] They concluded that the airplane's speed and freedom of movement necessitated restricting activity in Panamanian airspace to US government aircraft, a total surrender of a nation's aerial sovereignty that any government would find wholly unacceptable.[53] Tempered by the State Department's desire to "interfere as little as possible with the sovereignty of Panama," the secretaries of state, navy, and war agreed to aerial provisions that would establish prohibited areas under US control around the Canal Zone, place aviation in Panama under a joint board dominated by the United States, and require US and Panamanian aircraft to operate under the regulatory provisions of the 1919 convention.[54] This policy—a clear attempt to appropriate Panama's aerial sovereignty while at the same time ensuring compatibility with the convention's operational requirements— became part of a new bilateral treaty to replace the Taft Agreement.[55] Signed by representatives of both nations a mere two months before the passage of the 1926 Air Commerce Act, the Panamanian government viewed this proposed treaty as an afront to its sovereignty and would call for its complete revision three years later.[56] Nevertheless, its potential ratification profoundly shaped how US delegates would approach discussions over a hemispheric air convention at the 1928 Havana Conference.

A desire to conclude a multilateral air agreement, the second aspect of US aerial diplomacy, grew directly out of this first concern. Limiting foreign aerial access to the Canal Zone and greater Panama might address the immediate situation along the isthmus, but it did not solve the larger issue of European aerial activity in the hemisphere. With air mail contracts a requirement for the successful operation of nascent airlines, the establishment of US operators throughout the hemisphere could both expand US commerce and prestige and serve as a bulwark against European encroachment. Air agreements that awarded blanket overflight rights would go a long way toward the creation of a US-dominated aerial network. In 1923, the War Department, with the full support of Herbert Hoover's Commerce Department, recommended that the Department of State begin negotiations with Central American nations for aviation treaties that could "assist in [the] development of aerial services in Latin America by American interests," but the Navy Department cautioned that such agreements could hamper the future adoption of the 1919 convention.[57] A May 1925 communique from the Colombian government requesting permission for two SCADTA aircraft to land in the Canal Zone and Key West on a planned survey flight from Colombia to the United States prompted the Departments of War, Navy, Commerce, Post Office, and Treasury "to formulate a national policy in regard to American dominance of aerial commercial and postal activities in Central America" in early 1926.[58] With Leland Harrison representing the State Department, this Interdepartmental Aviation Conference approved SCADTA's pioneering effort but asserted that any regular airline through the region must be "substantially controlled by Americans."[59]

The Interdepartmental Aviation Conference also recommended that the State Department pursue ratification of the 1919 convention to secure the overflight rights necessary to establish US airlines. Since Colombia had not yet ratified the 1919 convention, ratification by both the United States and Panama could provide a means to bar SCADTA from the Canal Zone based on the convention's requirement that members close their airspace to non-members. But Panama's ratification raised a host of other issues. First, the convention's grant of innocent passage would allow *all* member states to fly over both the Canal Zone and Panama, including any states that joined in the future, such as Colombia and Germany. Second, if Panama exercised its right under the convention to establish a prohibited zone in its territory around the Canal Zone, such restrictions would apply to *all* aircraft, including those of Panama and the United States. Finally, the full application of the convention's declaration of national air sovereignty, combined with the 1903 Isthmian Canal Convention's silence on US rights over Canal Zone airspace, could potentially deny the United States a say in Panamanian aerial affairs, something completely unacceptable to the military and political leadership in Washington. And there was still the issue of the convention's connection to the League of Nations. Faced with the consensus of the Interdepartmental Aviation Conference and the threat from SCADTA, Harrison reluctantly worked to make the 1919 convention as amicable to US interests as possible. Under his guidance, the State Department drafted two new reservations in addition to the original three from the Wilson administration that explicitly stated US ratification did not create or imply a "legal relation . . . or . . . obligation" to either the League of Nations or the Permanent Court of International Justice. More than six years after Ambassador Hugh Wallace signed the 1919 convention on behalf of the United States, President Coolidge submitted it to the Senate on June 16, 1926, so that US citizens could fully enjoy "the benefits to be derived from international co-operation" in aeronautics.[60]

But the Senate would never debate the 1919 convention. Although the NACA, the interested executive departments, and industry had called for its ratification for more than half a decade, the nation's aerial situation had radically changed with the passage of the Air Commerce Act that May. As detailed in previous chapters, MacCracken fully recognized the need for regulatory compatibility with the 1919 convention's operational and registrational provisions and fought to ensure that this new federal law represented a compatible statement of US foreign aerial policy.[61] Having consulted with the nation's aeronautical interests that fall, MacCracken informed Harrison in early December 1926 that the nation's new Air Commerce Regulations, while generally compatible, "were not exactly identical with the regulations attached to the Air Convention" and hoped "it would be possible to postpone final action on the Air Convention until an opportunity had been afforded to see how the Commerce regulation[s] worked out."[62] Harrison, who continued to harbor his own reservations about the convention, readily agreed. The two assistant secretaries decided "they would do nothing about it, thinking that no action would be taken by the Senate Foreign Relations Committee if neither Department pushed it."[63] This strategy of indifference on the part of the two assistant secretaries proved successful, and the 1919 convention gathered dust in committee. But this did not mean the complete abandonment of a multilateral convention as a tool to secure US dominance over Latin American skies.[64] A properly tailored multilateral air convention would clearly benefit the United States, and by early 1923 Harrison and other State Department officials had concluded that they "prefer[red] to submit . . . a new convention wholly divorced from the League of Nations."[65] The first postwar Inter-American conference, scheduled for March 25 to May 3, 1923, in Santiago, Chile, provided the perfect opportunity to begin this process, and Secretary of State Hughes inserted a charge to discuss the "policy, laws, and regulations concerning commercial aircraft" into the conference agenda.[66] As they pursued a hemispheric air agreement tailored to meet their national security and economic priorities over the next five years, US policymakers would find their efforts repeatedly stymied as delegates from PAU member states used the principles and norms embedded within *two* existing multilateral air agreements to curtail Washington's hegemonic designs.

Checking US Ambitions

The assignment to lead the US delegation to the Fifth International Conference of American States in Santiago, Chile, had not lived up to Henry Prather Fletcher's expectations. A Pennsylvania lawyer who harbored presidential aspirations, Fletcher had joined Theodore Roosevelt's Rough Riders at age twenty-five and spent the next two decades in diplomatic service, with six years in Chile and one as undersecretary of state at the request of newly appointed secretary of state Charles Evans Hughes.[67] Now serving as ambassador to Belgium, Fletcher personally lobbied to lead the US delegation to Santiago, and his good friend President Harding happily secured his ap-

Henry Prather Fletcher upon his appointment as US ambassador to Mexico, ca. 1916. Photographs, box 20, Henry Prather Fletcher Papers, Library of Congress.

pointment.[68] Fletcher's familiarity with the host government gave him high hopes for the conference, but the world had drastically changed in the thirteen years since the last Inter-American gathering.[69] With the League of Nations now standing in stark contrast to the hegemonic power imbalance embedded within the Pan American Union, a clear division arose among attendees. On the one hand stood the United States' desire to maintain the PAU's limited focus on commercial exchange and its control over the organization. On the other stood numerous smaller Latin American states seeking to transform the Monroe Doctrine into a broad nonintervention treaty and the PAU into an American League of Nations. As chair of the PAU's Executive Committee in charge of the conference agenda, Hughes did his best to contain these demands but—as shown with the election of League of Nations president Agustin Edwards of Chile as conference president—he could not stem the rising tide of pro-league sentiment.[70] Although "a skillful diplomat," Fletcher faced the unenviable task of securing agreements on trademarks, copyrights, and other commercial aspects while also defending Washington's hegemonic policies.[71] He soon found himself in "open disagreement" with Edwards and other delegates on the Political Committee on numerous points, and his blunt assertion that the Monroe Doctrine would continue to be "the unilateral national policy of the United States" did not sit well with his contemporaries or the Latin American press.[72] Fletcher succeeded in postponing these thorny political questions to the next conference, but his negative experiences in Santiago prompted him to swear he "never would take on a similar job in the future."[73] At Santiago, the tension between postwar Wilsonian internationalism and decades of US interventionist policy set into sharp relief the inequities that undergirded Pan-Americanism.

Just as the League of Nations shaped political discussions at Santiago, so too did the 1919 convention shape the aeronautical dialogue. At the suggestion of US delegate and future secretary of state senator Frank B. Kellogg, members of the Commerce Committee supported the establishment of an Inter-American Commercial Aviation Committee (IACAC) that would study and draft "laws and regulations" for air navigation in anticipation of the next conference scheduled for Havana in 1928. With the State Department effectively controlling the PAU's operations, this would permit the creation of a separate hemispheric convention tailored to US economic and security desires. But Mr. Luis Barros Borgoño of Chile and other Latin American delegates—fully aware of their nations' strong economic ties to Europe— insisted that the proposed IACAC "take into consideration . . . the Conventions already existing . . . so far as possible."[74] While Fletcher and the US delegation secured support for a new air convention, this unexpected amendment showed that Latin American members of the PAU expected any hemispheric agreement to conform to accepted aerial principles and norms already established in international treaties.

The extent of deviation from the 1919 convention that Latin American governments would accept became apparent when Spain, Portugal, and twenty American republics signed a second multilateral air convention in 1926.[75] Generally dismissed as "practically [an] exact textual copy of the 1919

Convention," the Ibero-American Aerial Navigation Treaty—popularly known as the Madrid Convention—retained both the 1919 convention's universal applicability of prohibited zones and ban on international munitions transport.[76] But it also diverged significantly from its predecessor. Member states could enter into air agreements with nonmembers, it contained no customs provisions or reference to the League of Nations, and it allowed non-nationals to register their aircraft in member states, a particular desire of Latin American nations that would enable European-supported carriers such as SCADTA to operate as fully national entities.[77] In light of von Bauer's desire to expand SCADTA and Spain's increasing closeness to Germany, the Madrid Convention fed into the established narrative of German designs in the Western Hemisphere and accentuated the need for a US-led hemispheric agreement to uphold the Monroe Doctrine. Charles H. Cunningham, the US commercial attaché in Madrid, cautioned that German desires for an air service to South America could be "behind the whole movement," while Leighton W. Rogers of the Bureau of Foreign and Domestic Commerce had no doubt that Spain's push for Latin American aerial dominance occurred "with German technical help and financing."[78] Because the Santiago amendment required the IACAC to take existing international air agreements into account, the Madrid Convention only further complicated US designs for a hemispheric air convention.

As Spain worked to consolidate the Madrid Convention over the summer of 1926, efforts began in Washington to create a draft agenda to serve as a foundation for the work of the IACAC approved in Santiago.[79] Responsibility for the initial draft fell on PAU assistant director Esteban Gil Borges. A trained lawyer, former Venezuelan minister of foreign affairs, and firm believer in the power of international law to unite the American republics into a common polity, Borges sought to balance US reservations to the 1919 convention with the Fifth Conference's mandate to remain faithful to current international air agreements.[80] Like the Madrid Convention, Borges's draft recognized a member state's right to establish separate aerial treaties with nonmembers, eliminated all mention of the League of Nations, and contained no customs provisions, but it retained the universal applicability of restricted areas and the complete ban on munitions transport found in both previous international air agreements. Leighton W. Rogers found Borges's draft "inadequate [to] American conditions" and suggested that "an Interdepartmental Committee should prepare the Agenda."[81] Over the next three months, newly promoted chief of the Air Regulations Division Clarence Young led representatives from the Commerce and State Departments in the creation of a revised agenda.[82] Several provisions ensured that future US carriers would dominate Latin American skies, but three in particular simultaneously embodied the primary economic and security concerns of the United States and contrasted sharply with established international norms. The first allowed a nation's own aircraft access to its restricted areas, a provision lifted straight from the US reservations to the 1919 convention. The second, a blatant effort to prevent European-owned aircraft from registering in Latin American member states, limited registration to citizenship as required in the recently promulgated Air Commerce Regulations. The third permitted the aerial transportation of munitions at the discretion of the overflown states, something that would allow aircraft to resupply US military forces in Latin America.[83] As a result of these revisions, the PAU's draft agenda for the IACAC became an aerial expression of the Monroe Doctrine.

Just as they had pushed back against a unilateralist expression of the Monroe Doctrine at the Santiago conference, Latin American representatives opposed US attempts to elevate its own policy concerns over generally accepted aerial principles and norms. But unlike the seasoned diplomat Fletcher, head of the US delegation William P. MacCracken Jr. found himself outmatched and outmaneuvered when the IACAC met at the PAU's Washington headquarters in May 1927.[84] Unanimously elected chairman of the conference, no sooner had MacCracken announced that the PAU's working draft would serve as the basis of discussion than Enrique Olaya Herrera of Colombia and his technical advisor, Peter Paul von Bauer of SCADTA, presented their own project of convention. Based on "precepts and regulations approved at the Paris Conference of 1919 and the Madrid meeting of 1926," this rival proposal retained the universal applicability of restricted areas and allowed non-nationals to register their aircraft with member states.[85] Unwilling to stand by as the United States used a multilateral convention to restrict SCADTA's access to the Canal Zone, Olaya Herrera and von Bauer led a concerted effort over the next two weeks that forced the US delegation to capitulate on all three primary security provisions within the PAU draft agenda. MacCracken agreed to a compromise wherein a state's own aircraft could access its restricted zones for domestic and noncommercial international flight, but any restrictions on international commercial operations must apply equally to all members. If the United States wanted to allow its

Table 6.1. The Three Primary Security Provisions within the IACAC Draft Agenda Compared with Complementary Provisions in the 1919 Air Convention and the Madrid Convention

Key US Security Provisions in the IACAC Draft Agenda	1919 Convention	Madrid Convention
Members may "prohibit the aircraft of the other contracting States from flying over certain areas of its territory"	Restricted zones applicable to all aircraft	Restricted zones applicable to all aircraft
Aircraft's nationality based on the nationality of owner	Aircraft's nationality based on the nationality of owner	Aircraft's nationality based on place of registration
Transportation of munitions "forbidden in international navigation unless expressly authorized by each State through whose airspace the aircraft passes"	Transportation of munitions strictly forbidden	Transportation of munitions strictly forbidden

Source: Inter-American Commission on Commercial Aviation, "Data Prepared by the Pan American Union as a Contribution to the Work of the Commission," Inter-American Commission on Commercial Aviation 1927, box 8, John Lansing Callan Papers, LOC.

Note: These three provisions constituted Articles 5, 9, and 16 of the US-produced IACAC draft agenda.

own commercial carriers access to the Canal Zone, it would have to do so for SCADTA as well. When the IACAC voted overwhelmingly to leave the registration of foreign-owned aircraft up to member states as called for in the Madrid Convention, the US delegation abstained.[86] Due to the United States' history of armed intervention, the provision to allow the aerial transportation of munitions, even with consent, raised serious concerns among Latin American delegates, and the majority voted to incorporate the complete prohibition on arms transport in both the 1919 convention and the Madrid Convention.[87] At the signing of the IACAC's Final Act on May 19—one day prior to the start of Lindbergh's historic transatlantic flight—Secretary of State Kellogg commended the delegates for an accomplishment that "exceeds any other in the history of the Pan American Union."[88] While this may have been the case, Latin American opposition to US attempts to deviate from international principles and norms meant that the IACAC's final project of convention was no longer the aerial representation of the Monroe Doctrine that policymakers in Washington had envisioned.

Left out of the creation of the IACAC's deliberations due to the civilian nature of the proposed hemispheric air agreement, representatives of the War and Navy Departments took serious issue with MacCracken's failure to defend US security priorities. Assistant Secretary of the Navy for Aeronautics Edward P. Warner strongly opposed any restriction on the ability of a state's own aircraft to access its prohibited zones, while Assistant Secretary of War for Aeronautics F. Trubee Davison—concerned that Europeans could gain "backdoor" access to the hemisphere's airspace through the proposed convention's innocent passage clause—demanded the reinsertion of registration based on nationality.[89] To fully ensure Canal Zone security, Davison even recommended expanding the definition of state aircraft to include carriers engaged in contract air mail services, an extreme deviation from established practices explicitly designed to prevent SCADTA from taking advantage of the proposed convention's provisions.[90] But it was too late for revisions to make it into the convention prior to the Sixth International Conference of American States; any changes would have to occur at Havana. When Warner, Davison, and chief of the Air Corps Maj. Gen. Mason Patrick joined MacCracken and representatives of the State Department in October to recommend instructions for delegates to the upcoming conference, MacCracken convinced them that any attempt to reinsert the three security provisions "would be useless" and might even prompt revisions that could further "disadvantage . . . the United States."[91] As a last-ditch effort to protect the Canal Zone, all present agreed to forward Davison's proposed redefinition of state aircraft to the State Department.[92] If accepted, this change would, of course, also mean that US air mail carriers could not take advantage of the proposed convention's innocent passage provision, but this seemed like a necessary sacrifice in the name of security.

The difficult task of modifying the IACAC's proposed convention into a document more amicable to US interests at Havana fell to Henry Prather Fletcher. Having witnessed the challenges to US policy firsthand in Santiago and aware of growing resentment against US intervention in Nicaragua, Secretary of State Kellogg called on Fletcher, now ambassador to Italy, to lead a "competent" delegation to "suppress . . . elements going . . . purely for the purpose of making trouble."[93] True to the oath he had made in 1923, Fletcher initially refused. Only after his former boss Charles Evans Hughes— arguably the greatest American statesman of the era—agreed to lead the US delegation did Fletcher acquiesce to the assignment.[94] As he prepared for the conference, Fletcher found the proposed convention crafted by the

IACAC to be an "unsatisfactory" product of "technical experts" that failed to account for the aerial security stipulations in the still-unratified treaty with Panama.[95] After discussions with Hughes and Francis White, Harrison's successor as assistant secretary, Fletcher recommended the inclusion of the following italicized clause concerning separate agreements into the convention:

> The right of any of the contracting States to enter into any convention or special agreement with any other State or States concerning international aerial navigation is recognized, so long as such convention or special agreement shall not impair the rights or obligations of any of the States party to this Convention, acquired or imposed herein; *provided however that prohibited areas within their respective territories, and regulations pertaining thereto, may be agreed upon by two or more States for military reasons or in the interest of public safety. Such agreements, and all regulations pursuant thereto, shall be subject to the same conditions as those set forth in Article 5 of this Convention with respect to prohibited areas within the territory of a particular State.*[96]

Having lost the struggle to include the three primary security concerns of the United States in the IACAC's project of convention, this so-called Fletcher amendment—tailored specifically to account for the proposed 1926 US-Panama Treaty—and Davison's redefinition of state aircraft to include air mail carriers represented rearguard actions to restrict foreign access to the Canal Zone and greater Panama. Kellogg inserted both suggested amendments into the instructions for delegates alongside directions that they secure "a satisfactory international convention regarding Commercial Aviation," take "special care that this country is not compelled under any international convention to sanction procedure and practices which might jeopardize the safety of the Panama Canal," and distance themselves from any "arrangement . . . or resolution . . . which could possibly be interpreted as curtailing in any way the application by the United States of the Monroe Doctrine."[97]

As US Marines employed Vought O2U-1 Corsairs and De Havilands on aerial bombing and recognizance missions against Sandinistas in Nicaragua, Fletcher struggled to secure US interests on the Communications Committee in Havana. There he faced two individuals who had successfully resisted the blanket acceptance of the US draft agenda at the IACAC eight months earlier: SCATDA supporter Enrique Olaya Herrera of Colombia and Felipe A. Espil, Argentina's counsellor in Washington, who the State Department considered decidedly "pro-German."[98] The two vehemently denounced the US attempt to include air mail carriers within the definition of state aircraft as an obvious effort to restrict foreign access to the Canal Zone, and they compelled Fletcher to accept the definition of state aircraft in the 1919 convention.[99] The ability of member states to create separate air agreements provided for in the so-called Fletcher amendment generated particularly intense debate. Co-opting the essence of President Monroe's speech more than a hundred years earlier, Olaya Herrera worried that such a clause could lead to the creation of regional aerial fiefdoms that would undermine the Pan-American ideal and argued that the 1919 convention should "be considered as the maximum of what is admissible with respect to military precautions."[100] But Fletcher refused to capitulate on the last vestiges of Washington's security priorities. In a series of individual meetings, Fletcher and Olaya Herrera negotiated a compromise based on two potentially contradictory clauses: that any regulations established by agreement of two or more member states must "guarantee equality of treatment" to all convention adherents *and* that "nothing in this convention shall affect the rights and obligations established by existing treaties."[101] But Olaya Herrera desired further reassurances that regulations agreed on by the US and Panama would not prejudice non-US carriers, and Fletcher agreed to an additional ambiguous clause stating that any such "regulations shall in no case prevent the establishment and operation of practicable inter-American aerial lines and terminals."[102] With Olaya Herrera satisfied that the revised convention would not restrict access to the Canal Zone and assurances that the reference to "existing treaties" did not compel the Panamanian government to accept the unratified (and deeply disliked) US-Panama Treaty of 1926, the Communications Committee approved the Pan American Convention on Commercial Aviation.[103]

The hemispheric air agreement adopted by the Sixth International Conference of American States on February 8, 1928, looked decidedly different than the atmospheric extension of the Monroe Doctrine that had emerged from the executive branch a year earlier. On the one hand, the United States obtained a document that, for the most part, corresponded with its reservations to the 1919 convention: members could create air agreements with nonmembers; no connections existed to the League of Nations or the international court; and the absence of technical annexes meant no forced adherence to an externally mandated aerial customs regime and the freedom to use MacCracken's compatible albeit dissimilar air regulations. But only a single nebulous article open to interpretation remained out of the three distinct security provisions embedded within the earlier IACAC draft agenda.[104] Not only did the Havana Convention fail to safeguard US national security priorities, but the Communications Committee's desire to align it more closely with the 1919 and Madrid conventions had resulted in the inclusion of a provision requiring member states to ensure equality of landing rights that would necessitate the nationalization of airfields, something the NACA and Congress had explicitly rejected with the Air Commerce Act.[105] Armed with identical claims to aerial sovereignty and demanding an equal application of the established right of innocent passage, Latin American delegates—with Colombia's Olaya Herrera at the vanguard—successfully wielded both Pan-Americanism's stated commitment to hemispheric cooperation and the accepted principles embedded within existing international air agreements as powerful weapons to check the aerial ambitions of the United States. Constrained within the established boundaries of two prior international air agreements, US officials had no choice but to drastically curtail their expectations about the role this new hemispheric air convention would play in US policy.

Slipping into Irrelevancy

By March 1928, the United States was a signatory to two international air agreements, neither of which fully aligned with its aerial priorities: the 1919 convention—with its connection to the League of Nations, detailed technical annexes, customs regime, universally applicable restricted zones, and prohibition against separate air agreements with nonmembers—and the Havana Convention, a document that incorporated most of the State Department's reservations to the 1919 convention but did not fully guarantee Washington's ability to prevent foreign access to Canal Zone airspace. Nevertheless, MacCracken and others in the Aeronautics Branch believed the new hemispheric air agreement settled the question of US adherence to the 1919 convention, and Commerce Secretary Herbert Hoover supported the latter's withdrawal from the Senate.[106] Opinions in the State Department remained mixed. Some believed that submission of the Havana Convention should coincide with the 1919 convention's withdrawal to "arouse less antagonism" among advocates of US participation in that international air agreement, while others cautioned that ratification of the Havana Convention could "endanger" US freedom to regulate aviation in the Canal Zone and Panama in light of the uncertain state of the 1926 US-Panama Treaty.[107] Coolidge's recommendation to have the Senate's Foreign Affairs Committee "hold the treaty and see" settled the matter.[108] As a result, the 1919 convention continued to languish in the Senate as the State Department debated the fate of the Havana Convention.[109]

With the three security provisions in the IACAC draft agenda absent, prohibitions against international commercial service through restricted areas, and the Fletcher/Olaya Herrera compromise open to widely different interpretations, rapid ratification of the Havana Convention could actually hinder US desires. Under the Havana Convention, any grant of commercial access to Canal Zone airspace given to Pan Am had to extend to all member state aircraft, including any foreign-owned machines—such as SCADTA's—registered in those states. This was completely unacceptable. Bolstered by air mail contracts awarded under the Foreign Air Mail Act of 1928 and State Department assistance, Trippe's Pan Am raced to establish a US monopoly of service on the isthmus prior to the Havana Convention's submission to the Senate in order to create an aerial barrier against European encroachment in Central America and the Caribbean. On September 28, 1928, President Coolidge granted the secretary of state full authority "to receive and pass upon all applications for the privilege of operating commercial aircraft between the Canal Zone and foreign countries," and Kellogg readily granted Pan Am such permissions.[110] After receiving permission from the nations along his route, Pan Am's newest technical advisor Charles Lindbergh spent his twenty-seventh birthday inaugurating the airline's new Miami-to-Cristóbal route on February 4, 1929.[111] Unbound by the Havana Convention's requirement to ensure equality of treatment, the State Department granted Pan Am access to the Canal Zone without extending the same right to von Bauer's SCADTA.

But for Pan Am to secure a recently announced US air mail contract from the Canal Zone to Chile and connect with its subsidiary Pan American-Grace Airways (PANAGRA) operating in Peru, it needed overflight rights in Colombia, and quickly.[112] Ratification of the Havana Convention by both countries could take considerable time, and the issue of how that agreement affected Washington's ability to promulgate air regulations under the imperfect 1926 US-Panama Treaty remained unclear. A bilateral executive agreement tailored to the needs of this particular situation—one that did not require Senate approval—offered the best solution, and Trippe actively lobbied the

State Department to conclude a separate accord between the two nations.[113] Agreed on by an exchange of notes between Kellogg and Olaya Herrera on February 23, 1929, the United States' first ever international air agreement set the stipulations by which commercially registered aircraft of the United States and Colombia could access the other's airspace, including the Canal Zone.[114] Washington abandoned a policy of complete prohibition of foreign aircraft in the Canal Zone to secure Pan Am's expansion, but such access would remain heavily regulated. Proclaimed as the representatives of the US and Colombia negotiated their landmark air agreement, Executive Order 5047 restored the Canal Zone to its wartime designation as a "Military Airspace Reservation" and placed "control of civil aviation" under its military governor. Pilots had to strictly adhere to predetermined routes, aircraft underwent detailed inspections on landing and prior to takeoff, aerial photographs or sketches of the Canal Zone were forbidden, and "authorized Panama Canal Zone officials" could approve the aerial transportation of munitions.[115] With the US-Panama Treaty still in limbo, the airspace surrounding the Canal Zone became an aerial cordon sanitaire when a newly established joint Aviation Board dominated by US personnel promulgated regulations for Panama based on the Air Commerce Regulations of the United States two months later.[116] Three months after the Post Office advertised the new mail route to Chile, the United States had negotiated its first bilateral air agreement with Colombia, established provisions for commercial air operations in the Canal Zone, and erected a system of protective regulations in Panama. By the summer of 1929, PANAGRA extended Pan Am's operations along South America's Pacific coast, and Pan Am effectively controlled SCADTA as a result of a secret gentleman's agreement between Trippe and von Bauer.[117] A year later, Pan Am's incorporation of the New York, Rio, and Buenos Aires Line's former routes along the Atlantic coast of South America brought the vision of a single US-owned and -operated American air system into reality. Without the three security provisions within the IACAC draft agenda, the United States rejected the Havana Convention as a means to secure the Canal Zone and did so instead through unilateral action and bilateral agreements.

The Havana Convention became further diminished as a result of a 1929 attempt to revise the 1919 convention. Prompted by German calls to "reunite the entire civilized world" under a "universal public aerial law" instead of the three similar yet decidedly different multilateral agreements now in effect, the ICAN invited both member and nonmember states to participate in an Extraordinary Session.[118] With the blessing of the NACA, newly appointed secretary of state Henry L. Stimson turned to MacCracken to lead the US delegation.[119] Stimson charged him with revising the 1919 convention to reflect the provisions of the Havana Convention and stressed the need to incorporate a clause that guaranteed the revised 1919 convention would not conflict with the still-pending 1926 US-Panama Convention, the recently passed regulations for the Canal Zone and Panama, or any additional regulatory agreement between American states as provided for in the Havana Convention.[120] Stimson made clear that the United States Government had no intention of accepting just any modified agreement. If MacCracken could not secure the liberalization of the 1919 convention along the lines of the Havana Convention, the United States would not ratify the revised treaty. But MacCracken's diplomatic skills had not improved in the two years since the IACAC, and US designs wilted before the defense of established norms when the Extraordinary Session convened from June 10 to June 15 in Paris.[121] Tasked with bringing the 1919 convention in line with the Havana Convention, the US delegation succeeded only in securing arbitration outside of the Permanent International Court of Justice.[122] MacCracken signed this revised 1919 convention with almost the exact reservations that accompanied the previous version still languishing in the Senate.[123] The inability of the US delegation to convince representatives of twenty-seven sovereign states to accept the Havana Convention's provisions—several of them signatories— showed both the tenacity of the 1919 convention's principles and the limited appeal of the new hemispheric agreement. In the wake of the ICAN's Extraordinary Session, the 1919 convention quickly fell out of favor within the NACA.[124]

Even with the regulation of Canal Zone airspace and Pan Am entrenched throughout the hemisphere, a multilateral air agreement would still secure overflight rights for *all* US civil aircraft among its adherents. With the 1919 convention now an impossibility and the Canal Zone situation secured, the State Department officially supported ratification of the Havana Convention.[125] This drew immediate support from the US Chamber of Commerce and PAA president Juan Trippe, who viewed the Havana Convention as the best means to assist "private efforts to establish international commercial air lines." Whereas Assistant Secretary of State Harrison and Assistant Secretary of Commerce for Aeronautics MacCracken had worked in 1926 to corral the 1919 convention in the Senate Foreign Relations Committee, their respective successors Francis White and Clarence Young now called for the rapid ratification of the Havana Convention to open Latin American skies to all US aircraft.[126] On February 20, 1931, a multilateral air convention finally made its way out of the Senate. When Hoover proclaimed the Havana Convention's provisions in force on July 27, the United States joined Guatemala, Mexico, Nicaragua, and Panama as official members to an air agreement that looked nothing like the aerial extension of the Monroe Doctrine originally embodied in the IACAC's initial draft agenda.[127]

Devoid of the security provisions the United States had originally sought, the Havana Convention proved unable to even provide the access that Trippe, Young, and others hoped. This stemmed from three reasons. First, the lack of technical specifics in the Havana Convention—something that MacCracken expressly desired to ensure US regulatory freedom—complicated its application. MacCracken, Harrison, and others in the Commerce and State Departments always assumed that Latin American members would simply adopt US regulations, but as equally sovereign states they were under no obligation to do so.[128] Without clear technical annexes or a permanent organization to establish uniform regulations, individual member states remained free to adopt their own individual operational and registrational practices. To make matters worse, the convention required member states to ensure that their airworthiness certificates and pilots' licenses complied with the regulations of all states the aircraft would fly over, a stipulation that would require the Aeronautics Branch to apply foreign provisions to domestic certification and licensing that would undermine US sovereign regulatory authority.[129] Second, not all signatories to the Havana Convention agreed on the extent of its grant of innocent passage. As far as the State Department and the Aeronautics Branch were concerned, it negated the need for private aircraft to obtain special permission before entering a member state's airspace, as was required for Lindbergh's Central American and Caribbean goodwill tour. But the governments of Guatemala and Mexico vigorously disagreed and asserted that the innocent passage provision applied only to aircraft engaged in international commerce.[130] Finally, the Havana Convention would require a significant extension of federal authority to fully enact its provisions. The requirement that member states guarantee equal access to landing facilities would necessitate federal control over state and municipal airports, while the provision that members work to secure uniformity of regulations demanded an extension of federal regulatory control over intrastate aviation.[131] Both the NACA and Congress had explicitly excluded these aspects from the 1926 Air Commerce Act out of concerns over constitutionality and federal overreach. Focused on the Great Depression, unwilling to expend political capital for a flawed air agreement crafted under his Republican predecessors, and with Pan Am ensconced as the nation's "chosen instrument" in Latin America, newly elected president Franklin Delano Roosevelt let the Havana Convention fade into irrelevance.[132]

The Lessons of Havana

When one looks at the established narrative of US aviation in the 1920s and 1930s, the Havana Convention appears, if at all, as a sideshow, an outlier with no real significance. Juan Trippe established the Pan American Airways System through individual agreements with Latin American governments, and the United States settled the issue of Canal Zone aviation through a mixture of bilateral agreements and unilateral actions. But history is a study of how past human beings addressed situations as they existed in *their* present, and focusing solely on successful activities or policies results in a Whiggish view of the past that misses the reality of human experience. US policymakers believed that a hemispheric agreement represented the best way to limit air activity in the Canal Zone to US aircraft, prevent European competition, and secure unfettered access to the skies of Latin America. But when they modified the 1919 convention to meet US national security and economic priorities, representatives from fellow PAU member states—vested with equal air sovereignty—opposed attempts to deviate from established international norms. Devoid of its security provisions and further complicated by issues of interpretation and constitutionality, the Havana Convention became "an admirable makeshift agreement" instead of a fully formed regime to facilitate US aerial dominance in the Western Hemisphere.[133] When Assistant Secretary of State Adolf A. Berle Jr. sought to create another postwar air convention that would provide American aircraft unfettered access to foreign airspace on a global scale, he faced forces similar to those that MacCracken and Fletcher confronted more than a decade earlier. Like the Havana Convention, the Roosevelt administration's efforts at the 1944 Chicago Conference resulted in an air agreement "far less favorable to US interests than . . . Washington leaders had envisioned."[134]

Conclusion

On September 25, 2018, US President Donald J. Trump spoke before the United Nations General Assembly in New York City. In a speech that touched on trade, terrorism, immigration, and a host of other issues, he declared that America would "never surrender . . . sovereignty to an unelected, unaccountable, global bureaucracy"; rejected "the ideology of globalism"; and promised to "protect our cherished independence above all."[1] Elected on a platform of "America First," the speech unequivocally demonstrated the forty-fifth president's commitment to the pursuit of US interests above those of other nations or international institutions. A month later, a Boeing 737 Max 8 aircraft crashed ten minutes after taking off from Jakarta, Indonesia, killing all 189 people on board.[2] As the investigation into this incident continued, a second Max 8 operated by Ethiopian Airlines crashed shortly after takeoff on March 10, 2019, resulting in the deaths of 157 people.[3] Global reaction was swift and decisive. Within two days, China, the European Union, Australia, and other nations banned the use of the Max 8 in their airspace, and carriers around the world grounded the aircraft due to safety concerns.[4] With the Max a major commercial aviation program for Boeing, US regulators hesitated to take similar action, which could negatively impact a vital aerospace company with over 153,000 employees across the country.[5] But rising safety concerns prompted a direct presidential order, and the FAA grounded all Boeing 737 Max 8s and 9s on March 13.[6] Unable to force equally sovereign nations to allow the Max into their airspace and under increasing pressure at home to follow the international consensus, the United States was forced to adopt a regulatory policy that ran counter to its economic interests and undercut its global prestige.

This book shows that the interconnectivity among domestic and foreign considerations vividly demonstrated in the Boeing Max episode existed from the very beginning of US civil aviation policy. As Americans grappled with the airplane's proper place in the Constitution's federal system, they did so within a wider context. As Howard Coffin, Samuel Bradley, Maj. Gen. Mason Patrick, Joseph Ames, and others struggled to achieve federal legislation to safeguard public safety and promote commercial aviation, they recognized that the need for regulatory uniformity extended beyond the US border. While the postwar air convention affirmed the unrestricted aerial sovereignty of all nations, Canada's adoption of its provisions made clear the need to establish a domestic system compatible with standard international operational and registrational procedures. Voluntary regulatory attempts to do so proved unsuccessful without the power of the state to enforce them. It took a perceived national security crisis to prompt Congress to act, but representatives refused to accept the complete federal control that Bradley, MacCracken, and others viewed as essential to safe operations.[7] Nevertheless, the 1926 Air Commerce Act and Air Commerce Regulations established a regulatory framework for US aviation that vastly extended federal power *and* intentionally complemented the postwar international civil aviation regime. Washington's efforts to craft a hemispheric air agreement more amicable to its national security and economic interests, however, met fierce resistance from fellow Pan American Union members that used established international principles and norms to defend their own aerial sovereignty. Whether addressing the delineation between federal and state power, crafting specific regulations, or attempting to create an air convention for the Americas, the 1919 air convention drafted as part of the Versailles Conference served as both a guide for and a check on the initial creation of US civil aviation policy.

Although focused on the origins of US aviation in the 1920s, this book provides several implications for future studies. First, it illustrates the central role that technologies play in the reevaluation of state power within a federalist system. With arguments draped in technologically deterministic overtones, regulatory advocates argued that the airplane's capacity for rapid movement within three-dimensional space necessitated an unprecedented expansion of federal power that—with continued refinement based on experience—led to the current federal dominance over the atmosphere.[8] In other words, the device mattered. Second, this book highlights the importance of mezzo-level bureaucrats and officials in policy creation. Presidents Wilson, Harding, and Coolidge agreed to plans and signed off on legislation to regulate civil aviation, but they did not generate them. The difficult work of policy formulation occurred among members of the military and civilian departments who, in consultation with private sector stakeholders, carried their vision from one administration to the next. Civil aviation policy arose not as the result of the regulatory genius of any one individual but through a process of consensus creation and refinement among numerous aeronautical constituencies.[9] Finally, this book illustrates how international conventions, agreements, and regulatory regimes affect national policy formation even in the absence of official adherence. The airplane's capabilities blurred the line between the domestic and international spheres, and US policymakers and regulatory advocates found themselves forced to operate within an international regime that they had no direct ability to influence. The convention's "rules of the game" for aviation profoundly affected the development of US aerial policy, but in the absence of ratification Washington lacked a say in their modification at the hands of the International Commission for Air Navigation. As the nations of the world face the need to respond to nanotech, artificial intelligence, global health pandemics, and other transnational challenges, the creation of US aviation policy in the 1920s reminds us that active engagement in international regimes, even seemingly imperfect ones, can serve the national interest.

Notes

Abbreviations

AHC   American Heritage Center, University of Wyoming, Laramie,
               Wyoming
HHPL   Herbert Hoover Presidential Library, West Branch, Iowa
LAC   Library and Archives, Ottawa, Ontario, Canada
LOC   Library of Congress, Manuscript Division, Washington, DC
NARA I   National Archives and Records Administration, Washington, DC
NARA II   National Archives and Records Administration, College Park,
               Maryland
NASM   National Air and Space Museum Archives, Udvar-Hazy Center,
               Chantilly, Virginia
OAS   Organization of American States Archives, Washington, DC
PAA   Pan American World Airways, Inc. Records, University of Miami
               Special Collections, Miami, Florida
Smithsonian   Smithsonian Institution Archives, Washington, DC
UL Archives   Underwriters Laboratories Archive, Northbrook, Illinois

Introduction

1. "Harding Dedicates Lincoln Memorial; Blue and Grey Join," *New York Times*, May 31, 1922.

2. "Sherrill Asks Fahy Why He Disregarded Requests," *Washington Post*, June 1, 1922.

3. According to one account, Secretary of War John W. Weeks rushed to the edge of the Lincoln Memorial and "projected his bulk between two marble columns" in an attempt to wave Fahy away. "Harding Dedicates Lincoln Memorial; Blue and Grey Join."

4. "Killed in Plane Fall," *Washington Post*, July 26, 1922; Bradley to Hoover, Aug. 24, 1922, in Herbert Hoover Papers, Commerce Department, Bureau of Aeronautics, Legislation, 1921, box 121, Commerce Papers, HHPL.

5. "Weeks Urges Control of Civilian Aviation," *New York Times*, June 2, 1922.

6. "Drops Reserve Flier by Orders of Harding," *New York Times*, June 3 1922.

7. Transportation, 49, C.F.R. § 40103 (2011), https://www.gpo.gov/fdsys/pkg /USCODE-2011-title49/pdf/USCODE-2011-title49-subtitleVII-partA-subparti-chap401 -sec40103.pdf; George McKeegan and William Ranieri, *A Handbook of Aviation Law* (Chicago: American Bar Association, 2017), 55.

8. Federal Aviation Administration, "Air Traffic by the Numbers," https://www .faa.gov/air_traffic/by_the_numbers/media/Air_Traffic_by_the_Numbers_2019.pdf (accessed Jan. 1, 2020); Department of Transportation, Federal Aviation Administration,

Performance and Accountability Report 2019 https://www.faa.gov/about/plans_reports /media/2019_PAR.pdf (accessed Jan. 1, 2020).

9. The NAS divides US airspace into Classes A through E based on altitude and proximity to terminal facilities. US Department of Transportation, Federal Aviation Administration, *Pilot's Handbook of Aeronautical Knowledge*, 2016, chapter 15: Airspace, https://www.faa.gov/regulations_policies/handbooks_manuals/aviation/phak/media /17_phak_ch15.pdf (accessed Oct. 5, 2018).

10. Federal Aviation Administration, Special Awareness Training, Washington, DC, Special Flight Rules Area, https://www.faasafety.gov/files/gslac/courses/content/55 /707/SFRA%20Course%20Notes%20%20111130.pdf (accessed Oct. 5, 2018). The use of unmanned aerial vehicles, or drones, is also strictly prohibited in the Flight Restricted Zone. The FAA Reauthorization Act of 2018 tasked the FAA and various national security agencies with the creation of systems "to detect and mitigate potential risks posed by errant or hostile unmanned aircraft system operations" (FAA Reauthorization Act of 2018, Sec. 383, Pub. L. No. 115-254). As my colleague Alan Meyer pointed out, members of the military can receive special security authorization to undertake flights using visual flight rules in the Flight Restricted Zone.

11. For a detailed analysis of the frontier nature of early aviation and its "taming" over the twentieth century, see David Courtwright, *Sky as Frontier: Adventure, Aviation, and Empire* (College Station: Texas A&M University Press, 2005).

12. Up to this point, the literature directly addressing the development of interwar aviation policy has myopically focused on the Air Commerce Act through a strictly domestic lens. In the 1940s, Henry Ladd Smith concentrated on the activities of the 69th Congress and presented the Air Commerce Act as the inevitable by-product of the 1925 Air Mail Act's transfer of air mail carriage from the Post Office Department to private industry; Henry Ladd Smith, *Airways: The History of Commercial Aviation in the United States* (1942; repr., Washington, DC: Smithsonian Press, 1991). Two decades later, Donald R. Whitnah argued that the Air Commerce Act stemmed from a desire to support aviation within the context of America's traditional aversion to direct federal subsidies; Donald Whitnah, *Safer Skyways: Federal Control of Aviation, 1926–1966* (Ames: Iowa State University Press, 1966). In an exhaustive and widely cited dissertation from 1970, Thomas W. Walterman drew on the work of Robert H. Wiebe, Gabriel Kolko, and K. Austin Kerr to concluded that the Air Commerce Act—with its indirect support for industry at the cost of the individual freedom to fly at will—represented the ultimate victory of a postwar "quasi-progressivism" that emphasized the economic rationalization and industrial efficiency within traditional Progressive ideology at the expense of its social justice component. Thomas W. Walterman, "Airpower and Private Enterprise: Federal-Industrial Relations in the Aeronautics Field, 1918–1926" (PhD diss., Washington University, 1970). Nick Komons presented the Air Commerce Act as the result of special interest group advocacy and stressed the role of William P. MacCracken Jr., the first assistant secretary of commerce for aeronautics, in shaping the dialogue that resulted in its passage; Nick Komons, *Bonfires to Beacons: Federal Civil Aviation Policy under the Air Commerce Act, 1926– 1938* (Washington, DC: Smithsonian Institution Press, 1989). Most recently, M. Houston Johnson V—relying heavily on Komons—presented the Air Commerce Act as the sole

result of Secretary of Commerce Herbert Hoover's regulatory vision; M. Houston Johnson V, *Taking Flight: The Foundations of American Commercial Aviation, 1918–1938* (College
Station: Texas A&M University Press, 2019). While the Air Commerce Act has become
ubiquitous in the literature on American aviation in the twentieth century, no study has
placed the origins of US aviation policy within its proper historical and international
context, a deficiency this book seeks to rectify.

    13. Thomas J. Misa, "Retrieving Technological Change from Technological Deter-
minism," in *Does Technology Drive History? The Dilemma of Technological Determinism*,
ed. Merritt Roe Smith and Leo Marx (Cambridge, MA: MIT Press, 1994), 139; Daniel P.
Carpenter, *The Forging of Bureaucratic Autonomy: Reputations, Networks, and Policy Innovations in Executive Agencies, 1862–1928* (Princeton, NJ: Princeton University Press, 2001),
19. For examples of similar dynamics in railroad and automobile regulation, respectively,
see Steven W. Usselman, *Regulating Railroad Innovation: Business, Technology, and Politics
in America, 1840–1920* (Cambridge: Cambridge University Press, 2002), and Lee Vinsel,
Moving Violations: Automobiles, Experts, and Regulations in the United States (Baltimore:
Johns Hopkins University Press, 2019).

    14. In this respect, this book could be seen as a historical case study in support of the
process of mental model formation, shared mental model creation, and institutionalization put forth by Denzau and North. Arthur T. Denzau and Douglass C. North, "Shared
Mental Models: Ideologies and Institutions," *KYKLOS* 47 (Feb. 1994): 3–31.

    15. Tom D. Crouch, *A Dream of Wings: Americans and the Airplane, 1875–1905* (New
York: W. W. Norton, 1981), 303–304. Langdon Winner also argues that airplane flight
represented "something new" to the human experience in *The Whale and the Reactor*
(Chicago: University of Chicago Press, 1986), 13. For a history of the dream of flight in
the human experience, see Bayla Singer, *Like Sex with Gods: An Unorthodox History of
Flying* (College Station: Texas A&M University Press, 2003).

    16. The ability of technologies to prompt new ideas and reconfigure existing ones
has been aptly demonstrated in numerous studies. In his influential 1954 essay "The
Question concerning Technology," German philosopher Martin Heidegger proposed that
the essence of technology rests in its power to reshape how human beings view the world
around them. For Heidegger, modern technology results in people viewing the natural
world as "standing reserve," a collection of resources that constitute a means to an end
rather than something of value in and of itself. As the creation of the FAA's National
Airspace System shows, the airplane has transformed the atmosphere into a resource;
Martin Heidegger, "The Question concerning Technology," in *Philosophy of Technology:
The Technological Condition, An Anthology*, 2nd ed., ed. Robert C. Scharff and Val Dusek
(Oxford: Wiley-Blackwell, 2014). A decade later, Thomas Kuhn illustrated how devices
created to extend human sense perception, "constructed mainly for anticipated functions," can foster anomalies within the established framework of normal science that
precipitate shifts in fundamental scientific understanding; Thomas S. Kuhn, *The Structure
of Scientific Revolutions*, 3rd ed. (Chicago: University of Chicago Press, 1962, 1996), 64–65.
Evidence to support Kuhn's and Heidegger's argument that technological developments
both modify existing ideas and stimulate new ways of thinking can be found throughout
historical studies. A few of the myriad examples include Peter Norton's analysis of how

the proliferation of motor vehicles instigated a reconceptualization of the use and purpose of urban streets, Peter D. Norton, *Fighting Traffic: The Dawn of the Motor Age in the
American City* (Cambridge, MA: MIT Press, 2011); Stephen Skowronek's emphasis on the
centrality of the railroad industry in the creation of the modern American administrative
state, Stephen Skowronek, *Building a New American State: The Expansion of National Administrative Capacities, 1877–1920* (Cambridge: Cambridge University Press, 1982); Thomas P.
Hughes's illustration of how the economies of scale within electrical production precipitated a shift to ever-larger systems, Thomas P. Hughes, *Networks of Power: Electrification
in Western Society, 1880–1930* (Baltimore: Johns Hopkins University Press, 1983); and
Richard R. John's study of how the telegraph and telephone facilitated a shift from the
nineteenth-century's antimonopolistic, competitive political economy to one that advo-
cated government regulation of "natural" monopolies, Richard R. John, *Network Nation:
Inventing American Telecommunications* (Cambridge, MA: Harvard University Press, 2010).
In the field of aviation history, Michael Sherry examined how the airplane instigated a
radical reappraisal of warfare in the first half of the twentieth century in *The Rise of
American Air Power: The Creation of Armageddon* (New Haven, CT: Yale University Press,
1987), Stuart Banner detailed how the airplane prompted a preconception of the legal
standing of the atmosphere in *Who Owns the Sky? The Struggle to Control Airspace from
the Wright Brothers On* (Cambridge, MA: Harvard University Press, 2008), and Jason
Weems illustrated how the airplane reinforced and expanded spatial understandings of
the American Midwest in *Barnstorming the Prairies: How Aerial Vision Shaped the Midwest*
(Minneapolis: University of Minnesota Press, 2015).

17. Joseph J. Corn, *The Winged Gospel: America's Romance with Aviation*, 2nd ed.
(Baltimore: Johns Hopkins University Press, 2002), 30.

18. The simultaneous emergence of radio along with the airplane instigated a broader
reconceptualization of the atmosphere during the first decades of the twentieth century.
It is important, however, not to project this analytical connection backward into the
minds and motivations of actors at the time. Archival documents plainly show that—
other than nationality designations for aircraft engaged in international flight, as discussed in chapter 2—Americans grappling with the government's relationship to aviation
looked to contemporary regulatory regimes for waterway transportation, railroads, and
the automobile for guidance and not radio. For a holistic approach to atmospheric regulation, see Rikitianskaia et al., "The Mediation of the Air: Wireless Telegraphy and the
Origins of a Transnational Space of Communications, 1900–1910s," *Journal of Communications* 68, no. 4 (2018): 758–779. For studies of telegraphy and radio in the domestic
and international context during this period, see John, *Network Nation*; and Heidi J. S.
Tworek, *News from Germany: The Competition to Control World Communications, 1900–1945*
(Cambridge, MA: Harvard University Press, 2019).

19. Stephen D. Krasner, "Abiding Sovereignty," *International Political Science Review* 22,
no. 3 (July 2001): 229–251.

20. Crouch, *Wings*, 112–113.

21. For powered flight's ability to stir pilots and passengers toward new artistic expressions, see Laurence Goldstein, *The Flying Machine and Modern Literature* (Blooming-

ton: Indiana University Press, 1986); Robert Wohl, *A Passion for Wings: Aviation and the Western Imagination, 1908–1918* (New Haven, CT: Yale University Press, 1994); Robert Wohl, *The Spectacle of Flight: Aviation and the Western Imagination, 1920–1950* (New Haven, CT: Yale University Press, 2005); and Weems, *Barnstorming the Prairies.*

22. Richard P. Hallion, *Taking Flight: Inventing the Aerial Age from Antiquity through the First World War* (Oxford: Oxford University Press, 2003), 337.

23. In this respect, the airplane of the early twentieth century shares much in common with other emerging dual-use technologies such as nuclear power in the 1940s and 1950s and the internet of the 1980s and 1990s. In each of these cases, technological development occurred "at an exponential pace . . . outpacing regulatory processes" that in turn prompted the creation of new governmental structures to facilitate their commercial use and mitigate their potentially destabilizing applications. See Marc A. Saner and Gary E. Merchant, "Proactive International Regulatory Cooperation for Governance of Emerging Technologies," *Jurimetrics: The Journal of Law, Science and Technology* 55, no. 2 (2015): 147–178.

24. Lyria Bennett Moses argues that technology can prompt legal change in four ways: (1) the need for new laws to regulate new conduct made possible by the new technology, (2) uncertainty as to how the law relates to the new technology, (3) existing rules do not adequately incorporate the new technology, and (4) existing rules become obsolete due to the new technology. All four criteria applied to the airplane. Lyria Bennett Moses, "Towards a Theory of Law and Technology: Why Have a Theory of Law and Technological Change," *Minnesota Journal of Law, Science, and Technology* 8 (2007): 589–715.

25. G. Edward White, *Law in American History, Vol. 1, From the Colonial Years through the Civil War* (New York: Oxford University Press, 2012), 187. Stephen D. Krasner divides sovereignty into four elements: international legal sovereignty, Westphalian sovereignty, domestic sovereignty, and interdependent sovereignty. Krasner, "Sovereignty and Its Discontents," in *Power, the State, and Sovereignty*, ed. Stephen Krasner (Oxford: Routledge, 2009), 179–210.

26. Wartime aviator Edgar Gorrell pointed out that "a flyer flying from Washington, D.C., to Mineola, Long Island, passes over six states in about two hours" to make a technologically deterministic argument in favor of federal control and against state action. Edgar S. Gorrell, "Rules of the Air," *US Air Service* 4, no. 2 (Sept. 1920): 15.

27. Stephen Krasner, "Structural Causes and Regime Consequences: Regimes as Intervening Variables," in *International Regimes*, ed. Stephen Krasner (Ithaca, NY: Cornell University Press, 1983), 9.

28. The transnational turn has resulted in a plethora of studies showing the influence of international ideas and developments on the United States. See Ira Katznelson and Martin Shefter, eds., *Shaped by War and Trade: International Influences on American Political Development* (Princeton, NJ: Princeton University Press, 2002); Eric Rauchway, *Blessed among Nations: How the World Made America* (New York: Hill and Wang, 2006); Ian Tyrrell, *Transnational Nation: United States History in Global Perspective Since 1789* (New York: Palgrave Macmillan, 2007); Jay Sexton, *The Monroe Doctrine: Empire and Nation in Nineteenth-Century America* (New York: Hill and Wang, 2011); Daniel Gorman, *The Emergence of Inter-*

national Society in the 1920s (Cambridge: Cambridge University Press, 2012); and Benjamin Allen Coates, *Legalist Empire: International Law and American Foreign Relations in the Early Twentieth Century* (New York: Oxford University Press, 2016).

29. James Hansen, "Aviation History in the Wider View," *Technology and Culture* 30 (July 1989): 643–656.

Chapter 1. Where Does the Regulatory Power Lie?

1. Civil Aeronautics Act of 1923, H.R. 13715, 67th Cong. (1923), 18. The two lawyers hit it off so well that they went on to open a law practice together in 1930.

2. William P. MacCracken Jr., "Aeronautical Legal Problems," address delivered at the First National Air Institute under the auspices of the Detroit Aviation Society, Oct. 11, 1922, 31; *Law Memoranda upon Civil Aeronautics: Printed for the Use of the Committee on Interstate and Foreign Commerce House of Representatives* (Washington, DC: Government Publishing Office, 1923), 49.

3. *Law Memoranda upon Civil Aeronautics*, 48; US Const. art. I, § 8, cl. 3.

4. US Const. amend. X.

5. Regulatory historian Harry N. Scheiber refers to this dynamic as "the problem of areal-functional congruity." Harry N. Scheiber, "Federalism and the American Economic Order, 1789–1910," *Law and Society* 10, no. 1 (Oct. 1975): 70; William J. Novak, *The People's Welfare: Law and Regulation in Nineteenth-Century America* (Chapel Hill: University of North Carolina Press, 1996), 2.

6. According to a path-dependent theory of change, previous decisions heavily influence possible decisions in the present by affecting the various opportunity costs attached to them. Those options with lower opportunity costs as a result of previous decisions are therefore more likely to be chosen than those that cause widespread disruption. An extreme example, bordering on the absurd, would be if the people of the United States decided to abandon agriculture and return to a hunter-gatherer lifestyle. It technically could be done, but the costs would be extraordinarily high and the transition painfully difficult. For more on the effect of path-dependent change on technological development, see W. Brian Arthur, "Competing Technologies, Increasing Returns, and Lock-In by Historical Events," *Economic Journal* 99, no. 394 (Mar. 1989): 116–131. For a fascinating study of the importance of path-dependent change in understanding American political development, see Paul Pierson, *Politics in Time: History, Institutions, and Social Analysis* (Princeton, NJ: Princeton University Press, 2004).

7. Herbert A. Johnson, *Gibbons v. Ogden: John Marshall, Steamboats, and the Commerce Clause* (Lawrence: University Press of Kansas, 2010), 27–29; George R. Taylor, *The Transportation Revolution, 1815–1860* (New York: Rinehart and Co., 1951), 58–59.

8. Charles F. Hobson, *The Great Chief Justice: John Marshall and the Rule of Law* (Lawrence: University Press of Kansas, 1996), 1–4.

9. G. Edward White, *Law in American History, Vol. 1: From the Colonial Years through the Civil War* (New York: Oxford University Press, 2012), 158–159.

10. US Const. art. I, § 8, cl. 3 (the commerce clause) and US Const. art. I, § 8, cl. 18 (the Necessary and Proper Clause). Myriad questions concerning commercial exchange dominated the Constitutional Convention. See Walton Hale Hamilton and Douglass

Adair, *The Power to Govern: The Constitution, Then and Now* (New York: W. W. Norton, 1937).

11. "Chapter 2, 1 Congress, Session 1, An Act: For laying a duty on goods, wares, and merchandises imported into the United States," US Statutes at Large 1: 24–27; "Chapter 3, 1 Congress, Session 1, An Act: Imposing duties on tonnage," US Statutes at Large 1: 27–28; "Chapter 5, 1 Congress, Session 1, An Act: To regulate the collection of the duties imposed by law on the tonnage of ships or vessels, and on goods, wares, and merchandise," US Statutes at Large 1: 29–49; "Chapter 9, 1 Congress, Session 1, An Act: For the establishment and support of light-houses, beacons, buoys, and public piers," US Statutes at Large 1: 53–54. Congress's creation of a system of navigational aids calls into question the belief in a laissez-faire, nonexistent national government during the early Republic. For further evidence of an active and indispensable federal government during this period, see Richard R. John, *Spreading the News: The American Postal System from Franklin to Morse* (Cambridge, MA: Harvard University Press, 1995), and Brian Balogh, *A Government Out of Sight: The Mystery of National Authority in Nineteenth-Century America* (Cambridge: Cambridge University Press, 2009). For a detailed discussion of tariffs in the early Republic, see Gautham Rao, *National Duties: Custom Houses and the Making of the American State* (Chicago: University of Chicago Press, 2016).

12. This consensus on Congress's power to facilitate navigational aids for waterborne transport did not extend to the creation of roads and canals, collectively referred to as internal improvements. Although Jefferson's treasury secretary Albert Gallatin proposed a systematic program in 1808 and Senator Henry Clay of Kentucky adopted a similar plan as part of his "American System," states' rights supporters—concerned about the impact of expanded federal power on the continued existence of slavery and most at home in the Democratic Party of first Jefferson and then Jackson—viewed a constitutional amendment as a prerequisite to any comprehensive federal plan. See John Lauritz Larson, *Internal Improvement: National Public Works and the Promise of Popular Government in the Early United States* (Chapel Hill: University of North Carolina Press, 2001), and George R. Taylor, *The Transportation Revolution 1815–1860*, 3rd ed. (New York: M. E. Sharpe, 1977).

13. Grant S. Nelson and Robert J. Pushaw Jr., "Rethinking the Commerce Clause: Applying First Principles to Uphold Federal Commercial Regulations but Preserve State Control over Social Issues," *Iowa Law Review* 85, no. 1 (1999): 14. Much ink has been spilt over the years in an attempt to understand the original intent of the commerce clause. For analyses that support a limited interpretation, see Randy E. Barnett, "The Original Meaning of the Commerce Clause," *University of Chicago Law Review* 68, no. 1 (2001): 101–148, and Robert G. Natelson, "The Legal Meaning of 'Commerce' in the Commerce Clause," *St. John's Law Review* 80, no. 3 (2006): 789–848. For arguments in support of a broad interpretation, see Nelson and Pushaw, "Rethinking the Commerce Clause"; William Winslow Crosskey, *Politics and the Constitution in the History of the United States* (Chicago: University of Chicago Press, 1953); and Hamilton and Adair, *The Power to Govern*.

14. In *McCulloch v. Maryland* (1819), Marshall drew upon the Constitution's Supremacy Clause and the Necessary and Proper Clause to declare that the federal government, "though limited in its powers, is supreme; and its laws, when made in pursuance of the constitution, form the supreme law of the land." According to the doctrine of supremacy,

whenever a conflict exists between federal and state law, the federal law takes precedent. *McCulloch v. the State of Maryland et al.*, 17 US 316, 437 (4). Recently appointed Justice Smith Thompson recused himself from *Gibbons* because he had sat on the New York Court of Errors in its previous decisions on the case.

15. Hobson, *The Great Chief Justice*, 111–112; Felix Frankfurter, *The Commerce Clause under Marshall, Taney, and Waite* (Chapel Hill: University of North Carolina Press, 1937), 14.

16. *Gibbons, Appellant, v. Ogden, Respondent*, 22 US 1, 240 (9).

17. *Gibbons, Appellant, v. Ogden, Respondent*, 22 US 1, 240 (9); Marshall reaffirmed this pro-state position five years later in *Thompson Willson and Others, Plaintiffs in Error v. the Black Bird Creek Marsh Company, Defendants*, 27 US 245, 252 (2). For a detailed analysis of the public health argument in support of state police powers, see Novak, *The People's Welfare*.

18. *Cooley v. Board of Wardens*, 53 US 299.

19. Hamilton and Adair convincingly demonstrate that members of the Constitutional Congress held the same definition of the term "commerce" as Marshall. With *Gibbons*, this broader conception of the term became legal precedent subject to the principle of stare decisis. See Hamilton and Adair, *The Power to Govern*. Chief Justice Roberts drew heavily upon Marshall's definition of commerce in *Gibbons v. Ogden* to rule that the commerce clause did not support the Patient Protection and Affordable Care Act's mandate to purchase health insurance as argued by the Obama administration. Roberts' court ultimately upheld the mandate as a tax that fell under Congress's established constitutional authority. See *National Federation of Independent Business v. Sebelius* (2012).

20. Paul F. Paskoff, *Troubled Waters: Steamboat Disasters, River Improvements, and American Public Policy, 1821–1860* (Baton Rouge: Louisiana State University Press, 2007), 14.

21. For more on the shifting American economy during the antebellum period, see John Lauritz Larson, *The Market Revolution in America: Liberty, Ambition, and the Eclipse of the Common Good* (Cambridge: Cambridge University Press, 2010).

22. John G. Burke, "Bursting Boilers and the Federal Power," *Technology and Culture* 7, no. 1 (Winter 1966): 3–10.

23. Hunter, *Steamboats on the Western Rivers*, 523–524; Jerry L. Mashaw, *Creating the Administrative Constitution: The Lost One Hundred Years of American Administrative Law* (New Haven, CT: Yale University Press, 2012), 380n13.

24. Memorial of the Legislature of Mississippi, Feb. 15, 1838. An 1847 Federal Circuit Court ruling that the Louisiana law could not extend to steamboats engaged in interstate voyages cemented the position that steamboat safety regulation was a federal issue. *Halderman v. Beckwith*, 11 Federal Cases 174-175 (D. Ohio 1847) (No. S. 907), cited in Hunter, *Steamboats on the Western Rivers*, 524.

25. "Chapter 191, 25 Congress, Session 2, An Act: To provide for the better security of the lives of passengers on board of vessels propelled in whole or in part by steam," 5 Main Section Stat. 304, 306.

26. Andrew Jackson, *Fifth Annual Message to Congress*, Dec. 3, 1833; 23 Cong. Globe 49 (1834). In 1830, the Philadelphia-based Franklin Institute established a committee to study the steam boiler issue and conducted a series of experiments in the preceding years with funding from the Treasury Department; Burke, "Bursting Boilers and the Federal

Power." For more on the importance of the Franklin Institute in the subsequent development of the engineering profession and standard setting, see Andrew L. Russell, *Open
Standards and the Digital Age: History, Ideology, and Networks* (Cambridge: Cambridge University Press, 2014), and JoAnne Yates and Craig N. Murphy, *Engineering Rules: Global
Standard Setting since 1880* (Baltimore: Johns Hopkins University Press, 2019).

27. "Chapter 106, 32 Congress, Session 1, An Act: To amend an act entitled 'An act to
provide for the better security of the lives of passengers on board of vessels propelled in
whole or in part by steam,' and for other purposes," 10 Main Section Stat. 61, 75. In 1903,
the Steamboat Inspection Service was transferred from the Treasury Department to the
newly established Department of Commerce and Labor.

28. *Proceedings, Board of Supervising Inspectors of Steamboats, 1861*, 30; *Proceedings,
1863*, 59, cited in Hunter, *Steamboats on the Western Rivers*, 541. Hunter points out that
casualty figures prior to 1852 came largely from newspaper reports, which had a tendency
to exaggerate the death toll. Even if one assumes a 100 percent exaggeration rate and
adjusts the numbers of deaths to 577 from 1848 to 1852, it still represents a substantial
reduction.

29. Congress established the Bureau of Navigation in 1884 within the Treasury Department to oversee enforcement of the various navigation acts. Both it and the Steamboat Inspection Service were incorporated into the Department of Labor and Commerce
upon its creation in 1903 and then placed under the purview of the Department of Commerce ten years later. For more on the early history of the Steamboat Inspection Service
and the Bureau of Navigation, see Lloyd M. Short, *Steamboat Inspection Service: Its History,
Activities, and Organization* (New York: D. Appleton, 1922), and Lloyd M. Short, *The Bureau
of Navigation: Its History, Activities, and Organization* (Baltimore: Johns Hopkins University Press, 1923).

30. Steven W. Usselman, *Regulating Railroad Innovation: Business, Technology, and
Politics in America, 1840–1920* (Cambridge: Cambridge University Press, 2002), 61; Franz
Anton Ritter von Gerstner, *Early American Railroads*, ed. Fredrick C. Gamst, trans. David J.
Diephouse and John C. Decker (Stanford, CA: Stanford University Press, 1997), 295; Wil-
liam Z. Ripley, *Railroads: Rates and Regulations* (London: Longmans, Green, 1912), 34.

31. For more on the relationship between the growth of railroads and the expansion
of federal power, see Herbert Hovenkamp, "Regulatory Conflict in the Gilded Age: Federalism and the Railroad Problem," *The Yale Law Journal* 97, no. 6 (1988): 1017–1072, and
Steven Skowronek, *Building a New American State: The Expansion of National Administrative Capacities, 1877–1920* (Cambridge: Cambridge University Press, 1982).

32. The Maryland General Assembly's charter of the Baltimore & Ohio Railroad
Company, the first of its kind, vested the new corporation with "all the rights and powers
necessary" to construct, operate, and repair a railroad from Baltimore to "some suitable
point on the Ohio River"; exempted its capital from taxation; and granted access to the
state's power of eminent domain for the appropriation of lands and material with just
compensation if the owner objected. These broad powers were mirrored in the 1827 Virginia and 1828 Pennsylvania charters for the B&O and quickly became standard in state
charters throughout the nation. *An Act to Incorporate the Baltimore and Ohio Rail Road
Company, Passed at December Session, 1826* (Baltimore: William Wooddy, 1827); An Act

passed by the Virginia Legislature to confirm a law, passed at the present session of the General Assembly of Maryland, entitled, "An Act to incorporate the Baltimore and Ohio Rail Road Company," passed March 8, 1827; An Act, passed by the Legislature of Pennsylvania, to authorize the Baltimore and Ohio Rail Road Company to construct a Rail Road through Pennsylvania, in a direction from Baltimore to the Ohio River, passed Feb. 1828.

33. State charters were an established method of creating private entities vested with a public interest that had previously been used for insurance companies, benevolent and bible societies, turnpike and canal companies, and churches. According to historian Harvey H. Segal, state-financed regional projects between 1815 and 1834 resulted in 3,326 miles of canals at a cost of $125,000,000, imprisoning many states in debt just as the new technology of the railroad promised a means to more cheaply connect the nation. Charters therefore offered an economical way to facilitate railroad construction that neatly complemented a rising belief in the superiority of private enterprise and the anti-government overtones of the Jacksonian Era. Harvey H. Segal, "Cycles of Canal Construction," found in Carter Goodrich, ed., *Canals and American Economic Development* (New York: Columbia University Press, 1961), 179.

34. Novak, *The People's Welfare*, 121–123; Statement of W. M. Acworth, Esq., London, England, *Extension of Tenure of Government Control of Railroads: Hearings Before the Committee on Interstate Commerce* (Washington, DC: Government Publishing Office, 1919); Thomas M. Cooley, "State Regulation of Corporate Profits," *North American Review* 137, no. 322 (Sept. 1883): 205–217; An Act to Incorporate the Baltimore and Ohio Rail Road Company, Passed at December Session, 1826.

35. Alfred D. Chandler Jr., *The Visible Hand: The Managerial Revolution in American Business* (Cambridge, MA: Belknap Press of Harvard University Press, 1977), 81–82; James W. Ely, *Railroads and American Law* (Lawrence: University Press of Kansas, 2001), 73; Constitution of the State of Illinois, adopted and ratified in 1870. The idea of railroads as public highways "free to all persons" persisted in the minds of many throughout the nineteenth century, and in many ways this divergence between cultural expectations of transportation and the technical realities of railroading formed the basis for all subsequent calls for action to address the "railroad problem."

36. *Historical Statistics of the United States, Millennial Edition Online*, table Df874–881, "Railroad Mileage and Equipment: 1830–1890."

37. An Act to Incorporate the Pennsylvania Railroad Company, April 13, 1846. For more on the expansion of the Pennsylvania Railroad, see Albert J. Churella, *The Pennsylvania Railroad, Vol. 1: Building an Empire* (Philadelphia: University of Pennsylvania Press, 2013).

38. Adams firmly believed that the nature of railroads meant that the industry constituted a "natural monopoly" that precluded Adam Smith's invisible hand of market forces from serving as a check. Thomas K. McCraw, *Prophets of Regulation: Charles Francis Adams, Louis D. Brandeis, James M. Landis, Alfred E. Kahn* (Cambridge, MA: Belknap Press of Harvard University Press, 1984), 7–10. Adam Plaiss masterfully traces the development and application of natural monopoly ideology across the nineteenth and early twentieth centuries in "From Natural Monopoly to Public Utility: Technological Determinism and the Political Economy of Infrastructure in Progressive-Era America," *Technology and Culture* 57, no. 4 (Oct. 2016): 806–830.

39. The best biographical treatment of Adams remains McCraw, *Prophets of Regulation*. For an analysis of Adams's thoughts on the value of commissions and their connection to elitism, see Samuel DeCanio, *Democracy and the Origins of the American Regulatory State* (New Haven, CT: Yale University Press, 2015).

40. Hovenkamp, "Regulatory Conflict in the Gilded Age," 1033.

41. Richard C. Cortner, *The Iron Horse and the Constitution: The Railroads and the Transformation of the Fourteenth Amendment* (Westport, CT: Greenwood Press, 1993), 3–4.

42. Charles Francis Adams, "The Government and the Railroad Corporations," *North American Review* 112, no. 230 (Jan. 1871), 6–7.

43. *Proceedings of the Seventh Session of the National Grange of the Patrons of Husbandry* (New York: S. W. Green, 1874), 15. Dudley traced his lineage to Henry Adams, the founder of the famed Adams family of Massachusetts that included Charles Adams Jr. Harriet M. Bryant, "Dudley Warren Adams, Pioneer," *Florida Historical Quarterly* 35, no. 3 (Jan. 1957): 241.

44. Quoted in Schonberger, *Transportation to the Seaboard: The "Communications Revolution" and American Foreign Policy, 1860–1900* (Westport, CT: Greenwood Press, 1971), 41.

45. *Munn v. Illinois*, 94 US 113, 154 (1877); *Chicago, Burlington & Quincy Railroad Company v. Iowa*, 94 US 155, 163 (1877).

46. *Case of the State Freight Tax* 82 US 15 Wall. 232 232 (1872); *Pensacola Telegraph Company v. Western Union Telegraph Company*, 96 US 1, 23 (1878). See also James E. Ely Jr., "The Railroad System Has Burst through State Limits: Railroads and Interstate Commerce, 1830–1920," *Arkansas Law Review* 55, no. 4 (Winter 2003), 940.

47. From 1850 to 1871, grants of federal land totaling nearly two hundred million acres made with the condition that the government could transport troops and mails at a preferred rate along subsequent rail lines gave the federal government an indirect stake in the operation of the nation's railroads; *Historical Statistics of the United States, Millennial Edition Online*, table Df890, "Federal Land Grants Used by Railroads, by State: 1850–1871." Such grants were usually made to states, which then dispersed them to railroad corporations.

48. "Chapter 15, 37 Congress, Session 2, An Act: To authorize the President of the United States in certain cases to take possession of railroad and telegraph lines, and for other purposes," 12 Main Section Stat. 334, 335. This grant of authority expressly forbade the construction of federal railroads within sovereign states. For detailed analysis of the importance of railroads in the Civil War, see John E. Clark, *Railroads in the Civil War* (Baton Rouge: Louisiana State University Press, 2001), and William G. Thomas, *The Iron Way: Railroads, the Civil War, and the Making of Modern America* (New Haven, CT: Yale University Press, 2011).

49. "Chapter 120, 37 Congress, Session 2, An Act: To aid in the construction of a railroad and telegraph line from the Missouri River to the Pacific Ocean, and to secure to the government the use of the same for postal, military, and other purposes," 12 Main Section Stat. 489, 489. The same act also authorized the Central Pacific Railroad, chartered in California, to extend eastward to meet up with the Union Pacific Railroad. Zachary Callen argues that a lack of resources and the high level of competition among states undermined the creation of a transnational railroad through state action, and that this failure

compelled the elevation of the issue to the federal level. See *Railroads and American Political Development: Infrastructure, Federalism, and State Building* (Lawrence: University Press of Kansas, 2016).

50. "Chapter 127, 39 Congress, Session 1, An Act: To facilitate commercial, postal, and military Communication among the several States," 14 Main Section Stat. 124; Lewis Henry Haney, "A Congressional History of Railways in the United States, 1850–1887," *Bulletin of the University of Wisconsin*, no. 842, Economics and Political Science Series 6, no. 1 (Jan. 1910): 215–227. New Jersey's continued support of the Camden and Amboy's monopoly on rail transportation within that state and Pennsylvania's cancellation of an Ohio company's charter to protect the Pennsylvania Railroad compelled congressional action. Congress did recognize, however, that the construction of any railway through a state required its approval. That same year, Congress also passed the 1866 Telegraph Act, which allowed the construction of lines over federal-controlled lands, gave government transmissions priority, and provided a means for possible government purchase of lines at a fair price. For more on the telegraph's place in the political economy of the United States, see Richard R. John, *Network Nation: Inventing American Telecommunications* (Cambridge, MA: Belknap Press of Harvard University Press, 2010).

51. 41 Cong. Globe 423, 448 (1871); "Chapter 252, 42 Congress, Session 3, An Act: To prevent cruelty to animals while in transit by railroad or other means of transportation within the United States," 17 Main Section Stat. 584, 585; Haney, "A Congressional History of Railways," 260–262. This 1873 law arose not only out of public health concerns about the prolonged exposure of cattle to their dead brethren within unventilated railcars but also an increased interest in the humane treatment of animals. The American Society for the Prevention of Cruelty to Animals (ASPCA), founded in 1866, drew upon deep-seated Christian notions of stewardship in its push for legal remedies for animal cruelty. See Janet M. Davis, *The Gospel of Kindness: Animal Welfare and the Making of Modern America* (Oxford: Oxford University Press, 2016). For more on the relationship between eastern markets and the western hinterland, as well as the central role of Chicago and new technologies in that relationship, see William Cronin, *Nature's Metropolis: Chicago and the Great West* (New York: W. W. Norton, 1991).

52. *Select Committee on Transportation Routes to the Seaboard*, S. Rep. 43-307, pt. 1 (Washington, DC: Government Publishing Office, 1874), 93. This investigative committee was popularly known as the Windom Committee after its chairperson, Republican senator William Windom of Minnesota.

53. Chester A. Arthur, *Third Annual Message to Congress*, Dec. 4, 1883; Ely, *Railroads and American Law*, 90.

54. The House passed the stricter Reagan bill while the Senate passed the more moderate Cullom bill.

55. Report of the Senate Select Committee on Interstate Commerce, 49th Cong., 1st Sess., submitted to the Senate Jan. 18, 1886 (Washington, DC: Government Publishing Office, 1886), 38–39.

56. *Wabash, St. Louis & Pacific Railway Company v. Illinois*. 118 US 557, 595 (1886). The Supreme Court had first applied the concept of "national character" to waterway transportation in *Cooley v. Board of Wardens* (1851) and then to railroads in the *Case of the State*

Freight Tax (1872) mentioned previously. For a discussion of commerce clause precedents leading up to the *Wabash* decision, see David P. Currie, *The Constitution in the Supreme Court: The First Hundred Years, 1789–1888* (Chicago: University of Chicago Press, 1985), 403–413.

57. Grover Cleveland, *Second Annual Message to Congress* (first term), Dec. 6, 1886.

58. "Chapter 104, 49 Congress, Session 2, An Act: To regulate commerce," 24 Main Section Stat. 379, 387 (1883–1887).

59. Skowronek, *Building a New American State*, 145.

60. Roosevelt shepherded the passage of the 1903 Elkins Act and the 1906 Hepburn Act, while Taft signed the 1910 Mann-Elkins Act. The Supreme Court tacitly affirmed the Elkins Act's expansion of the ICC's authority in 1910 when it upheld the commission's decision against a railroad company–supported rate increase in the Eastern Rate Case. For a detailed analysis of the Eastern Rate Case and Louis Brandeis's use of Fredrick Winslow Taylor's notions of scientific management, see Usselman, *Regulating Railroad Innovation*, and Gerald Berk, *Louis D. Brandeis and the Making of Regulated Competition, 1900–1932* (Cambridge: Cambridge University Press, 2009).

61. *Gibbons, Appellant, v. Ogden, Respondent*, 22 US 1, 240 (9); Hoovenkamp, "Regulatory Conflict in the Gilded Age: Federalism and the Railroad Problem," 1065.

62. For an overview of Hughes's career, see McCraw, *Prophets of Regulation*, and Paul Brickner, "Different Styles and Similar Values: The Reformer Roles of Charles Evans Hughes and Louis Dembitz Brandeis in Gas, Electric, and Insurance Regulation," *Indiana Law Review* 33, no. 3 (2000): 893–920. For a personal account, see Charles Evans Hughes, *The Autobiographical Notes of Charles Evans Hughes*, ed. David J. Danelski and Joseph S. Tulchin (Cambridge, MA: Harvard University Press, 1973).

63. The two cases in questions are the so-called Minnesota Rate Case, 230 US 352 (1913) and *Houston E. & W. Tex. Ry. Co. v. United States*, 234 US 342 (1914). The quote comes from Hughes's opinion in the latter.

64. Hughes, *The Autobiographical Notes of Charles Evans Hughes*, ed. David J. Danelski and Joseph S. Tulchin, 210.

65. "Chapter 196, 52 Congress, Session 2, An Act: To promote the safety of employees and travelers upon railroads by compelling common carriers engaged in interstate commerce to equip their cars with automatic couplers and continuous brakes and their locomotives with driving-wheel brakes, and for other purposes," 27 Main Section Stat. 531, 532. With the 1893 Safety Appliance Act, Congress empowered the ICC to mandate automatic brakes and couplers, initiating the commission's formal relationship with railroad safety some forty years after the establishment of the Steamboat Inspection Service. For an in-depth analysis of railroad safety devices and the ICC's role in their development and application, see Mark Aldrich, *Death Rode the Rails: American Railroad Accidents and Safety, 1828–1965* (Baltimore: Johns Hopkins University Press, 2006); Arwen P. Mohun, *Risk: Negotiating Safety in American Society* (Baltimore: Johns Hopkins University Press, 2013); and Steven W. Usselman, "Air Brakes for Freight Trains: Technological Innovation in the American Railroad Industry, 1869–1900," *Business History Review* 58, no. 1 (Spring 1984): 30–50.

66. On December 26, 1917, Wilson issued a proclamation wherein the federal gov-

ernment, through Secretary of War Newton D. Baker, took possession and control of "each and every system of transportation . . . within the boundaries of the continental United States and consisting of railroads, and owned or controlled systems of coastwise and inland transportation, engaged in general transportation, whether operated by steam or by electric power" and all "equipment and appurtenances commonly used upon or operated as a part of such rail or combined rail and water systems of transportation." The subsequent Federal Railroad Control Act, signed by Wilson on March 21, 1918, called for the creation of government–railroad contracts with compensation, provided $500,000,000 to cover necessary maintenance and improvements, and stipulated that federal control must end no later than seventeen months after the ratification of a peace treaty. The Transportation Act of 1920 returned the railroads to private control within a heavily regulated system that recognized the federal government's vital role in fostering a rational transportation network. Wilson, Proclamation 1419; Federal Railroad Control Act [as amended], 40 Stat. 451 (1918). Also see K. Austin Kerr, "Decision for Federal Control: Wilson, McAdoo, and the Railroads, 1917," *Journal of American History* 54 (Dec. 1967): 550–560; Rogers MacVeagh, *The Transportation Act, 1920: Its Sources, History, and Text, Together with Its Amendments to the Interstate Commerce Act* (New York: Henry Holt, 1923); and Richard Saunders, *Merging Lines: American Railroads, 1900–1970* (DeKalb: Northern Illinois University Press, 2001).

67. Harlan F. Stone, "Charles Thaddeus Terry," *Columbia Law Review* 23, no. 5 (May 1923): 415–417; Charles Thaddeus Terry, "Opening Address," *Proceedings of the First Annual Good Roads and Legislative Convention, Held at Buffalo, NY, July 7–8, 1908* (New York: American Automobile Association, 1908), 34. The AAA was established in 1902 with a charge to promote "rational [automobile] legislation."

68. Charles Thaddeus Terry, "Memorandum of Arguments and Authorities in Support of House Bill 428" in "Automobiles Engaged in Interstate Commerce," in H.R. Rep. No. 61-2270 (1911); "Full Text of Automobile Bill," *Automobile Topics*, Mar. 9, 1907.

69. Julius Goebel, *A History of the School of Law, Columbia University* (New York: Columbia University Press, 1955), 192. Terry served as NCCUSL secretary from 1906 to 1912 and president from 1912 to 1915.

70. "Full Text of Automobile Bill."

71. Committee on Interstate and Foreign Commerce, United States House, 61st Cong., 2nd sess., *Providing for Regulation, Identification, and Registration of Motor Vehicles Engaged in Interstate Travel* (Washington, DC: Government Publishing Office, 1910).

72. 46 Cong. Rec. H 4175 (Mar. 3, 1911). The revised bill that made it out of committee, H.R. 32570, included a federal driver's license at the request of motorists alongside an explicit affirmation of a state's right to establish its own rules of the road and speed restrictions. An accompanying minority report issued by democratic committee members William C. Adamson and Charles L. Bartlett of Georgia, Thetus W. Sims of Tennessee, and William R. Smith of Texas ruthlessly attacked the bill as an unlawful federal appropriation of sovereign state authority. See Automobiles Engaged in Interstate Commerce, H.R. Rep. No. 61-2270 (1911).

73. Adamson became chair of the House Committee on Interstate and Foreign

Commerce after the election of 1910 put Democrats in control of the House for the first time since 1895.

74. The 1935 Motor Carrier Act placed interstate trucking under the jurisdiction of the ICC. See Shane Hamilton, *Trucking Country: The Road to America's Wal-Mart Economy* (Princeton, NJ: Princeton University Press, 2008). For a detailed study of how the federal government expanded its regulatory authority over safety and environmental aspects of the automobile in response to various constituencies during the twentieth century, see Lee Vinsel, *Moving Violations: Automobiles, Experts, and Regulations in the United States* (Baltimore: Johns Hopkins University Press, 2019). For more on the voluntary adoption of uniform rules of the road in the 1920s under the auspices of Commerce Secretary Herbert Hoover, see Peter Norton, *Fighting Traffic: The Dawn of the Motor Age in the American City* (Cambridge, MA: MIT Press, 2008), and Vinsel, *Moving Violations*.

75. *Providing for Additional Security in States' Identification Act of 2009*, S. Rep. 111-104 (Washington, DC: Government Publishing Office, 2009).

76. *The 9/11 Commission Report*, 220, 231, https://www.9-11commission.gov/report /911Report.pdf (accessed July 17, 2019). Instead of establishing a national form of identification, Congress instituted stricter parameters for state-issued identifications under the REAL ID Act of 2005, partially in response to concerns that a national identification would negatively affect civil liberties (a concern that almost comes off as quaint in a world of corporate-controlled artificial intelligence assistants, facial recognition software, and foreign election meddling via social media). For analyses of the REAL ID Act, see Anna Ya Ni and Alfred Tat-Kei Ho, "A Quiet Revolution or a Flashy Blip? The Real ID Act and US National Identification System Reform," *Public Administration Review* 687, no. 6 (Nov.–Dec. 2008): 1063–1078, and Priscilla M. Regan and Christopher J. Deering, "State Opposition to REAL ID," *Publius* 39, no. 3 (Summer 2009): 476–505. For a sample of the widespread concern over a national ID even in the immediate wake of the 9/11 terrorist attacks, see *Does America Need a National Identifier?*, H. HRG. 107-118 (Washington, DC: Government Publishing Office, 2002).

77. "Memoranda in Re Section 221 of the Proposed Civil Aeronautics Act," in *Law Memoranda upon Civil Aeronautics: Printed for the Use of the Committee on Interstate and Foreign Commerce House of Representatives* (Washington, DC: Government Publishing Office, 1923), 48.

78. The first horseless carriages were powered by electric, steam, or internal combustion engines, but gas-powered machines promised the fastest speeds. In 1895, J. Frank Duryea averaged eight miles per hour when he won the first automobile race in the United States sponsored by the *Chicago Times-Herald*. By 1908, Henry Ford's Model T could achieve a maximum speed of roughly forty-five miles per hour. See James J. Flink, *America Adopts the Automobile, 1895–1910* (Cambridge, MA: MIT Press, 1970), 23. The relatively delicate nature of early automobiles and the horrendous state of the nation's rural roadways ensured that automobiles would initially develop as an urban phenomenon, and motorists quickly joined farmers in their call for improved roads. For an in-depth discussion of the relationship between roadways and automobile development, see John B. Rae, *The Road and the Car in American Life* (Cambridge, MA: MIT Press, 1971), and Clay McShane, *Down*

the Asphalt Path: The Automobile and the American City (New York: Columbia University Press, 1994).

79. "Runaway Due to Automobile: Mrs. Charles Welsh Is Injured in an Accident in Lincoln Park," *Chicago Daily Tribune*, June 23, 1899.

80. "Auto Causes Accident," *New York Tribune*, Nov. 15, 1902.

81. "Death Due to Automobile," *New York Tribune*, June 8, 1902.

82. "Killed in Carriage Accident," *New York Tribune*, Sept. 13, 1902.

83. For more on the public reaction to automobiles and how these new devices shaped roadway usage, see Norton, *Fighting Traffic*; Brian Ladd, *Autophobia: Love and Hate in the Automobile Age* (Chicago: University of Chicago Press, 2008); and Vinsel, *Moving Violations*.

84. While some states passed rules of the road after the Civil War, urban incorporation laws generally affirmed the authority of municipal governments to license common carriers and set rules of the road within their jurisdictions. See *The Municipal Code of Chicago: Comprising the Laws of Illinois Relating to the City of Chicago, and the Ordinances of the City Council; Codified and Revised* (Chicago: Beach, Barnard, 1881), 279–293; *The Revised Statutes of the State of New York as Altered by the Legislature* (Albany, NY: Packard and Van Benthuysen, 1836), 692–693; *The Statutes of Illinois, 1818 to 1868* (Chicago: E. B. Myers and Co., 1868), 663–664; *The Kentucky Statutes* (Louisville, KY: Courier-Journal Job Printing Company, 1899), 1182.

85. Because the Constitution vests all powers not expressly given to the federal government with the states, a municipality's regulatory power ultimately rests on a grant of state authority. Over the course of the nineteenth century, both state and federal courts recognized two primary means of such delegation: (1) a clear and specific grant of power within a township's charter or subsequent act of the legislature, or (2) an implied power required to carry out an explicitly granted power; C. P. Berry, *A Treatise on the Law Relating to the Automobile* (Chicago: Callaghan and Company, 1909), 52–53; Charles J. Babbitt, *The Law Applied to Motor Vehicles* (Washington, DC: John Byrne and Co., 1911), 176–177.

86. "Regulate Automobile Speed," *Chicago Daily Tribune*, June 24, 1899; *Proceedings of the City Council of the City of Chicago for the Municipal Year 1899–1900* (Chicago: John F. Higgins, 1900), 944–947. Approved on July 6, 1899—exactly two weeks after the Welsh's incident in Lincoln Park—the Chicago licensing ordinance required drivers to obtain a three-dollar license; instituted a five-to-twenty-five-dollar fine per offense; established a Board of Examiners of Operators of Automobiles to develop licensing examination criteria, administer examinations, and revoke licenses for cause; set an eight-mile speed limit for automobiles within the city; mandated the use of a four-inch bell or gong at intersections to notify pedestrians; and required all automobiles to be equipped with lamps and brakes capable of bringing the vehicle to a full stop within ten feet. Over the course of its first year, the board issued an additional 376 licenses, ten of which went to women. In June 1902, the City Council required the registration of all automobiles in the city, and a year later owners had to display their assigned registration number on a tag prominently affixed to their vehicle. See "Report of Chicago Automobile Examiners," *Western Electrician* 28, no. 7 (Feb. 16, 1901): 111; "Examinations for Automobile Operators," *Railroad Gazette* 32, no. 8 (Feb. 23, 1900): 125; and "Chicagoans Fight Tags," *Motor Age*, April 14, 1904.

87. *Washington Elec. Vehicle Co. v. District of Col.*, 19 App. D.C. 462, 1902; *People v. Schneider*, 139 Mich. 673, 676, 1905; *Commonwealth v. Hawkins*, 14 Pa. Dist. Ct. Reps. 592, 594, 1905.

88. *Chicago v. Banker*, 112 Ill. App. 94, 1904. In ruling against the Chicago ordinance's driver's license provision, the court accepted the Chicago Automobile Club's position that it represented an unfair restriction on one class of travelers in direct violation of the Fourteenth Amendment's equal protection provisions. The impression among early adopters that they were under siege by "class legislation" provided an important unifying force in the early years of automobility, particularly among the newly established Chicago Automobile Club as it struggled with "apathy, factionalism, and ineffective leadership" (Flink, *America Adopts the Automobile*, 149).

89. "Numbering Case Complex," *Motor Age*, April 28, 1904; Resolution adopted October 17, 1904, in "Club Will Use Numbers," *Motor Age*, Oct. 20, 1904.

90. *Charter and Revised Ordinances of the City of Joliet, Will County, State of Illinois* (Joliet, IL: Joliet Republican Printing Company, 1902), 264; *Revised Ordinances of the City of Waukegan, County of Lake, and State of Illinois* (Waukegan, IL: Waukegan Sun Press, 1905), 29–30.

91. "Evils of Local Ordinances," *Horseless Age*, Feb. 13, 1907.

92. The bill set speed limits at eight miles per hour in populated areas and fifteen miles per hour in other municipal areas. "Clubs Advocate a Uniform Speed Law," *Horseless Age*, Feb. 27, 1901; *The Revised Statutes, Codes and General Laws of the State of New York* (New York: Baker, Voorhis, 1901), 1640, 1645–1646. The ACA included such wealthy luminaries as Alfred G. Vanderbilt, Henry Clay Frick, and John W. Gates. The Hill-Cocks Bill, approved May 3, 1904, in *Automobile Laws of All the States of the United States* (New York: Automobile Club of America, 1905). Long Island Representative William W. Cocks was elected to Congress in 1905 and sponsored Terry's federal automobile bill in the 59th, 60th, and 61st Congresses.

93. US Department of Commerce, *Highway Statistics: Summary to 1955* (Washington, DC: Government Publishing Office, 1957), table MV-230, available at https://www.fhwa.dot.gov/ohim/summary95/mv230.pdf, and DL-230, available at https://www.fhwa.dot.gov/ohim/summary95/dl230.pdf, both accessed July 22, 2019. The District of Columbia and the territory of Hawaii also required state registrations by 1910, while New York's 1910 driver's license provision applied only to chauffeurs.

94. See *Radnor Township v. Bell*, 27 Pa. Super. Ct. 1, 5, 1904; *Commonwealth v. Boyd*, 188 Mass., 79, 1905; *Brazier v. Philadelphia*, 215 Pa. St. 297, 1906; *People ex rel. Hainer v. Prison Keeper*, 190 N.Y. 315, 1907; *State v. Thurston*, 28 R. I. 265, 1907; and *Fletcher v. Dixon*, 107 Md. 420, 426, 1908.

95. Flink, *America Adopts the Automobile*, 166–168; "Oppressive Regulations Proposed in New Jersey," *Horseless Age*, Jan. 28, 1903.

96. Xenophon P. Huddy, "Proposed Federal Legislation," *Horseless Age*, Feb. 6, 1907.

97. "Federal and Uniform Regulation, *Horseless Age*, Jan. 23, 1907; US Const. art. I, § 8, cl. 7.

98. "The Springfield Good Roads and Legislative Convention," *Good Roads Magazine*, Oct. 1907.

99. Xenophon P. Huddy, "Interstate Commerce," *Horseless Age*, Jan. 23, 1907; Xenophon P. Huddy, "Proposed Federal Automobile Law," *Horseless Age*, Feb. 6, 1907. Huddy regularly contributed articles on legal matters to *The Horseless Age*, an early auto enthusiast publication, before joining its staff in June 1907.

100. New York's 1904 law explicitly recognized out-of-state registrations whereas New Jersey's 1903 law did not. As a result, New York drivers were forced to purchase an additional New Jersey registration if they wanted to drive legally in that state, a source of serious contention for motorists as well as an auto industry that wanted to promote sales.

101. "New Jersey Automobile Law Upheld—A Decision in the Test Case," *Horseless Age*, Jan. 31, 1906; *Unwen v. State*, 73 N.J. Law, 529, 1906; Charles Thaddeus Terry, "Opening Address," *Proceedings of the First Annual Good Roads and Legislative Convention, Held at Buffalo, NY, July 7–8, 1908* (New York: American Automobile Association, 1908), 29–31.

102. *Hendrick v. State of Maryland*, 235 US 610, 624 (1915). Transcript of Record, filed Dec. 16, 1912.

103. *Kane v. State of New Jersey*, 242 U.S. 160, 168 (1916). Brief for Plaintiff in Error. The Illinois Supreme Court had also adopted a position that registration and licensing fees represented usage fees rather than a tax and therefore could not be discriminatory when it upheld Chicago's municipal ordinance in *Chicago v. Banker*, 112 Ill. App. 94, 1904.

104. *Highway Statistics*, 18; James W. Martin, "The Motor Vehicle Registration License," *Bulletin of the National Tax Association* 12, no. 7 (April 1927): 206–207. Model T sales jumped from 5,986 in 1908 to 577,036 in 1916. See David Hounshell, *From American System to Mass Production, 1800–1932: The Development of Manufacturing Technology in the United States* (Baltimore: Johns Hopkins University Press, 1984), 224.

105. This ad hoc system of reciprocity proved extraordinarily difficult to enforce and offered, in reality, "virtually . . . complete freedom to tourists" ("Features of the Touring Season," *New York Times*, Sept. 11, 1910). Only the meticulous keeping of ferry records allowed the New York Automobile Bureau's New York City office to collect $30,976 in fees from New Jersey motorists in 1916, but it would be prohibitively expensive to apply the same level of scrutiny along every mile of a state's border ("Making Reciprocity Pay," *New York Times*, Jan. 28, 1917). A map displaying the disparate provisions for recognition of out-of-state auto licenses in the United States and Canada can be found in *Motor Age* 3, no. 14 (Apr. 4, 1918): 77.

106. For more on the relationship between roads and the states, see Novak, *The People's Welfare*.

107. For more on the Wrights' systematic approach to achieving powered flight and how they drew upon a network of other pioneering aeronautical minds, see Tom D. Crouch, *A Dream of Wings: Americans and the Airplane, 1875–1905* (New York: W. W. Norton, 1981). Bayla Singer provides a wonderfully inciteful analysis of the dream of flight throughout human history and across cultures in *Like Sex with Gods: An Unorthodox History of Flying* (College Station: Texas A&M University Press, 2003).

108. Joseph J. Corn details the American fascination with the airplane in *The Winged Gospel: America's Romance with Aviation*, 2nd ed. (Baltimore: Johns Hopkins University Press, 2002), 30. This vision of the airplane as a redemptive device also existed in Europe but was severely tempered by the use of airpower in World War I and visions of even more

destructive power to come. See Robert Wohl, *A Passion for Wings: Aviation and the Western Imagination, 1908–1918* (New Haven, CT: Yale University Press, 1994), and Richard P. Hallion, *Taking Flight: Inventing the Aerial Age from Antiquity through the First World War* (Oxford: Oxford University Press, 2003).

109. Tom D. Crouch, *Wings: A History of Aviation from Kites to the Space Age* (New York: W. W. Norton, 2003), 132.

110. For an excellent discussion of the early legal debate over freedom of the air and aerial sovereignty as well as the tenacity of the common law maxim, see Stuart Banner, *Who Owns the Sky? The Struggle to Control Airspace from the Wright Brothers On* (Cambridge, MA: Harvard University Press, 2008).

111. Banner, *Who Owns the Sky?*, 45–52; Alan Dobson, *A History of International Civil Aviation: From Its Origins through Transformative Evolution* (London: Routledge, 2017), 7–8.

112. The nations represented included Austria-Hungary, Belgium, Bulgaria, Denmark, France, Germany, Great Britain, Italy, Luxembourg, Monaco, the Netherlands, Portugal, Romania, Russia, Spain, Sweden, Switzerland, and Turkey. Although its participants adjourned without a formal agreement, the 1910 Paris Convention cemented an air sovereignty stance within Great Britain and, by extension, its empire that only hardened after the war. For an overview of the debate over air sovereignty and ownership occurring during the period, as well as the views of major international thinkers, see Blewett Lee, "Sovereignty of the Air," *American Journal of International Law* 7 (July 1913): 470–496, and Banner, *Who Owns the Sky?*, 42–56. Dorthe Gert Simonsen provides a detailed analysis of the aerial freedom versus air sovereignty debate in Europe and Britain's response to the 1910 Conference in "Island in the Air: Powered Aircraft and the Early Formation of British Airspace," *Technology and Culture* 59, no. 3 (July 2018): 597–607. Further details on the 1910 Conference can be found in John C. Cooper, "The International Air Navigation Conference, Paris 1910," *Journal of Air Law and Commerce* 19 (Spring 1952): 127–143; Alfred M. Gollin, *The Impact of Air Power on the British People and Their Government, 1909–14* (London: Macmillan, 1989), 134–158; and Christer Jönsson, *International Aviation and the Politics of Regime Change* (London: Frances Pinter, 1987), 77–83.

113. Frederick H. Jackson, *Simeon Eben Baldwin: Lawyer, Social Scientist, Statesman* (New York: Columbia University King's Crown Press, 1955), 75–86.

114. Simeon E. Baldwin, "Annual Address of the Director of the Bureau of Comparative Law," in *Report of the Thirty-Third Annual Meeting of the American Bar Association Held at Chattanooga, Tennessee August 30 and 31, and September 1, 1910* (Baltimore: Lord Baltimore Press, 1910), 902.

115. Ostensibly passed so the town of fewer than 2,200 people would "not be caught napping when the new means of locomotion shall burst upon an astonished atmosphere," the Kissimmee ordinance extended municipal boundaries twenty miles into the air, forbade the dropping of any material from an aircraft or balloon, levied license taxes for aeronautical devices, and required the use of "bells, whistles, or horns" similar to those for the automobile. "Airship Ordinance Suggested," *Kissimmee Valley Gazette*, July 7, 1908, and "An Ordinance Regulating the Status and Employment of Aircraft within the Town of Kissimmee City, FL., 1908," Call #248.211-83K, IRIS # 160079, AFHRA, Maxwell AFB, AL.

116. Simeon Baldwin, "The Law of the Air-Ship," *American Journal of International Law* 4 (Jan. 1910): 96, 98; Banner, *Who Owns the Sky?*, 112.

117. Baldwin, "The Law of the Air-Ship," 106. The creation of international conventions was also endorsed by Baldwin's fellow ABA members Arthur K. Kuhn of Columbia Law School—who saw them as the best means to ensure uniformity in "rules of the road, symbols, and ceremony"—and Dean of Iowa State University's College of Law Charles Noble Gregory. Arthur K. Kuhn, "The Beginnings of an Aerial Law," *American Journal of International Law* 4 (Jan. 1910): 131; Charles N. Gregory, "Aviation as Affecting the Judicial Settlement of Disputes," in *Proceedings of International Conference under the Auspices of American Society for Judicial Settlement of International Disputes, December 15–17, 1910, Washington, D.C.* (Baltimore: Waverly Press, 1911), 174.

118. Baldwin's bill asserted federal sovereignty over the air, placed regulatory power over all foreign and interstate flights within the federal sphere, required the visible marking of flags and numbers, and necessitated registration of all aircraft with the office of the Collector of Internal Revenue in the owner's district of residence along with a surety bond of no less than $1,000. It also called for the licensing of pilots through a certificate of competency issued "by the District Attorney of the United States for any Judicial District," prohibited the licensing of minors, set specific fees for licenses and registration, and stipulated that any violation of the act's provisions constituted a misdemeanor "punishable by a fine not exceeding $1000 or by imprisonment for not exceeding thirty days, or by both, at the discretion of the court." Simeon E. Baldwin, "Liability for Accidents in Aerial Navigation," *Michigan Law Review* 9 (Nov. 1910): 20, 25–27.

119. "Report of the Committee on Jurisprudence and Law Reform," in *Transactions of the Thirty-Fourth Annual Meeting of the American Bar Association Held at Boston, Massachusetts, August 29, 30 and 31, 1911* (Baltimore: Lord Baltimore Press, 1911), 380–384.

120. "Laws to Regulate Aviation," *Virginia Law Register* 16 (Feb. 1911): 778.

121. Juxta analyses of the sections for reciprocity, licensing, and registration in the 1909 Connecticut auto law (An Act concerning the Registration, Numbering, Use, and Speed of Motor Vehicles and the Licensing of Operators of Such Vehicles, Conn. Public Acts c. 211, approved Aug. 10, 1909) and the 1911 Connecticut airship law (An Act Governing the Registration, Numbering, and Use of Air Ships, and the Licensing of Operators thereof, Conn. Public Acts c. 86, approved June 8, 1911) clearly show that Baldwin and Connecticut legislators relied heavily on the existing regulatory framework for the automobile in their state aviation bill. Fully interactive versions of textual comparisons can be found at https://drive.google.com/file/d/1Q4kks-eIn6CMITadofgG2bqKi8RjUgsw /view?usp=sharing (licensing), https://drive.google.com/file/d/1-jo3byE-wVhX-oFUpo CbPz9GICJ1F_So/view (reciprocity), and https://drive.google.com/file/d/1NgFoOZy -lP4nQHy_gD7aTtLjCqLo_LxD/view (registration). The unhighlighted sections represent verbatim transfers of terminology from the 1909 auto law to the 1911 airship law. Highlighted areas denote a divergence in exact wording, but even in these sections the airship law retained the spirit of its automotive predecessor. This analogous application of the existing regulatory framework for the automobile to the airplane in this early period represents yet another example of how human beings draw upon previous experiences when confronting something new to formulate mental models, what

cognitive psychologist P. N. Johnson-Laird defined as a "representation that corresponds to a set of situations . . . that [have] a structure and content that captures what is common to these situations." P. N. Johnson-Laird, "Formal Rules versus Mental Models in Reasoning," in *The Nature of Cognition*, ed. Robert J. Sternberg (Cambridge, MA: MIT Press, 1999), 60. For the past thirty years, cognitive scientists, organizational psychologists, and political scientists have conducted extensive research on mental models and the use (and misuse) of analogy in their formulation. For more on the creation of mental models, see P. N. Johnson-Laird and Eugenia Goldvarg-Steingold, "Models of Cause and Effect," in *The Mental Models Theory of Reasoning: Refinements and Extensions*, ed. Walter Schaeken et al. (Mahwah, NJ: Lawrence Erlbaum Associates, 2007). For an analysis of how human beings use analogy to approach novel situations, see Keith J. Holyoak and Paul Thagard, *Mental Leaps: Analogy in Creative Thought* (Cambridge, MA: MIT Press, 1995), and Gavetti et al., "Strategy Making in Novel and Complex Worlds: The Power of Analogy," *Strategic Management Journal* 26 (2005): 691–712.

122. Howard Mansfield, *Skylark: The Life, Lies, and Inventions of Harry Atwood* (Hanover, NH: University Press of New England, 1999), 7.

123. Atwood learned to fly alongside future head of the World War II Army Air Force Lieutenant Henry H. "Hap" Arnold and Cal Rogers, who would pilot the "Vin Fiz" in the first cross-country flight later that year. Mansfield, *Skylark*, 8–10; Henry Harley Arnold, *Global Mission* (New York: Harper and Brothers, 1949), 17. With his lively account of Harry Atwood's heroic qualities and shortcomings, Mansfield provides insight into the risk-taking, publicity-seeking personalities of many Early Bird aviators.

124. Mansfield, *Skylark*, 27–29.

125. "Young Aviator Tours Wall Street District," *Washington Herald*, July 2, 1911; "Atwood Away To-Day in Flight to Washington," *New York Times*, July 4, 1911.

126. Conflicting reports exist as to the identity of the second Army aviator. The *New York Times* reported it as being Lieutenant Roy C. Kirtland while the *Washington Post* stated that it was Lieutenant Thomas D. Milling.

127. "Plays Hoax in Air," *Washington Post*, July 11, 1911; "Aviators' Hoax Mystifies Many in Washington"; and "Aviators Reach City To-Day," *Washington Herald*, July 11, 1911.

128. "Aviator Atwood Makes Thrilling Flight over City," *Washington Herald*, July 14, 1911; "Taft Greets Atwood after Rainy Flight," *New York Times*, July 15, 1911. Those in attendance at Atwood's brief award ceremony included Signal Corps Brigadier General James Allen, Lieutenants Arnold and Kirtland, and Catholic University of America professor of mechanics Dr. Alfred F. Zahm of the Washington Aero Club.

129. "Flier Violates Law," *Washington Herald*, July 2, 1911; "Atwood is at Goal," *Washington Post*, Aug. 28, 1911. Atwood's flight took him through Missouri, Illinois, Indiana, Ohio, Pennsylvania, New Jersey, and New York.

130. "The Connecticut Statute for the Regulation of Aerial Navigation," *Bench and Bar* 26 (July 1911), 10.

131. Not only did the Massachusetts air law carry over much of the licensing and registration particulars from the state's automobile law, but it also tasked the State Highway Commission with the licensing of pilots and registration of aircraft. 1913 Mass. Acts c. 663 (approved May 16, 1913).

132. "Hawaii Aviation, Aeronautic Laws, 1915–1942," accessed May 20, 2015, http://aviation.hawaii.gov/aeronautic-laws/1915–1942/.

133. "An Ordinance to Prevent Persons from Alighting or Dropping Objects from Aircraft into the Streets, Trees, and Parks of the Town of Nutley, New Jersey," enclosed in Emil Diebitsch to Director of Air Service, September 27, 1919, 248.211-83K, 1916, Call #248.211-83K, IRIS #160079, AFHRA, Maxwell AFB, AL.

134. "Want Aeronauts Licensed," *New York Times*, Sept. 13, 1908; Henry A. Wise Wood, "The Status and Regulation of the Marine Aeroplane," *Flying* 3 (April 1914): 84–85. In 1909, the ACA began to issue aircraft pilot's licenses based on the limited testing criteria of the Paris-based Fédération Aéronautique Internationale, but no legal requirement existed to compel aviators to obtain an ACA license and enforcement relied solely on "excluding the offender from record attempts and sanctioned competitions—events in which only a relatively small number of pilots participated." Bill Robie, *For Greatest Achievement: A History of the Aero Club of America and the National Aeronautic Association* (Washington, DC: Smithsonian Institution Press, 1993), 57.

Chapter 2. World War I and the Internationalization of American Aviation Policy

1. "Howard Earle Coffin," in *Who's Who in American Aeronautics* (New York: Gardner, Moffat, 1922), 36; Memorandum re: Personnel of the Aircraft Board, folder 84, box 2, Aircraft Board Correspondence Files, 1917–1918, RG 18, NARA II. Coffin served as chair of the Aircraft Board and the renamed Aircraft Production Board from May 1917 through April 1918, when John D. Ryan of the Anaconda Copper Mining Company succeeded him in response to charges of mismanagement.

2. Coffin to House, Feb. 19, 1919, folder 4, box 485, MAA papers, AHC; Janice Cheryl Beaver, CRS Report for Congress: "US International Borders: Brief Facts," Nov. 9, 2006, accessed June 7, 2018, http://www.fas.org/sgp/crs/misc/RS21729.pdf.

3. Coffin to Wilson, Feb. 14, 1919, folder 7, box 122, MAA papers, AHC. Coffin proposed a delegation composed of representatives from the Army, the Navy, the National Advisory Committee for Aeronautics, the Society of Automotive Engineers, the Interstate Commerce Commission, and the State Department. To denote the inclusion of aeronautics within the SAE, Coffin successfully promoted an organizational name change from the Society of Automobile Engineers to the Society of Automotive Engineers to denote the inclusion of both motor-powered vehicles within the society's mandate. Tom D. Crouch, *Rocketeers and Gentlemen Engineers: A History of the American Institute of Aeronautics and Astronautics . . . and What Came Before* (Reston, VA: American Institute of Aeronautics and Astronautics, 2006), 12.

4. Privy Council Order 2389, Sept. 17, 1914, enclosed in Foster to Bryan, Sept. 21, 1914, Records of the State Department, RG 59 (National Archives Microfilm Publication M51, roll 1435). Wartime contraband included explosives, firearms, photographic apparatus, carrier or homing pigeons, and mail.

5. "Aeroplane is Fired upon by Canadian Troops at Montreal," *Niagara Falls Gazette*, Sept. 17, 1914, 1. For more on Canadian concerns of an aerial attack from pro-German elements in the United States, see Robert Bartholomew, "Phantom German Air Raids

on Canada: War Hysteria in Quebec and Ontario during the First World War," *Canadian Military History* 7, no. 4 (1998): 29–36.

6. American Consul Mosher to Bryan, April 6, 1915, and Barclay to Bryan, Oct. 7, 1914, both in Records of the State Department, RG 59 (National Archives Microfilm Publication M51, roll 1435).

7. Lansing to Governors, Aug. 5, 1915, Records of the State Department, RG 59 (National Archives Microfilm Publication M51, roll 1435).

8. Richard P. Hallion, *Taking Flight: Inventing the Aerial Age from Antiquity through the First World War* (Oxford: Oxford University Press, 2003), 337.

9. H. G. Wells, *The War in the Air* (New York: Grosset and Dunlap, 1907). For a discussion of the war's place in the larger history of aviation, see Tom D. Crouch, *Wings: A History of Aviation from Kites to the Space Age* (New York: W.W. Norton, 2003), and Hallion, *Taking Flight*. In *The Winged Gospel: America's Romance with Aviation* (Baltimore: Johns Hopkins University Press, 1983) and *A Passion for Wings: Aviation and the Western Imagination, 1908–1918*, Joseph Corn and Robert Wohl, respectively, illustrate how World War I shaped cultural connotations of aviation, strengthening positive prewar conceptions in the American case and painting the airplane with more ominous tones in Europe. For more on the development of strategic bombing doctrine in WWI, see Tami Davis Biddle, *Rhetoric and Reality in Air Warfare: The Evolution of British and American Ideas about Strategic Bombing, 1914–1945* (Princeton, NJ: Princeton University Press, 2002), and Michael Sherry, *The Rise of American Air Power: The Creation of Armageddon* (New Haven, CT: Yale University Press 1987).

10. Stephen Krasner, "Structural Causes and Regime Consequences: Regimes as Intervening Variables," in *International Regimes*, ed. Stephen Krasner (Ithaca, NY: Cornell University Press, 1983), 9.

11. Wilson's indifference to aeronautical questions can be seen in the complete lack of any designated aeronautical advisors as part of the American Mission, whereas Great Britain, France, Italy, Japan, Belgium, and Portugal all included aviation experts within their official delegations. *Papers Relating to the Foreign Relations of the United States: The Paris Peace Conference 1919*, vol. 1 (Washington, DC: Government Publishing Office, 1942), 155–156 (hereafter referred to as *FRUS*).

12. For examples of studies that effectively merge the domestic and international aspects of US policy, see Jay Sexton, *The Monroe Doctrine: Empire and Nation in Nineteenth-Century America* (New York: Hill and Wang, 2011), and Benjamin Allen Coates, *Legalist Empire: International Law and American Foreign Relations in the Early Twentieth Century* (New York: Oxford University Press, 2016).

13. "Ruth Law Lands Here from Chicago in Record Flight," *New York Times*, Nov. 21, 1916.

14. "Signal by President Bathes Liberty Statue in Flood of Light," *New York Times*, Dec. 3, 1916.

15. "Ruth Law to Enlist Men," *New York Times*, Apr. 30, 1917; "Ruth Law, in Liberty Loan Flight, to Drop Bombs on Western Cities," *Washington Post*, June 4, 1917; "A Fiery Plea from the Sky," *Chicago Daily Tribune*, July 26, 1917.

16. According to Law, President Wilson and Chief Signal Officer Maj. Gen. George O. Squier both supported her receiving a commission, but New York representative George M. Hulbert's bill providing for female officers at the president's discretion failed to gain traction in Congress. "Ruth Law Applies for Commission in Aviation Corps; Champion Flier," *New York Times*, Nov. 23, 1917; "Congress Asked to Pass Ruth Law as Army Flier," *New York Times*, Dec. 7, 1917; "Ruth Law Still Seeks Commission in Army," *New York Tribune*, Nov. 29, 1917.

17. Minutes of the Joint Army and Navy Board on Aeronautic Cognizance, Aug. 1, 1918, box 90, Aeronautical Board Agenda and Minutes of Meetings 1916–48, RG 334, NARA II.

18. US Bureau of the Census, *Historical Statistics of the United States, Colonial Times to 1970, Bicentennial Edition, Part 1* (Washington, DC: Government Publishing Office, 1975), 116, 118; Pub. L. No. 24, 40 Stat. 217 (1917). For discussions of the experience of German immigrants and their descendants in the United States during World War I, see Matthew D. Tippens, *Turning Germans into Texans: World War I and the Assimilation and Survival of German Culture in Texas, 1900–1930* (Austin, TX: Kleingarten Press, 2010), and Petra DeWitt, *Degrees of Allegiance: Harassment and Loyalty in Missouri's German-American Community during World War I* (Athens: Ohio University Press, 2012). Nancy Mitchell masterfully details the historical roots of US fears about German imperial expansion into the Western Hemisphere in *The Danger of Dreams: German and American Imperialism in Latin America* (Chapel Hill: University of North Carolina Press, 1999), while Robert Bartholomew discusses Canadian fears over air raids by German Americans in "Phantom German Air Raids on Canada."

19. "Ruth Law Denies Tale That She Is a Spy," *New York Times*, June 8, 1918.

20. Aircraft Production Board to Aeromarine Engineering and Sales Co., Aug. 17, 1917, folder 51, box 1, Aircraft Board Corr. Files, 1917–1918, 1–71, RG 18, NARA II; "Second Pan-American Aeronautic Expo Postponed," *Flying* 6, no. 12 (Jan. 1918): 1055.

21. Fullam to Swanson, July 6, 1917, Report of Truman H. Newberry, Aug. 28, 1917, and Foulois to Squier, Sept. 19, 1917, all in folder 518-2, box 9, Interservice Agencies, Aeronautical Board Subject-Numerical Correspondence File 1916–45, RG 334, NARA II. For a history of this interservice body in the first half of the twentieth century, see Adrian O. Van Wyen, *The Aeronautical Board, 1916–1947* (Washington, DC: Government Printing Office, 1947). The Joint Army and Navy Board on Aeronautic Cognizance represented one of a multitude of such boards established in the years between the Spanish-American War and World War II to coordinate issues of shared interest between the Army and Navy. Such joint boards were permanently replaced by the creation of the Joint Chiefs of Staff with the passage of the National Security Act of 1947. For more information, see Robert G. Albion, *Makers of Naval Policy, 1798–1947* (Annapolis, MD: United Naval Institute Press, 1980). On June 24, 1919, the Joint Army and Navy Board on Aeronautical Cognizance was renamed the Joint Army Navy Board on Aeronautics.

22. Presidential Proclamation No. 1432, Feb. 28, 1918.

23. Dean to Bane, Sept. 19, 1918, folder 518-1, box 9, Interservice Agencies, Aeronautical Board Subject-Numerical Correspondence File 1916–45, RG 334, NARA II.

24. List of Applicants, enclosed in Whitehead to Irwin, May 9, 1918, folder 518-2,

box 9, Interservice Agencies, Aeronautical Board Subject-Numerical Correspondence File 1916–45, RG 334, NARA II. Of the 108 applications received up to that point where the Joint Board issued a final decision, 89 were denied and 19 were approved.

25. List of Joint Board Licenses, undated, folder 518-2, box 9, Interservice Agencies, Aeronautical Board Subject-Numerical Correspondence File 1916–45, RG 334, NARA II. For Atwood's wartime naval contract, see Howard Mansfield, *Skylark: The Life, Lies, and Inventions of Harry Atwood* (Hanover, NH: University Press of New England, 1999), 96–99.

26. Wartime federal control of the railroads via legislation offers an interesting contrast with restrictions on aviation through executive order. With the Federal Control Act of 1918, Congress empowered a new United States Railroad Administration under Wilson's Treasury Secretary and son-in-law William Gibbs McAdoo to operate the nation's railroads, with compensation, for the duration of the wartime emergency. Federal Control Act (as amended), 40 Stat. 451 (1918). Whereas this bill and the dominance of the railroads in American life ensured that Congress would address the nation's rail system after the war—as it did with the 1920 Transportation Act—the passage and repeal of an executive order on wartime flying did not precipitate anywhere near the same level of congressional interest in aviation regulation; Transportation Act of 1920, in effect March 1, 1920 (Cambridge: Cambridge University Press, 1920), 45. For more on the creation and ramifications of the Transportation Act of 1920, see Rogers MacVeagh, *The Transportation Act, 1920: Its Sources, History, and Text, Together with Its Amendments to the Interstate Commerce Act* (New York: Henry Holt, 1923); Morton Keller, *Regulating a New Economy: Public Policy and Economic Change in America, 1900–1933* (Cambridge, MA: Harvard University Press, 1990); and Mark H. Rose, Bruce E. Seely, and Paul F. Barrett, *The Best Transportation System in the World: Railroads, Trucks, Airlines, and American Public Policy in the Twentieth Century* (Philadelphia: University of Pennsylvania Press, 2010).

27. Public Law 271, 63d Cong., 3d sess., passed March 3, 1915 (38 Stat. 930); testimony of Joseph Ames, in *Hearings Before the President's Aircraft Board, vol. 1* (Washington, DC: Government Publishing Office, 1925), 343. The best in-depth history of the NACA remains Alex Roland, *Model Research: The National Advisory Committee for Aeronautics, 1915–1958,* vol. 1 (Washington, DC: NASA Scientific and Technical Information Branch, 1985). Over four decades, the NACA served as the preeminent governmental organization for the promotion of aeronautics before its transformation into the National Aeronautics and Space Administration (NASA) under President Dwight Eisenhower. For the role of the NACA and NASA in the development of aviation and aerospace technologies over the course of the twentieth century, see James R. Hansen, *The Bird Is on the Wing: Aerodynamics and the Progress of the American Airplane* (College Station: Texas A&M University Press, 2004).

28. Robert C. Hilldale, "History of Development of Aircraft Production during the War of 1917," box 1, entry 107, Office of the Chief of the Air Service, Information Group, General Records, RG 18, NARA II. The NACA was composed of two representatives from the War Department, two from the Navy, one from the Smithsonian, one from the Weather Bureau, one from the Bureau of Standards, and five additional individuals "acquainted with the needs of aeronautical science" who all served without compensation. Aeronautical activity within the US government prior to the NACA remained diffuse

and uncoordinated, but several notable activities occurred, such as the creation of the Aeronautical Division in the US Army in 1907, Eugene Ely's first ship takeoff and landing in 1910 and 1911, respectively, and the Post Office's experimentations with air mail service at several aviation meets.

29. For a biographical treatment of Walcott, see Ellis Leon Yochelson, *Smithsonian Institution Secretary, Charles Doolittle Walcott* (Kent, OH: Kent State University Press, 2001).

30. For more on Stratton's work at the Bureau of Standards, see Lee Vinsel, "Virtue via Association: The National Bureau of Standards, Automobiles, and Political Economy, 1919–1940," *Enterprise and Society* 17, no. 4 (Dec. 2016): 809–838, and Rexmond C. Cochrane, *Measure for Progress: A History of the National Bureau of Standards* (Washington, DC: National Bureau of Standards, US Department of Commerce, 1966). President Taft signed legislation in 1913 that separated the Department of Commerce and Labor, established ten years earlier, into two separate cabinet-level departments.

31. "Joseph Ames," in *Who's Who in American Aeronautics* (New York: Gardner, Moffat, 1922), 20.

32. Ames "served as Chairman of the Executive Committee of the NACA from 1919 to 1936, [and] as Chairman of the main committee from 1927 to 1939." Henry Crew, "Biographical Memoir of Joseph Sweetman Ames, 1864–1943," National Academy of Sciences, *Biographical Memoirs*, vol. 23 (1944): 194, accessed Aug. 2, 2019, http://www .nasonline.org/publications/biographical-memoirs/memoir-pdfs/ames-joseph-s.pdf.

33. Established in 1916, the Industrial Preparedness Committee advised the president on industrial policy as the United States became increasingly drawn into the war. Congress authorized the creation of the Council of National Defense, a more formal structure composed of the Secretaries of War, Navy, Interior, Agriculture, Commerce, and Labor who drew upon the expertise of civilian advisors, in August 1916, but it did not begin operations until early March 1917. *First Annual Report of the Council of National Defense* (Washington, DC: Government Publishing Office, 1917), 6. For more on the industrial preparedness movement, Coffin's role in it, and the development of a public/private partnership that laid the foundation for the military-industrial complex, see Paul A. C. Koistinen, *Mobilizing for Modern War: The Political Economy of American Warfare, 1865–1919* (Lawrence: University Press of Kansas, 1997), and Hugh Rockoff, *America's Economic Way of War: War and the US Economy from the Spanish-American War to the Persian Gulf War* (Cambridge: Cambridge University Press, 2012).

34. "Sidney Dunn Waldon," in *Who's Who in American Aeronautics* (New York: Gardner, Moffat, 1922), 102. Membership in the AMA was comprised of "about 80 per cent of the industry" and included the Wright-Martin Aircraft Corporation, the Curtiss Aeroplane and Motor Corporation, the Aeromarine Plane and Motor Company, and the Burgess Company. Frank Russell, "A Comparison of the Aircraft Industry in England and [the] United States," *Flying* 6, no. 8 (Sept. 1917): 682.

35. Wilbur and Orville Wright. 1906. Flying Machine. US Patent No. 821, 393, filed Mar. 23, 1903, and issued May 22, 1906.

36. Minutes of Meeting of Executive Committee, Mar. 22, 1917, folder 8, box 94, Office of the Secretary Records 1903–1924, RU 45, Smithsonian.

37. Roland, *Model Research*, 39. According to Roland, Crisp had recently represented Coffin's Hudson Motor Car Company in a patent case dealing with crankshafts. For more of Crisp's role in the automobile patent dispute, see William Greenleaf, *Monopoly on Wheels: Henry Ford and the Selden Automobile Patent* (Detroit, MI: Wayne State University Press, 2011).

38. Minutes of Meeting of Executive Committee, May 10, 1917, folder 8, box 94, Office of the Secretary Records 1903–1924, RU 45, Smithsonian; *Aeronautics: Third Annual Report of the National Advisory Committee for Aeronautics, 1917* (Washington, DC: Government Publishing Office, 1918), 15.

39. Cross-License Agreement, in Report of F. D. Schnacke on Various Charges of James V. Martin, folder 400.11, box 1366, "Sarah Clark" Correspondence Files, Entry P 26, RG 342, NARA II.

40. The Aircraft Production Board was established by order of the Council of National Defense on May 16, 1917, and held its first meeting the next day. In addition to Coffin as chair, it included Chief Signal Officer George Squier, Rear Adm. David W. Taylor of the Navy's Bureau of Construction, past president of both the Packard Motor Car Company and Cadillac Motor Car Company Sydney Walton, manufacturer Edward A. Deeds, and Philadelphia banker Robert L. Montgomery. *First Annual Report of the Council of National Defense* (Washington, DC: Government Publishing Office, 1917), 10–11, 261.

41. Roland, *Model Research*, 38.

42. Pub. L. No. 29, 40 Stat. 243 (1917); Meeting of Incorporators of the Manufacturers Aircraft Association, Inc., July 24, 1917, folder 2, box 2, MAA papers, AHC.

43. First Meeting of the Manufacturers Aircraft Association, July 24, 1917, folder 2, box 2, MAA papers, AHC.

44. Edward J. Roach, *The Wright Company: From Invention to Industry* (Athens: Ohio University Press, 2013), 130, 140; "Frank Henry Russell," in *Who's Who in American Aeronautics* (New York: Gardner, Moffat, 1922), 90.

45. MAA Press Release, Samuel S. Bradley of Brooklyn Appointed General Manager, Aug. 12, 1917, folder 5, box 62, MAA papers, AHC.

46. List of Members of Manufacturers Aircraft Association Owning Airplane Patents Licensed under the Cross License Agreement, to June 6, 1922, folder 3, box 10, MAA papers, AHC.

47. Report of the President, Annual Meeting of Stockholders, Manufacturers Aircraft Association, July 10, 1918, folder 2, box 2, MAA papers, AHC. Bradley first suggested that the MAA draft a federal bill in May 1918, but the press of the war precluded any action at that time; Minutes of a Special Meeting of the Board of Directors of the Manufacturers Aircraft Association, May 8, 1918, folder 2, box 2, MAA papers, AHC.

48. Minutes of a Special Meeting of the NACA Executive Committee, Sept. 23, 1918, in folder 9, box 94, Office of the Secretary Records, 1903–1924, RU 45, Smithsonian.

49. Memorandum, enclosed in Bradley to Ames, Nov. 15, 1918, folder 9, box 47, MAA papers, AHC.

50. Bradley to Coffin, Nov. 23, 1918, folder 4, box 71, MAA papers, AHC; *Aeronautics: Fourth Annual Report of the National Advisory Committee for Aeronautics, 1918* (Washington, DC: Government Publishing Office, 1920), 27.

51. In addition to Walcott and Stratton, this subcommittee also included Rear Adm. William R. Shoemaker and Cmdr. John H. Towers of the Navy, Col. Arthur Woods and Col. F. R. Kenny of the War Department, and second assistant postmaster Otto Praeger. Minutes of Conference of Representatives of War, Navy, Commerce, and Post Office Departments, to Consider Legislation concerning Aerial Navigation, Dec. 6, 1918, folder 15.02 NACA, box 44, Records of the Post Office Department, RG 28, NARA I; Minutes of Meeting of the Subcommittee on Aerial Legislation of the Interdepartmental Conference on Aerial Navigation, Dec. 31, 1918, folder 15.02 NACA, box 44, Records of the Post Office Department, RG 28, NARA I.

52. Thurman to Redfield, Jan. 25, 1919, folder 72965 [1 of 3], box 369, General Records of the Department of Commerce, RG 40, NARA II; Roland, *Model Research*, 52.

53. Memorandum, Jan. 20, 1919, and Proceedings of the Joint Army Navy Board on Aeronautic Cognizance, Feb. 12, 1919, both in Aeronautical Board Minutes, Aug. 1, 1918–April 9, 1924, box 90, Aeronautical Board, Subject-Numeric Correspondence Files, 1916–45, RG 334, NARA II.

54. Minutes of Regular Meeting of Executive Committee, Feb. 17, 1919, folder 10, box 94, Office of the Secretary Records, 1903–1924, RU 45, Smithsonian (emphasis added); Minutes of Special Meeting of Executive Committee, Feb. 21, 1919, folder 10, box 94, Office of the Secretary Records, 1903–1924, RU 45, Smithsonian. At this time, the membership of the Joint Board consisted of Maj. Gen. William L. Kenly, Rear Adm. William R. Shoemaker, Capt. N. E. Irwin, Capt. K. M. Bennett, Col. Henry H. ("Hap") Arnold, and Col. Claude E. Brigham.

55. Minutes of Regular Meeting of Executive Committee, Dec. 14, 1918, folder 9, box 94, Office of the Secretary Records, 1903–1924, RU 45, Smithsonian; Walcott to Wilson, Dec. 18, 1918, folder 2, box 96, Office of the Secretary Records, 1903–1924, RU 45, Smithsonian.

56. Minutes of Regular Meeting of Executive Committee, Dec. 14, 1918, folder 9, box 94, Office of the Secretary Records, 1903–1924, RU 45, Smithsonian.

57. Phillips to Baker, Dec. 26, 1918, and Phillips to Daniels, Dec. 26, 1918, both in box 5613, Records of the State Department, RG 59, NARA II; Margaret MacMillan, *Paris 1919: Six Months That Changed the World* (New York: Random House, 2003), 3. Wilson left for Paris on Dec. 4, 1918.

58. "Polk, Frank Lyon," in *Who's Who in America, vol. 11, 1920–1921* (Chicago: A. N. Marquis, 1920), 2269.

59. Daniels to State Department, undated, Baker to State Dept., Dec. 30, 1919, and Polk to de Chambrun, Jan. 10, 1919, all in box 5613, Records of the State Department, RG 59, NARA II.

60. Clemenceau to George, Wilson, and Orlando, Jan. 24, 1919, box 5613, Records of the State Department, RG 59, NARA II.

61. *Interim and Final Reports of the Civil Aerial Transport Committee with Appendices* (London: Air Ministry, July 1918), Air Ministry, Aerial Navigation after the War: recommendations of the Civil Aerial Transport Committee, T 1/12271/1967, National Archives, London. The Civil Aerial Transport Committee came into being on May 22, 1917, under the chairmanship of newspaper magnate and aviation advocate Lord Alfred Northcliffe to

develop a comprehensive postwar aerial policy for Great Britain and its empire rooted in the concept of air sovereignty. Its membership included Maj. Gen. William S. Brancker of the Air Ministry, High Commissioner to the United Kingdom for Canada George H. Perley, and representatives from the Home Office, Foreign Office, Colonial Office, Board of Customs, the Post Office, the Dominions, the Admiralty, the Meteorological Service, and industry.

62. Breckenridge Long, handwritten memo, Jan. 23, 1919, box 5613, Records of the State Department, RG 59, NARA II. Polk had forwarded the French request of December 11 to the American Mission in Paris on January 2 for guidance, but it was not received until January 19; Polk to the Ammission, Jan. 2, 1919, *FRUS: The Paris Peace Conference 1919*, vol. 1 (Washington, DC: Government Publishing Office, 1942), 550.

63. Polk to Ammission (telegram), Jan. 24, 1919, box 5613, Records of the State Department, RG 59, NARA II.

64. Woodrow Wilson to Georges Clemenceau, Feb. 7, 1919, box 5613, Records of the State Department, RG 59, NARA II. American Mission secretary Joseph C. Grew expressed Wilson's sentiment in response to Polk's request for guidance on January 2, but it was not received at the State Department until January 25, two days after Polk committed the United States to participation in the aviation commission planned for February 10; Polk to the Ammission, Jan. 2, 1919, *FRUS: The Paris Peace Conference 1919*, vol. 1, 551. According to Lansing, the president's desire to address postwar aeronautical questions only after the peace conference had been "superseded by the actions of the State Department" in Washington; Minutes of the Daily Meetings of the Commissioners Plenipotentiary, Feb. 13, 1919, *FRUS: The Paris Peace Conference 1919*, vol. 11 (Washington, DC: Government Printing Office, 1946), 35.

65. Minutes of the Daily Meetings of the Commissioners Plenipotentiary, Jan. 31, 1919, *FRUS: The Paris Peace Conference 1919*, vol. 11, 5; Paraphrase of Cable to Department of State, undated, and Grew to Knapp, Mar. 5, 1919, both in box 5613, Records of the State Department, RG 59, NARA II.

66. In addition to Knapp and Patrick, the American delegation included Capt. Luke McNamee, Cmdr. E. R. R. Pollock, Lt. Cmdr. John Lansing ("Lanny") Callan, and Lieut. Ralph Kiely of the Navy as well as Capt. C. I. Morton, Brig. Gen. Benjamin D. Foulois, Lt. Col. A. D. Butterfield, and Capt. H. S. Bacon of the Army Air Service. Walcott had recommended the inclusion of NACA member William F. Durand to the State Department, but the latter's recent return to the United States after a year in Europe precluded his participation in the US delegation.

67. Patrick, "Telegram No. Z14 P, June 25, 1918," Call #167.403-229, IRIS # 1145014, AFHRA, Maxwell AFB, AL. The Inter-Allied Aviation Committee agreed that the United States would concentrate on the production of Liberty engines and training aircraft while the Europeans would "manufacture . . . all fighting machines." It also addressed a range of issues such as the pairing of engines with cellules (complete airplanes lacking engines); the harvesting, shipment, and quality of American woods; the shipment of castor seeds and manufacture of lubricating castor oil; and the production and distribution of fabrics for both aircraft and hangars. Minutes of Inter-Allied Aviation Committee Meeting, Dec. 4, 1917, enclosed in Coffin to Daniels, Jan. 16, 1918, Subject File Aviation, Jan.–Apr. 1918,

reel 26, Josephus Daniels Papers, Library of Congress, Washington, DC. The minutes of the Inter-Allied Aviation Committee can be found in Col. Edgar S. Gorrell, "History of the American Expeditionary Forces Air Service, 1917–1919," series A, vol. 2, Records of the Records of the American Expeditionary Forces, RG 120 (National Archives Microfilm Publication M990, roll 1).

68. *Interim and Final Reports of the Civil Aerial Transport Committee with Appendices* (London: Air Ministry, July 1918), Air Ministry, Aerial Navigation after the War: recommendations of the Civil Aerial Transport Committee, T 1/12271/1967, National Archives, London; Conference held on Dec. 9, 1918, between the Air Council and the Society of British Aircraft Constructors, Records of the Air Ministry, AIR 1/2423/305/18/33, National Archives, London. Whereas the United States sent eight commissioned military officers to support Knapp and Patrick, the governments of Great Britain, France, and Italy included within their delegations many of the same legal minds that had addressed the issue in 1910.

69. Bradley to Menoher, Apr. 5, 1919, folder 4, box 221, MAA papers, AHC.

70. "Report of the United States Delegates to the Aeronautical Commission of the Peace Conference" enclosed in Knapp and Patrick to Lansing, July 10, 1919, Records of the State Department, RG 59 (National Archives Microfilm Publication M367, roll 0435), 2–5 (hereafter referred to as Report of Delegates). Although the commission included one delegate each from Belgium, Brazil, Cuba, Greece, Portugal, and Serbia, representatives from these so-called lesser powers did not join the IAAC plenary sessions until the end of March and had a very limited say in its operations and decisions.

71. Those in attendance at the first meeting on March 6 included Knapp and Patrick (United States); Brid. Gen. Groves, General Sykes, and Captain Tindal Atkinson (Great Britain); Col. Dhé and Lt. Col. Pujo (France); and Major Guidoni and Lt. Sauda (Italy). Capt. Albert Roper (France) served as interpreter and secretary. General Tanaka and Mr. Yamakawa (Japan) joined the commission at its second meeting on March 14. The Military Subcommission's recommendation to completely ban aviation in Germany due to the airplane's dual-use nature elicited the strongest (and possibly only) reaction from President Wilson regarding the commission's work when he insisted that enemy states be allowed to develop civilian aviation at the March 17 meeting of the Supreme Council. As a result, restrictions on German aviation in the Versailles Treaty were set to expire on Jan. 1, 1923; Minutes of the Military Sub-Commission of the Inter-Allied Aeronautical Commission, Mar. 20, 1919, *Report of Delegates*, appendix D-2, 1; *FRUS: The Paris Peace Conference 1919*, vol. 13 (Washington, DC: Government Publishing Office, 1947), 644; John C. Cooper, *Right to Fly: A Study in Air Power* (New York: Henry Holt and Company, 1947), 79.

72. Minutes of the Aeronautical Commission of the Peace Conference, Mar. 17, 1919, Report of Delegates, appendix B-3.

73. When the delegations circulated their initial draft conventions, Knapp and Patrick submitted a hastily composed US version that amounted to "nothing more than a redraft of the British proposal." A second US draft, created after more thought on the matter, included several differences from the British version, most notably concerning access to restricted areas; John C. Cooper, "United States Participation in Drafting Paris Convention 1919," *Journal of Air Law and Commerce* 18 (Summer 1951): 269. The full text of these five draft conventions can be found in Report of Delegates, appendix C-1 to C-6.

74. Minutes of the Military Sub-Commission of the Inter-Allied Aeronautical Commission, Mar. 21, 1919, Report of Delegates, appendix D-3; Legal, Commercial, and Financial Sub-Committee of the Aeronautical Commission of the Peace Conference, Report of Delegates, appendix H-1.

75. For the many proposed explanations of the "N" designation, see "What's in an N-Number?" American Aviation Historical Society, accessed Aug. 7, 2019, http://www.aahs-online.org/articles/N-number.php.

76. Progress Report of the Technical Subcommission of the Inter-Allied Aviation Commission, Mar. 29, 1919, Report of Delegates, appendix E-1. "John Lansing ("Lanny") Callan," in *Who's Who in American Aeronautics* (New York: Garner, Moffat, 1922), 31–32; Knapp to Callan, Apr. 1, 1919, 1919–1920 folder, box 4, John Lansing Callan Papers, Library of Congress, Washington, DC. For a thorough history of Callan's wartime activities in France and Italy, see Geoffrey L. Rossano, *Stalking the U-Boat: US Naval Aviation in Europe during World War I* (Gainesville: University Press of Florida, 2010).

77. Minutes of the Technical Subcommission of the Inter-Allied Aeronautical Commission, May 17, 1919, all in Report of Delegates, appendix D-3; Kiely to Craven, May 8, 1919, Subject File, Final Report of Aeronautical Commission of Peace Conference, box 9, John Lansing Callan Papers, Library of Congress, Washington, DC.

78. Two versions of the "official" 1919 convention exist within the source material, one composed of forty-five articles and another with forty-three. According to the Report of Delegates, the IAAC submitted the Drafting Committee's forty-five-article version to the Supreme Council and it was printed for both the US Senate (S. Doc. No. 66-91 [1919]) and British Parliament (Convention Relating to International Air Navigation [London: His Majesty's Stationary Office, 1919]). The forty-three-article version can be found in the International Commission for Air Navigation, *Official Bulletin* 7 (Dec. 1924): 49–57, and the League of Nations Treaty Series vol. 11 (1922): 190–198. This shorter version does not include Article 23's requirement that aircraft and passengers adhere to the laws of the state flown over, nor Article 42's criteria for bringing the convention into force. Both versions include eight technical annexes that provide uniform criteria for aircraft markings, certificate of airworthiness, log books, rules of the air, minimum qualifications for pilots and navigators, maps and ground markings, metrological information, and aerial customs provisions. I have found no explanation for the discrepancy among these two versions, but contemporary concerns over the validity of the Senate printing and the inclusion of the forty-three-article rendition within the ICAN's official publication support the belief that the shorter one constituted the true final form of the convention.

79. Jönsson posits that the 1919 convention established an international air regime based on "unrestricted sovereignty" that continued until 1944. Christer Jönsson, *International Aviation and the Politics of Regime Change* (London: Frances Pinter, 1987), 24, 29.

80. This resulted from British plenipotentiary Robert Cecil pointing out that Article 24 of the recently approved League of Nations Covenant placed all current and future "bureaux and . . . commissions for the regulation of matters of international interest . . . under the direction of the League." Art. 24, The Covenant of the League of Nations; Cecil to House, Paris, May 24, 1919, *FRUS: The Paris Peace Conference 1919*, vol. 6 (Washington, DC: Government Publishing Office, 1943), 33–34; handwritten note on memorandum for

Admiral Knapp, May 27, 1919, folder 1919–1920, box 4, John Lansing Callan Papers, Library of Congress, Washington, DC; Knapp to Pujo, May 29, 1919, Report of Delegates, appendix M-1b.

81. Knapp to Pujo, May 29, 1919, Report of Delegates, appendix M-1b.

82. They submitted a total of six reservations on May 29, 1919, one of which pertained to an unnecessary and unclear paragraph in Article 15.

83. Knapp and Patrick to Pujo, Paris, May 29, 1919, Report of Delegates, appendix K-1b. This included a complete reservation to Annex H and its sole focus on uniform aerial customs procedures.

84. In fact, Knapp submitted a memorandum to Lansing nearly two months after the International commission's last meeting explaining the reasons behind the US delegation's reservations to the convention. Such a report would not be necessary if coordination between the US delegation and the State Department existed at the time of the creation and submission of the former's reservations. Knapp to Lansing, July 12, 1919, box 5613, Records of the State Department, RG 59, NARA II.

85. "Benedict Crowell," in *Who's Who in American Aeronautics* (New York: Garner, Moffat, 1922), 38. For more on how Baker's progressivism shaped the US approach to war, see Douglas B. Craig, *Progressives at War: William G. McAdoo and Newton D. Baker, 1863–1941* (Baltimore: Johns Hopkins University Press, 2013).

86. "George Harrison Houston," in *Who's Who in American Aeronautics* (New York: Garner, Moffat, 1922), 58; "Clement Melville Keys," in *Who's Who in American Aeronautics* (New York: Garner, Moffat, 1922), 64.

87. Seven out of the eight members of the AAM departed Hoboken, New Jersey, aboard the USS *Mt. Vernon* on May 22, 1919, the same day the International Commission held its last meeting in Paris. Minutes of a Special Meeting of the Board of Directors of the Manufacturers Aircraft Association, May 14, 1919, folder 10, box 2; Minutes of the First Formal Meeting of the American Aviation Mission, May 26, 1919, folder 3, box 485; Bradley to Houston, Apr. 18, 1919, folder 4, box 10, all in MAA papers, AHC.

88. Crowell to State Department, May 20, 1919, box 7245, Records of the Department of State, RG 59, NARA II (emphasis added).

89. All members of the AMA except Captain Mustin toured the front from June 7 to June 9, and Bradley could find "no words" to describe "the extent of the devastation." Diary of Capt. Henry C. Mustin, American Aviation Mission, box 3, Henry C. Mustin Papers, LOC; Bradley, Report of the Tour of the Front, June 10, 1919, folder 3, box 485 and Bradley to Bell, June 13, 1919, folder 5, box 62, both in MAA papers, AHC.

90. Bradley to Bell, June 19, 1919, folder 5, box 62, MAA papers, AHC.

91. Partial List of Men and Officials Met by the American Aviation Mission in France, England, and Italy, June and July 1919, folder 3, box 485, MAA papers, AHC.

92. Committee on Commercial Development to Crowell, July 17, 1919, folder 3, box 485, MAA papers, AHC.

93. Legal Committee to Crowell, July 17, 1919, folder 3, box 485, MAA papers, AHC.

94. Committee on Organization Report to Crowell, July 16, 1919, folder 3, box 485, MAA papers, AHC. In addition to Coffin, this committee also included Colonel Dunwoody,

Lieutenant Colonel Blair, and Captain Mustin. The first half of this committee's report consisted of a verbatim articulation of Coffin's February letter to Wilson.

95. Final Report of the American Aviation Mission, July 19, 1919, box 5613, Records of the State Department, RG 59, NARA II.

96. William F. Trimble, *Admiral William A. Moffett: Architect of Naval Aviation* (Washington, DC: Smithsonian Institute Press, 1994), 48.

97. Mustin, Memorandum No. 1 and Memorandum No. 2, July 19, 1919, both in folder 3, box 485, MAA papers, AHC.

98. Meeting of the Heads of Delegations of the Five Great Powers, July 9, 1919, and Meeting of the Heads of Delegations of the Five Great Powers, July 10, 1919, both in *FRUS: The Paris Peace Conference 1919*, vol. 7 (Washington, DC: Government Printing Office, 1943), 62, 90–92.

99. Minutes of the Eighth Meeting of the American Aviation Mission, July 19, 1919, folder 3, box 485, and Bradley to Morse, July 24, 1919, folder 7, box 122, MAA papers, AHC.

100. Memorandum Regarding a Special Conference, undated, and Minutes of a Special Meeting of the Board of Directors of the Manufacturers Aircraft Association, July 23, 1919, folder 10, box 2, Menoher to Bradley, Aug. 4, 1919, folder 7, box 221, Foss to Bradley, Aug. 15, 1919, folder 3, box 29, all in MAA papers, AHC. The MAA-ICC distributed fifty confidential copies of the convention for comment to interested individuals in government and industry.

101. Report of the International Convention Committee of the Manufacturers Aircraft Association, Inc., Sept. 15, 1919, folder 3, box 29, MAA papers, AHC.

102. Crowell to Lansing, in folder 9, box 221, MAA papers, AHC.

103. The convention was forwarded to the State Department at Patrick's request by his chief of staff Col. Edgar S. Gorrell, an MIT-trained aeronautical engineer who had served in the First Aero Squadron assigned to Pershing's Punitive Expedition against Francisco ("Pancho") Villa in 1916; Gorrell to State Department, July 21, 1919, box 7695, Records of the Department of State, RG 59, NARA II. For a concise history of the First Aero Squadron's activities in Mexico, see Roger G. Miller, *A Preliminary to War: The 1st Aero Squadron and the Mexican Punitive Expedition of 1916* (Washington, DC: Air Force History and Museums Program, 2003).

104. This Special Subcommittee on International Air Navigation met on August 20 and 26, 1919, under the chairmanship of the Weather Bureau's Charles Marvin and included Gorrell of the Air Service, Callan representing the Navy, James C. Edgerton of the Post Office, the Geological Survey's Dr. George Otis Smith, Commissioner of Patents James T. Newton, and Joseph D. Nevius of the Customs Service. Marvin invited both Crowell and Bradley to serve as members as well, but Bradley's schedule and Crowell's recent return to Europe prevented their participation in the subcommittee's two meetings on August 20 and 26, 1919. Bradley to Marvin, Aug. 19, 1919, and Dunwoody to Marvin, Aug. 12, 1919, both in folder 32-6, box 177, Records of the National Aeronautics and Space Administration, RG 255, NARA II.

105. Minutes of Meeting of Special Subcommittee on International Air Navigation, Aug. 20, 1919, folder 32-6, box 177, Records of the National Aeronautics and Space Admin-

istration, RG 255, NARA II. The NACA Executive Committee unanimously approved the subcommittee's report on September 12, and Ames forwarded it to Secretary Lansing six days later; Minutes of Regular Meeting of Executive Committee, Sept. 12, 1919, folder 10, box 94, Office of the Secretary Records, 1903–1924, RU 45, Smithsonian Institution Archives, Washington, DC; Ames to Lansing, Sept. 15, 1919, box 5613, Records of the Department of State, RG 59, NARA II.

106. The NACA Subcommittee recognized that if the federal government did not have the power to legislate equal access to airfields, then the United States would have to reserve on that point, but expressed the hope that legislation could be passed to secure this power. Minutes of Meeting of Special Subcommittee on International Air Navigation, Aug. 26, 1919, folder 32-6, box 177, Records of the National Aeronautics and Space Administration, RG 255, NARA II.

107. Telegram, Lindsay to Governor General Cavendish, Aug. 14, 1919, Robert Borden Papers, MG 26, vol. 429, LAC.

108. "Biggar, Lieut.-Col. Oliver Mowat, B.A., K.C.," in *Who's Who and Why, 1919–1920* (Toronto: International Press Limited, 1919), 188; Report from Biggar concerning the convention, Nov. 22, 1919, Nov. 24, 1919, reel C-4317, Borden Papers, LAC.

109. "Carpenter, Walter Clayton," in *Register of the Department of State, December 23, 1918* (Washington, DC: Government Publishing Office, 1919), 93.

110. Memorandum of Conference among James Garfield of Mr. Long's Office, Mr. W. C. Carpenter, Assistant Solicitor, on Behalf of the Department of State, and Colonel Bigger [sic] of the Air Board of Canada, Held at the Department of State, Jan. 20, 1920, box 7695, Records of the State Department, RG 59, NARA II.

111. Quoted in John Herd Thompson, "Canada and the 'Third British Empire,' 1901–1939," found in Phillip Alfred Buckner, ed., *Canada and the British Empire* (Oxford: Oxford University Press, 2008), 97–99. For more on the impact of World War I on Canada's metamorphosis, see David MacKenzie, ed., *Canada and the First World War: Essays in Honor of Robert Craig Brown* (Toronto: University of Toronto Press, 2005).

112. For more on the settlement of US–Canada border issues, see Edward P. Kohn, *This Kindred People: Canadian-American Relations and the Anglo-Saxon Idea, 1895–1903* (Montreal: McGill-Queens University Press, 2004), and Ted L. McDorman, *Salt Water Neighbors: International Ocean Law Relations between the United States and Canada* (Oxford: Oxford University Press, 2009).

113. Doherty to Borden, Jan. 7, 1919, and Certified Copy of a Report of the Committee of the Privy Council Approved by His Excellency the Governor General on the 30th January, 1919, both in Robert Borden Papers, MG 26, vol. 429, LAC.

114. "Sifton, Hon. Sir Clifford, K.C.M.G., K.C.," in *Who's Who and Why, 1921* (Toronto: International Press Limited, 1921), 886.

115. Sifton to Borden, May 10, 1919, Robert Borden Papers, MG 26, vol. 429, LAC.

116. In late November 1919, the Council of Ambassadors modified the convention's exclusionary clause to allow the aircraft of nonmember states to access the airspace of convention members through "special and temporary authorization," but such permission required the unanimous approval of convention adherents. This did not fully resolve the

exclusionary clause issue for the US and Canadian governments since—if both nations did not take the same course of action in regard to the convention—it would make aerial relations between the North American neighbors subject to the decisions of a European-dominated body; Paraphrase of Excerpt of Cablegram from Wallace, April 15, 1920, enclosed in Adee to the NACA, May 1, 1920, folder 32-6, box 177, Records of the National Aeronautics and Space Administration, RG 255, NARA II.

117. Memorandum by the Canadian Plenipotentiaries Respecting the Draft Convention on International Air Navigation, May 10, 1919, Robert Borden Papers, MG 26, vol. 429, LAC.

118. Orders in Council no. 1379, July 7, 1919, *Canadian Gazette* 53 (July 19, 1919): 184.

119. The Governor in Council approved the new Canadian regulation on December 31, 1919. Wilson to Cull, May 2, 1919, reel 10779, John A. Wilson Fonds, LAC. In addition to Sifton, Biggar, and Wilson, the Privy Council also appointed the following individuals to the Air Board on June 23: Canadian Minister of Militia and Defense Maj. Gen. Sydney C. Mewburn, Minister of the Naval Service Charles C. Ballantyne, Deputy Postmaster Robert M. Coulter, and Chief Inspector of the Department of Customs and Inland Revenue E. S. Busby.

120. Air Regulations, 1920, Approved by the Governor in Council on the 31st Day of December, 1919 (Ottawa: J. de LaBroquerie Taché, 1920); O. M. Biggar, The Air Board Submission accompanying the Air Regulations of 1920, Dec. 22, 1919, microfilm reel 10783, John Armistead Wilson Fonds, LAC. In a comparative analysis of the convention and the Canadian Air Regulations, Chief of the US Air Service Information Group Col. Horace M. Hickam found "no inconsistency" between the two documents; Hickam, "Memorandum Comparing the International Air Convention and the Canadian Air Regulations of 1920," Jan. 19, 1920, folder 6, box, 411, MAA Papers, AHC.

121. Memorandum in Regards to Reservations to be Attached to Signature of Convention for the Regulation of International Air Navigation, April 5, 1920, box 5614, RG 59, NARA II. In addition to Carpenter, Crowell, and Bradley, this March 26 meeting also included Brig. Gen. Benjamin D. Foulois, Chief of the Air Service Information Group Col. Horace M. Hickam, American Aviation Mission member Lt. Col. Blair, member of the US delegation to create the air convention Capt. H. S. Bacon, and MAA member J. P. Tarbox. At this meeting, those present agreed that the United States did not need to adopt the numerous technical reservations proposed by the Canadian government.

122. Manning to Carpenter, Aug. 26, 1919, box 5613, Records of the Department of State, RG 59, NARA II.

123. Secretary of State to Wilson (telegram), April 7, 1920, box 5614, Records of the State Department, RG 59, NARA II.

124. High Commissioner to the United Kingdom for Canada George H. Perley signed on behalf of the Canadian government, while Ambassador Hugh Wallace signed for the United States; Colby to Wallace, Apr. 21, 1920, box 5614, Records of the State Department, RG 59, and Resolution of the Special Subcommittee on International Air Navigation, May 28, 1920, folder 25-30, box 153, Records of the National Aeronautics and Space Administration, RG 255, both at NARA II.

Chapter 3. Debating the Administrative Framework for Federal Control

1. "11 Killed, 27 Hurt in Blazing Blimp's Fall in Chicago," *New York Times*, July 22, 1919.

2. Ironically, the *Wingfoot* underwent this daylong flight to determine its fitness for a possible passenger route. "Injured in Balloon Crash" and "Blimp Burns, Kills 11," both in *Chicago Daily Tribune*, July 22, 1919.

3. "Airplane Kills Two Children," *Washington Post*, June 24, 1919; "Plane Crashes Into Crowd on Take-Off; Kills a Girl," *New York Tribune*, Aug. 4, 1919; "Three Children Killed by Aero," *Washington Post*, Aug. 15, 1919; "Dinner Interrupted by Crash of Airplane in Yard," *Chicago Daily Tribune*, Sept. 28, 1919.

4. Presidential Proclamation No. 1532, July 31, 1919.

5. Luther K. Bell, "The Need for Federal Aerial Legislation," *American City Magazine* 21 (1919): 415; Tom D. Crouch, *Wings: A History of Aviation from Kites to the Space Age* (New York: W. W. Norton, 2003), 235.

6. Ordinance No. 743, An Ordinance of the City of Venice Regulating Traffic and Travel in the Air, Aug. 4, 1919, folder 6, box 417, MAA papers, AHC.

7. Report of the Aviation Commission of the State of New York, 1920, Gov. A. E. Smith Subject and Correspondence Files, Box 96, New York State Archives, Albany, NY.

8. Official statistics show that New York and Illinois ranked first and second in the nation, respectively, in population and total income due predominantly to the exploding metropolises of New York City and Chicago. 1920 population: New York, 10,385,000; Illinois, 6,485,000. 1929 total income (the earliest date available): New York, $14,171,000; Illinois, $7,291,000. *US Bureau of the Census, Historical Statistics of the United States, Colonial Times to 1970, Bicentennial Edition, Parts 1 & 2* (Washington, DC: Government Publishing Office, 1975), 24–36, 243–245.

9. Bradley to Menoher, July 26, 1919, folder 4, box 221, MAA papers, AHC.

10. The quote cited comes from Col. Oscar Westover, but Information Group Chief Col. Horace M. Hickam, Capt. Laurence Miller, and Col. Thurman Bane, chief of the Technical Branch at McCook Field, were in full agreement and offered similar assessments. Testimony of Col. Oscar Westover, Jan. 28, 1920, in Report of the Proceedings of the Public Hearings Held by the Aviation Commission of the State of New York, Jan. 17–Jan. 31, 1920, Gov. A. E. Smith Subject and Correspondence Files, Box 96, New York State Archives, Albany, NY (hereafter referred to as NYAC Hearings).

11. Letter from Howard Coffin to the NY Aviation Commission, Jan. 29, 1920, NYAC Hearings; Testimony of George Houston, Jan. 31, 1920, NYAC Hearings.

12. This, and the fact that New York City served as the MAA's hub, probably explains why New York's members of Congress submitted so many aviation bills between 1921 and 1925. Minutes of the Fourth Meeting of the Sub-Committee on Commercial Aviation, Feb. 5, 1920, folder 13–3, box 41, Records of the National Aeronautics and Space Administration, RG 255, NARA II; Report of the Aviation Commission of the State of New York to His Excellency Alfred E. Smith, Governor of the State of New York, Governor Alfred E. Smith, Subject and Correspondence Files (Series 13682-53A), New York State Archives, Albany, NY.

13. L. Fuller, Memorandum for Executive, Nov. 3, 1920, folder 187, box 10, Records of

the Office of the Chief of Air Service, 1917–21, RG 18, NARA II. When Major Henry "Hap" Arnold consulted on a possible California state bill "as a stop-gap measure" while stationed in San Francisco, airpower advocate Brig. Gen. William "Billy" Mitchell ordered him to cease and desist so as to prevent "the same performance that had occurred with the automobiles, except . . . ten times worse"; Arnold to Mitchell, Jan. 3, 1921, and Mitchell to Arnold, Jan. 11, 1921, both in General Correspondence, 1921, box 9, William Mitchell Papers, LOC. The New York City and Chicago ordinances banned trick flying within city limits and established a two-thousand-foot minimum altitude. The New York City Board of Aldermen passed the ordinance at the insistence of its president, Fiorello La Guardia, on February 15, 1921, and Chicago adopted the exact same ordinance two months later. Bradley to Whitney, Dec. 31, 1920, folder 5, box 78, MAA papers, AHC; *Proceedings of the Board of Aldermen of the City of New York from Jan. 3 to Mar. 39, 1921*, vol. 1, 386–387; *Journal of the Proceedings of the City Council of the City of Chicago for the Council Year 1920–21*, 2230.

14. US Const. art. VI, § 8, cl. 2 (the Supremacy Clause); US Const. art. I, § 8, cl. 18 (the Necessary and Proper Clause); Baker to Lansing, Mar. 16, 1920, box 5614, Records of the State Department, RG 59, NARA II. The US Supreme Court affirmed Congress's authority to legislate in response to a treaty in the 1920 case *State of Missouri v. Holland*. The Migratory Bird Treaty, signed August 16, 1916, and proclaimed in effect four months later, sought to preserve game foul that flew freely across the US-Canada border. Over the objections of sportsmen and states' rights advocates that the Constitution did not explicitly grant the federal government such authority, Congress passed enabling legislation in 1918. The State of Missouri challenged US Game Warden Ray P. Holland's enforcement of the law, and the Supreme Court heard the case on March 2, 1920. In a 7–2 decision delivered seven weeks later, Justice Oliver Wendel Holmes declared that migratory birds constituted "a national interest" and that their transient nature made their preservation possible "only by national action in concert with that of another power." The Supreme Court ruled that the limited jurisdiction of states precluded them from sufficiently regulating a vital natural resource that could take to the skies and travel outside their borders at any moment. With unmistakable parallels between migratory birds and airplanes, *State of Missouri v. Holland* made ratification of the new air convention a possible means to secure federal control over aviation in peacetime. Convention between the United States and Great Britain for the Protection of Migratory Birds, *FRUS: 1916* (Washington, DC: Government Publishing Office, 1925), 279–282; *State of Missouri v. Holland, United States Game Warden*, 252 US 416 (1920); Kurkpatrick Dorsey, *Conservation Diplomacy: US-Canadian Wildlife Protection Treaties in the Progressive Era* (Seattle: University of Washington Press, 1998).

15. Minutes of a Special Meeting of the Board of Directors of the Manufacturers Aircraft Association, Inc., July 23, 1919, folder 10, box 2, MAA papers, AHC.

16. Bradley to Coffin (telegram), Aug. 2, 1919, folder 4, box 71, MAA Papers, AHC.

17. Woolsey to Crisp, Sept. 8, 1919, folder 8, box 122, MAA Papers, AHC.

18. Bradley to Woolsey, Sept. 9, 1919, and Flint to Woolsey, Sept. 9, 1919, both in folder 8, box 122, MAA Papers, AHC.

19. H.R. 13838, 64th Cong. (1916), introduced Mar. 28, 1916.

20. H.R. 3, introduced Apr. 2, 1917. Texas Democrat Morris Sheppard presented a complementary bill in the Senate two days later as S. 80.

21. S. 80: A Bill to Establish a Department of Aeronautics and for Other Purposes, Hearings Before a Subcommittee of the Committee on Military Affairs, 65th Cong. (1917).

22. For Mitchell's thoughts on air power in his own words, see William Mitchell, *Our Air Force: The Keystone of National Defense* (New York: E. P. Dutton, 1921), and William Mitchell, *Winged Defense* (New York: G. P. Putnam's Sons, 1925). Numerous studies of Mitchell and his doctrinal ideas exist, most notably Thomas Wildenberg, *Billy Mitchell's War with the Navy: The Interwar Rivalry over Air Power* (Annapolis, MD: Naval Institute Press, 2013); James J. Cooke, *Billy Mitchell* (Boulder, CO: Lynne Rienner, 2002); and Alfred F. Hurley, *Billy Mitchell: Crusader for Air Power* (Bloomington: Indiana University Press, 1964, 1975).

23. Col. Townsend F. Dodd, "Recommendations concerning the Establishment of a "Department of Aeronautics" Prepared by Direction of Brig. Gen. William Mitchell, Apr. 17, 1919, folder 5, box 457, MAA papers, AHC; Mitchell to Patrick, June 7, 1919, folder 5, box 485, MAA papers, AHC.

24. The bills in question were H.R. 7925, submitted July 28, 1919, and S. 2693, submitted July 31, 1919.

25. Neither Curry nor New consulted the MAA on their bills. Head of the MAA's Information Department Luther K. Bell first read about Curry's bill in the press, and Bradley informed Coffin that "New's bill [was] not sufficient" due to its lack of detail. Bell to Curry, Aug. 2, 1919, folder 2, box 226, MAA papers, AHC; Bradley to Coffin (telegram), 1919, folder 4, box 71, MAA papers, AHC.

26. Bradley, Crisp, and Woolsey were united in their belief that "it would be premature to draft and submit" a specific bill at this time, and therefore they did not draw up an MAA bill as called for in their charge. They did not need to, as sufficient candidates already existed in Congress. Report of the Subcommittee on the Legislation Which May Be Necessary to Make the Convention Effective, in Report of the International Convention Committee of the Manufacturers Aircraft Association, Sept. 15, 1919, folder 3, box 29, MAA Papers, AHC.

27. Walcott to Squier, May 19, 1917, folder 25–7, box 138, Records of the National Aeronautics and Space Administration, RG 255, NARA II. In 1917, members of the NACA from the military branches included Maj. Gen. George O. Squier, chief of the Signal Corp and his subordinate Lt. Col. Virginius E. Clark, as well as Chief Constructor Rear Adm. David W. Taylor and Lt. Comm. John H. Towers of the Navy.

28. Responses to Chief of the wartime Air Service Maj. Gen. Mason M. Patrick's request for feedback from his officers as to "the best organization under which developments of aviation may proceed most smoothly and most effectively" ran the gamut. Mitchell and several field officers supported a unified department; Brid. Gen. Benjamin Foulois suggested the inclusion of various assistant secretaries; wartime commander of the Air Service, 2nd Army, Col. Frank P. Lahm called for the continued separation of aviation between the services; and Air Service Col. Edgar Gorrell proposed the creation of a civil aviation regulatory body under an existing department. Statement of Benjamin D. Foulois, undated, folder 5, box 485, MAA papers, AHC; Lahm to Patrick, June 5, 1919,

folder 5, box 485, MAA papers, AHC; Gorrell to Patrick, June 6, 1919, Subject File, Aeronautics, Separate Department, Recommendations, box 31, William Mitchell Papers, LOC.

29. Patrick Memorandum, "Civil and Military Aviation in the United States," June 21, 1919, Subject File, Aeronautics, Separate Department, Recommendations, box 31, William Mitchell Papers, LOC; James P. Tate, *The Army and Its Air Corps: Army Policy towards Aviation 1919–1941* (Maxwell, AL: Air University Press, 1998), 6.

30. Statement of Secretary Baker upon Release of the Crowell Commission's Report, Aug. 11, 1919, reprinted in *Reorganization of the Army, Hearings Before the Subcommittee of the Committee on Military Affairs*, US Senate, 66th Cong., 1st sess., 195–196 (1919).

31. Adjutant General of the Army to Menoher, Aug. 8, 1919, folder 8, box 134, MAA papers, AHC. Maj. Gen. Frank W. Coe, Maj. Gen. William G. Haan, and Maj. Gen. William J. Snow, all nonflyers, served with Menoher on this board. Report of a Board of Officers Convened at Washington, DC, Oct. 27, 1919, folder 8, box 134, MAA papers, AHC.

32. Statement of Franklin D. Roosevelt, Sept. 12, 1919, *Reorganization of the Army, Hearings Before the Subcommittee of the Committee on Military Affairs*, US Senate, 66th Cong., 1st sess., 727 (1919); Wildenberg, *Billy Mitchell's War with the Navy*, 41.

33. Menoher to the Joint Army and Navy Board on Aeronautic Cognizance, May 19, 1919, and Joint Army and Navy Board on Aeronautics to the Secretary of War and the Secretary of the Navy, July 24, 1919, both in folder 518, box 9, Aeronautical Board Subject-Numerical Correspondence File 1916–45, RG 334, NARA II.

34. Baker to Menoher, Aug. 9, 1919, folder 518-1, box 9, Aeronautical Board Subject-Numerical Correspondence File 1916–45, RG 334, NARA II.

35. Letter of Appointment, Newton D. Baker, Oct. 10, 1919, folder 13, box 2, Records of the Air Service Advisory Board, RG 18, NARA II.

36. As Air Service Chief of Staff for the 2nd Army during World War I, Curry's postwar evaluation had focused on the need to closely coordinate aerial observation with ground forces, Callan had influenced Mustin in his reservations against the American Aviation Mission's call for a unified department, and the three NACA members had already come out in support of a clear separation between military and civil aviation. For more on Curry's report, see James J. Cooke, *The US Air Service in the Great War, 1917–1919* (Westport, CT: Praeger Publishers, 1996), 222.

37. Minutes of Meeting of Committee on Aerial Navigation, Oct. 20, 1919, folder 13, box 2, Records of the Air Service Advisory Board, RG 18, NARA II. Callan, Parker, and Ames were unable to attend this first meeting, and Maj. Henry S. Bacon, also a member of the US delegation during the convention's creation, advised on the convention in Callan's absence.

38. Minutes of Meeting of Committee on Aerial Navigation, Oct. 20, 1919, folder 13, box 2, Records of the Air Service Advisory Board, RG 18, NARA II; W. Clayton Carpenter, "Memorandum of First Meeting of Board to Draft Act Covering Air Navigation and Civil Aviation in the United States," Oct. 21, 1919, box 7695, Records of the State Department, RG 59, NARA II.

39. Statement of General Menoher, Aug. 20, 1919, *Reorganization of the Army, Hearings Before the Subcommittee of the Committee on Military Affairs*, US Senate, 66th Cong., 1st sess., 279 (1919); William F. Durand, *Civil Aerial Transport*, enclosed in Victory to Walcott,

Aug. 4, 1919, folder 3, box 96, Office of the Secretary Records, 1903–1924, RU 45, Smithsonian; Minutes of Regular Meeting of Executive Committee, July 25, 1919, folder 10, box 94, Office of the Secretary Records, 1903–1924, RU 45, Smithsonian.

40. W. Clayton Carpenter, Memorandum of First Meeting of Board to Draft Act Covering Air Navigation and Civil Aviation in the United States, Oct. 21, 1919, box 7695, Records of the State Department, RG 59, NARA II; Meeting of a Board to Draft a Proposed Act Covering Air Navigation, Oct. 21, 1919, folder 13, box 2, Records of the Air Service Advisory Board, RG 18, NARA II.

41. Carpenter, Memorandum of First Meeting of Sub-Committee of Board to Draft Act Covering Air Navigation and Civil Aviation in the United States, Nov. 4, 1919, box 7695, Records of the State Department, RG 59, NARA II.

42. At the War Department's request, Curry also submitted a revised draft of the sub-committee's bill that made the secretary of war the executor of regulations rather than the secretary of commerce, but the civilian majority on the Interdepartmental Board flatly rejected it. Carpenter, "Memorandum of Second Meeting of Board to Draft Act Covering Air Navigation and Civil Aviation in the United States," Nov. 14, 1919, box 7695, Records of the State Department, RG 59, NARA II.

43. Meeting of Full Committee on Aerial Navigation, Nov. 14, 1919, folder 13, box 2, Records of the Air Service Advisory Board, RG 18, NARA II; Carpenter, "Memorandum of Second Meeting of Board to Draft Act Covering Air Navigation and Civil Aviation in the United States," Nov. 14, 1919, box 7695, Records of the State Department, RG 59, NARA II.

44. Curry to Carpenter, Nov. 20, 1919, box 7695, Records of the State Department, RG 59, NARA II.

45. Minutes of Meeting of Committee on Aerial Navigation, Nov. 26, 1919, folder 13, box 2, Records of the Air Service Advisory Board, RG 18, NARA II. Those present at the November 26 meeting included Curry, Milling, Callan, Praeger, and Carpenter. Parker, Stratton, Lennon, Ames, and Marvin were absent.

46. Both NACA member Charles Marvin and Secretary of the Navy Josephus Daniels echoed Stratton's and Carpenter's earlier concern about the constitutionality of federally regulated intrastate flight, and Carpenter placed an invalidation clause in the bill so that if the courts declared any part of the act unconstitutional, "such judgment shall not affect, impair, or invalidate the remainder thereof." Final draft of bill enclosed in Curry to Carpenter, Dec. 19, 1919, box 7695, Records of the State Department, RG 59, NARA II.

47. The Interdepartmental Board accepted Carpenter's revisions on December 6 and approved the final draft of the bill, with minor revisions, twelve days later. Those present at the December 6 meeting included Curry, Milling, Callan, Praeger, Carpenter, Parker, Stratton, Marvin, and Lewis (for Ames), with Lennon absent. Those present at the December 18 meeting included Curry, Milling, Parker, Callan, Carpenter, Marvin, Lewis (for Ames), and Brown (for Stratton). Lennon and Praeger were absent, but the latter had submitted his approval of the draft legislation to Curry on December 17. Praeger to Curry, Dec. 17, 1919, folder 13, box 2, Records of the Air Service Advisory Board, RG 18 and Carpenter to Sec. of State, Nov. 12, 1919, box 7695, RG 59, both at NARA II.

48. The Senate rejected the Treaty of Versailles even with the reservations of Foreign Relations Committee chairman Henry Cabot Lodge on November 19, 1919, and again five

months later. The League of Nations debate in the US Senate has remained a subject of intense interest to diplomatic and political historians. See Frank Barth, *The Lost Peace, A Chronology: The League of Nations and the United States Senate, 1918–1921* (New York: Woodrow Wilson Foundation, 1945); John Milton Cooper Jr., *Breaking the Heart of the World: Woodrow Wilson and the Fight for the League of Nations* (Cambridge: Cambridge University Press, 2001); Herbert F. Margulies, *The Mild Reservationists and the League of Nations Controversy in the Senate* (Columbia: University of Missouri Press, 1989); Ralph A. Stone, ed., *Wilson and the League of Nations: Why America's Rejection?* New York: Holt, Rinehart and Winston, 1967); Ralph A. Stone, *The Irreconcilables: The Fight against the League of Nations* (Lexington: University of Kentucky Press, 1970); and William C. Widenor, *Henry Cabot Lodge and the Search for an American Foreign Policy* (Berkeley: University of California Press, 1980).

49. Minutes of Regular Meeting of Executive Committee, Nov. 25, 1919, folder 10, box 94, Office of the Secretary Records, 1903–1924, RU 45, Smithsonian.

50. Ames to Praeger, Dec. 11, 1919; Ames to Praeger, Dec. 12, 1919; both in box 18, Records of the Post Office Department, RG 28, NARA I.

51. Minutes of Meeting of Executive Committee, Feb. 12, 1920, folder 11, box 94, Office of the Secretary Records, 1903–1924, RU 45, Smithsonian.

52. To Create a Bureau of Aeronautics in the Department of Commerce, and Providing for the Organization and Administration Thereof, and Walcott to Hicks, May 19, 1920, both in folder 4, box 96, Office of the Secretary Records, 1903–1924, Smithsonian.

53. *Cong. Rec.* S 2185 (Jan. 29, 1920). Curry's unified department bill never made it out of the House Military Affairs Committee.

54. With the passage of the National Defense Act of 1920 on June 4, 1920, Congress tacitly supported the continued separation of military aeronautics when it elevated the Army Air Service to a combat arm equal to the Infantry, Cavalry, and Artillery. As a result, the Office of the Director of Air Service was renamed the Office of Chief of Air Service, and Mitchell's dream of a centralized department drifted further into the realm of improbability.

55. Baker to Esch, May 22, 1920, folder HR 66A-D14 HR 14061-14158, box 128, Records of the House of Representatives, RG 233, NARA I. The Interdepartmental Board draft bill was submitted by Californian Republican Julius Kahn as H.R. 14061 on May 13 and Republican James Wadsworth of New York as S. 4470 sixteen days later.

56. This was introduced by New York Republican Frederick C. Hicks as H.R. 14137 on May 20.

57. Minutes of Special Meeting of Executive Committee, Mar. 1, 1920, folder 11, box 94, Office of the Secretary Records, 1903–1924, RU 45, Smithsonian; Letter from Walcott to Wilson, Mar. 5, 1920, folder 4, box 96, Office of the Secretary Records, 1903–1924, RU 45, Smithsonian. By late February, both Secretary of War Baker and Secretary of the Navy Daniels had agreed that "federal control and regulation of civilian aviation should not be exercised through either the Army or the Navy. There should be a separate body provided for this purpose, preferably in the form of a bureau in the Department of Commerce." Memorandum of Points Enclosed in Craven to Menoher, Feb. 26, 1920, folder 518, box 9, Aeronautical Board Subject-Numerical Correspondence File 1916–45, RG 334, NARA II.

58. Ames's revised versions of H.R. 14061 (the Walcott, Taylor, Craven bill) and H.R. 14137 (the Interdepartmental Board's bill) discussed in this paragraph can be found in the NACA's Sixth Annual Report, 14–18.

59. Ames first floated the idea of replacing the joint board with the NACA at the Executive Committee's June 11 meeting. At a special meeting two weeks later, Ames discussed and further modified his revised version of the Walcott, Taylor, and Craven bill with Stratton, Marvin, and Menoher. Minutes of Regular Meeting of Executive Committee, June 11, 1920 and Minutes of Special Meeting of Executive Committee, June 28, 1920, both in folder 11, box 94, RU 45, Office of the Secretary Records, 1903–1924, Smithsonian.

60. Roland, *Model Research*, 55.

61. *Aeronautics: Sixth Annual Report of the National Advisory Committee for Aeronautics, 1920* (Washington, DC: Government Publishing Office, 1920), 11.

62. Bane to Ames, July 8, 1920, and Hayford to Ames, Nov. 4, 1920, both quoted in Roland, *Model Research*, 58.

63. Coffin to Thompson, Jan. 29, 1920, in Report of the Aviation Commission of the State of New York, 1920, Gov. A. E. Smith Subject and Correspondence Files, Box 96, New York State Archives, Albany, NY.

64. Rentschler to Bradley, July 30, 1920, folder 5, box 239, MAA Papers, AHC.

65. Bradley to Russell, Oct. 30, 1920, folder 3, box 78, MAA Papers, AHC.

66. Warren G. Harding, Readjustment Speech, recorded June 29, 1920, accessed Aug. 27, 2019, https://cdn.loc.gov//service/mbrs/nforum/9000056.pdf.

67. Warren G. Harding, Inaugural Address, delivered Mar. 4, 1921, Washington, DC, accessed Aug. 27, 2019, https://www.presidency.ucsb.edu/documents/inaugural-address-49.

68. Crowell to New, Oct. 7, 1920, and Crowell to Bradley, Oct. 25, 1920, both in folder 3, box 76, MAA Papers, AHC.

69. Address of President Warren G. Harding, delivered at a joint session of the two houses of Congress, Apr. 12, 1921 (Washington, DC: Government Publishing Office, 1921).

70. Robert K. Murray, *The Politics of Normalcy: Government Theory and Practice in the Harding-Coolidge Era* (New York: W. W. Norton, 1973), 23.

71. Daniel P. Carpenter, *The Forging of Bureaucratic Autonomy: Reputations, Networks, and Policy Innovation in Executive Agencies, 1862–1928* (Princeton, NJ: Princeton University Press, 2001), 20–23. Carpenter divides Washington's bureaucracy into three tiers: executive-level political appointees, mezzo-level career bureaucrats at the division chief and bureau head levels, and an operational level of lesser-ranked bureaucrats. The creation, expansion, and ultimate institutionalization of the bureau model for civil aviation regulation further attests to his argument that mezzo-level bureaucrats wield considerable power due to their relative permanence and control of administrative operations.

72. Minutes of Regular Meeting of Executive Committee, Mar. 10, 1921, folder 12, box 94, Office of the Secretary Records, 1903–1924, RU 45, Smithsonian.

73. Walcott to Denby, Mar. 23, 1921, folder 25–24, Navy Department, General, 1921, box 143, Records of the National Aeronautics and Space Administration, RG 255, NARA II; Minutes of Special Meeting of Executive Committee, Apr. 4, 1921, folder 12, box 94, Office of the Secretary Records, 1903–1924, RU 45, Smithsonian.

74. Menoher to Weeks, Apr. 1, 1921, National Advisory Committee for Aeronautics 1921–1926, box 427, Herbert Hoover Papers, HHPL.

75. Menoher to Weeks, Apr. 1, 1921, National Advisory Committee for Aeronautics 1921–1926, box 427, Herbert Hoover Papers, HHPL (emphasis added).

76. Minutes of First Meeting of Subcommittee on Federal Regulation of Air Navigation, Apr. 5, 1921, folder 5, box 159, MAA Papers, AHC. The Special Subcommittee on Federal Regulation of Air Navigation convened from April 5 to April 8, and Harding delivered his address to Congress on April 12. Moffett, the other Navy representative on the NACA, sat in for Taylor at its first meeting. NACA Executive Committee chair Joseph Ames was present at the subcommittee's second meeting, NACA executive officer George Lewis attended all but the first meeting, and the NACA's permanent secretary John F. Victory served as subcommittee secretary. Of the representatives from "civil life," only Frank Russell attended the subcommittee's first meeting, Glenn Martin joined him for the second, and all three were gathered together only at the third. Of the industry representatives, only Waldon attended the fourth and final meeting.

77. Warren G. Harding, Inaugural Address, delivered Mar. 4, 1921, Washington, DC, accessed Aug. 27, 2019, https://www.presidency.ucsb.edu/documents/inaugural-address-4.

78. Minutes of Third Meeting of Subcommittee on Federal Regulation of Air Navigation, Apr. 7, 1921, folder 5, box 159, MAA Papers, AHC.

79. Minutes of Fourth Meeting of Subcommittee on Federal Regulation of Air Navigation, Apr. 8, 1921, folder 5, box 159, MAA Papers, AHC; Martin to Gardner, May 1, 1921, and Kilner to Mitchell, Aug. 3, 1921, both in General Correspondence, 1921, box 9, William Mitchell Papers, LOC.

80. Martin to Gardner, May 1, 1921, General Correspondence, 1921, box 9, William Mitchell Papers, LOC.

81. This letter, attributed to Russell, Martin, Kilner, and Waldon but signed only by the last two, was submitted to Walcott to accompany the subcommittee's final report. Kilner and Waldon to Walcott, Apr. 8, 1921, folder 5, box 159, MAA Papers, AHC. The minority's letter, submitted to Harding on April 9, never made it into the Congressional Record alongside the Special Subcommittee's report, and some confusion exists as to whether Harding ever saw it. There is no reason to believe that it would have prompted him to adopt an approach that contradicted the subcommittee's report. Lewis to Bradley, Apr. 23, 1921, folder 5, box 159, MAA Papers, AHC; Report of National Advisory Committee for Aeronautics, H.R. Rep. No. 17 (1921).

82. Representative Kahn sponsored H.R. 201, a bill based on the Interdepartmental Board's bill of December 1919. Representative Hicks introduced two bills. H.R. 271 was Ames's revised bill that required the NACA to approve any regulations drafted by the commissioner of aeronautics, while H.R. 281, "prepared by Mr. Hicks personally," made the commissioner of aeronautics chair of an Aeronautical Board composed of one representative from the Air Service, the recently proposed Navy Bureau of Aeronautics, and the NACA, which would act in a purely advisory capacity. H.R. 201, 67th Cong. (1921); Analysis of H.R. 201, 67th Cong., 1st Sess., folder 282, box 14, Records of the Air Service Advisory Board, RG 18, NARA II; H.R. 271, 67th Cong. (1921); Minutes of Regular Meeting of Executive Committee, Apr. 14, 1921, folder 12, box 94, Office of the Secretary Records,

1903–1924, RU 45, Smithsonian; Bradley to Coffin, June 6, 1921, folder 5, box 71, MAA Papers, AHC; H.R. 281, 67th Cong. (1921).

83. Notes for Meeting of the Air Craft Men Monday, July 16, 1921, Aviation Correspondence, 1921–1921, box 39, Commerce Papers, HHPL. Submitted to Postmaster Hays for forwarding to Harding, the brief's proposed board would consist of representatives from the Aero Club of America, the MAA, the recently established National Aircraft Underwriters Association, the Society of Automotive Engineers, and the NACA. Minutes of a Special Meeting of the Board of Directors of the Manufacturers Aircraft Association, May 26, 1921, folder 1, box 3, MAA Papers, AHC; A Brief for the Advancement of Civilian Aviation in the United States, folder 4, box 406, MAA Papers, AHC.

84. Hoover coordinated food aid to Europe first through the Commission for Relief in Belgium and then the US Food Administration, where he became the de facto head of US wartime food policy.

85. David M. Hart, *Forged Consensus: Science, Technology, and Economic Policy in the United States, 1921–1953* (Princeton, NJ: Princeton University Press, 1998), 30, 40.

86. Ellis Hawley, "Three Facets of Hooverian Associationalism: Lumber, Aviation, and Movies, 1921–1930," in *Regulation in Perspective: Historical Essays*, ed. Thomas K. McCraw (Cambridge, MA: Harvard University Press, 1981), 99; Ellis Hawley, "Herbert Hoover, the Commerce Secretariat, and the Vision of an 'Associative State,' 1921–1928," *Journal of American History* 61 (June 1974): 116–140; Brian Balogh, *The Associational State: American Governance in the Twentieth Century* (Philadelphia: University of Pennsylvania Press, 2015).

87. Bradley to Lewis, June 8, 1921, folder 5, box 159, MAA Papers, AHC.

88. Borah submitted this resolution on June 23, 1921, most likely in consultation with Billy Mitchell. "Abolishing the organization known as the National Advisory Committee for Aeronautics and transferring its property and duties to existing governmental agencies," S. J. Res. 77, 67th Cong. (1921); "Senator Borah Wants N.A.C.A. Abolished," *Aviation and Aircraft Journal*, no. 26 (June 27, 1921): 800; Ames to Bradley, July 9, 1921, folder 9, box 47, MAA Papers, AHC.

89. Bradley also assured Ames that the MAA had nothing to do with Borah's resolution calling for the NACA's dissolution. Bradley to Ames, July 11, 1921, both in folder 9, box 47, MAA Papers, AHC.

90. An Act Making Appropriations for the Naval Service for the Fiscal Year Ending June 30, 1922, and for other purposes, Pub. L. No. 325, 42 Stat. 121–141; Trimble, *Admiral William A. Moffett*, 80.

91. Memorandum on Hoover Conference, July 7, 1921, folder 4, box 406, MAA Papers, AHC; Notes for Meeting of the Air Craft Men Monday, July 16, 1921, Aviation Correspondence, 1921–1921, box 39, Commerce Papers, HHPL.

92. The role of Small, Ely, and their respective organizations in the development of US aviation policy are detailed in chapter 5. Hoover to Hughes, July 8, 1921, National Advisory Committee for Aeronautics, 1921–1926, box 427, Commerce Papers, HHPL; Nick A. Komons, *Bonfires to Beacons: Federal Civil Aviation Policy under the Air Commerce Act, 1926–1938* (Washington, DC: Smithsonian Institution Press, 1989), 46.

93. Klein to Feiker, Oct. 23, 1923, and Klein to Winslow, Oct. 23, 1923, both in

Aviation—General, 1923–1924, box 2497, Records of the Bureau of Foreign and Domestic Commerce, RG 151, NARA II. As the Director of the Bureau of Foreign and Domestic Commerce within the Commerce Department, Julius Klein interacted extensively with Hoover and even served as acting secretary on occasion.

94. Several studies have perpetuated Hoover's assertion that "no one in the government was interested in developing commercial aviation" prior to his arrival at the Commerce Department. M. Houston Johnson V provides the most extreme example of this Hoover-centric narrative, one that reduces the complexity of the regulatory process to something akin to the dated genius-inventor archetype within the history of technology. He begins his study with the July 18 meeting and attributes unsigned notes taken prior to it to Hoover himself as evidence of the Secretary's "already firm grasp" of the problems facing aviation. But the author of these July 16 notes writes that they had just spent two days with Coffin, Bradley, Keys, and other industry representatives observing Mitchell's naval bombing tests, while Hoover's schedule shows him in Chicago at that time. Hoover Presidential Library Archivist Spencer Howard remains "quite certain that the memo was written for Hoover, not by him." Rather than a presentation of Hoover's thoughts prior to the meeting, the memo instead represents a statement of industry's newfound position prior to their July 18 meeting for Hoover's consumption; Herbert Hoover, *The Memoirs of Herbert Hoover: The Cabinet and Presidency, 1920–1933* (New York: Macmillan, 1952), 132; M. Houston Johnson V, *Taking Flight: The Foundations of American Commercial Aviation, 1918–1938* (College Station: Texas A&M University Press, 2019); Spencer Howard (Archives Technician, Herbert Hoover Presidential Library-Museum), email correspondence with author, Aug. 30, 2019. Other studies that elevate Hoover to a primary player in the development of American aviation policy include Ellis W. Hawley, "Three Facets of Hooverian Associationalism: Lumber, Aviation, and Movies, 1921–1930," in *Business and Government in America since 1870*, ed. Roberts F. Himmelberg (New York: Garland, 1994): 213–255; David D. Lee, "Herbert Hoover and the Development of Commercial Aviation, 1921–1926," *Business History Review* 58 (Spring 1964): 78–102; David D. Lee, "Herbert Hoover and the Golden Age of Aviation," in *Aviation's Golden Age: Portraits from the 1920s and 1930s*, ed. William M. Leary (Iowa City: University of Iowa Press, 1989): 127–147; and Randy Johnson, "Aviation, Herbert Hoover and his 'American Plan,'" *Journal of Aviation/Aerospace Education and Research* 10 (Spring 2001): 35–59.

95. Wright Aeronautical President George Houston had already began this process weeks earlier when he removed all mention of the proposed Aeronautical Board in H.R. 281, a revision that Bradley and Crowell, now the president of the Aero Club of America, fully supported. Bradley to Crowell, July 1, 1921, folder 3, box 76, MAA Papers, AHC.

Chapter 4. The Struggle for Legislation

1. Bureau of the Census, the census of 1920, District ED 92, Sheet 7B, GS Film Number 1821042, Image Number 00837, accessed Sept. 4, 2018, https://www.familysearch .org/search/collection/results?count=20&query=%2Bgivenname%3ACharles~%20%2B surname%3ANoeding~&collection_id=1488411&collectionNameFilter=true.

2. "3 Killed in Crash of 'Taxi' Airplane," *New York Times*, Aug. 25, 1924; "2 Official Probes of Plane Crash in Which 3 Died," *Brooklyn Daily Eagle*, Aug. 25, 1924.

3. With the MAA's membership limited to manufacturers of entire aircraft, its Board of Directors supported the creation of the ACCA in the fall of 1921 to represent manufactures engaged in all facets of the aviation industry. Untainted by persistent accusations of a wartime "aircraft trust," the ACCA constituted, in many ways, a rebranding of the MAA. With nine MAA companies among the ACCA's members and Bradley overseeing both organizations, the two industry groups remained indistinguishable in their support for federal regulation. The nine firms that held memberships in both the MAA and the ACCA were: Aeromarine Plane and Motor Company, Curtiss Aeroplane and Motor Corporation, Dayton Wright Company, Gallaudet Aircraft Corporation, L. W. F. Engineering Company, Lewis and Vought Corporation, Packard Motor Car Company, Thomas-Morse Aircraft Corporation, and Wright Aeronautical Corporation. Minutes of Special Meeting of the Board of Directors on the Manufacturers Aircraft Association, Oct. 14, 1921, folder 1, box 3, MAA Papers, AHC.

4. Report to the Department of Commerce from the Aeronautical Chamber of Commerce of America, Inc., "Statement of Urgency of Aerial Law and Survey of Safety in Flight," Apr. 10, 1922, folder 8, box 416, MAA Papers, AHC. The ACCA compiled this report under Bradley's leadership based solely on corporate records and newspaper stories, the only data available in the absence of official figures.

5. Of the 419 itinerant crashes from 1921 to 1923, the ACCA attributed 182 to pilot error, 137 to equipment failure, and 98 to stunt flying as compared to 51 fixed-base operator accidents that resulted in 25 deaths and 40 injuries. Aeronautical Chamber of Commerce of America, *Aircraft Year Book, 1924* (New York: Aeronautical Chamber of Commerce of America, 1924), 105, 108.

6. Aeronautical Chamber of Commerce of America, Inc., *Aircraft Yearbook 1925* (New York: Aeronautical Chamber of Commerce of America, 1925), 115–116.

7. William P. MacCracken Jr., interview by Charles E. Planck, Mar. 31, 1962, transcript, CM-008000–01, NASM.

8. For more on the hearings into Republican air mail contracts in the 1930s under Senator Hugo Black, see Nick Komons, *Bonfires to Beacons: Federal Civil Aviation Policy under the Air Commerce Act, 1926–1938* (Washington, DC: Smithsonian Institution Press, 1989), and F. Robert van der Linden, *Airlines and Air Mail: The Post Office and the Birth of the Commercial Aviation Industry* (Lexington: University Press of Kentucky, 2002).

9. George Bogert, "Problems in Aviation Law," *Cornell Law Quarterly* 6, no. 3 (Mar. 1921): 271–309.

10. Bradley to Cowles, June 28, 1920, folder 1, box 122, MAA Papers, AHC.

11. The following discussion draws from and expands on Stuart Banner's analysis in *Who Owns the Sky? The Struggle to Control Airspace from the Wright Brothers On* (Cambridge, MA: Harvard University Press, 2008), 150–158.

12. US Const. art. III, § 2, cl. 1.

13. Banner, *Who Owns the Sky*, 152.

14. "Conference of Delegates of State and Local Bar Associations, Boston, Mass., September 2, 1919," *American Bar Association Journal* 6 (Jan. 1920): 42–43; Banner, *Who Owns the Sky*, 152.

15. According to Banner, Rooker personally mailed this letter addressed to his fellow

committee members to numerous individuals in government and industry throughout the country. It gained even more exposure when it became part of the Congressional Record. Rooker to Members of the Special Committee Appointed by the American Bar Association to Study the Question of Aerial Legislation, Jan. 5, 1920, printed in 66 Cong. Rec. S 2189-2191 (Jan. 29, 1920).

16. Secretary of War Baker forwarded the Interdepartmental Board bill with an amendment that placed all navigable airspace and aircraft "within the admiralty jurisdiction of the Federal Courts" to the chairs of the House and Senate Military Affairs Committees "by direction of the President," in May 1920. Baker to Wadsworth, May 11, 1920, and May 22, 1920, both in folder S. 4470 66th Cong., box 59, Records of the US Senate, RG 46, NARA I.

17. Bradley to Crisp and Bradley to Woolsey, Feb. 14, 1920, Bradley to Rooker, Feb. 18, 1920, all in folder 7, box 134, MAA Papers, AHC.

18. Banner, *Who Owns the Sky*, 153–154. According to Banner, the cases in question were *The Crawford Bros. No. 2*, 215 F. 269, 271 (W.D. Wash. 1914) and *Reinhardt v. Newport Flying Service Corp.*, 232 N.Y. 115, 118 (1921).

19. Cowles to Bradley, Mar. 19, 1920, folder 7, box 134, MAA Papers, AHC; Cowles to Bradley, June 23, 1920, and Bradley to Cowles, June 28, 1920, both in folder 1, box 122, MAA Papers, AHC.

20. MAA Press Release, undated (but after Jan. 5, 1920), folder 3, box 29, MAA Papers, AHC.

21. "Report of the Committee on Admiralty and Maritime Law," *American Bar Association Journal*, 6 (July 1920), 415–416.

22. Miller to Mitchell, July 23, 1920, Reports, 1919, box 49, William Mitchell Papers, LOC.

23. Johnson to Director of Air Service, Enclosed in Fickel to Fuller, Oct. 8, 1920, folder 187, box 10, Records of the Air Service Advisory Board, Records of the Office of the Chief of the Air Service, 1917–21, RG 18, NARA II. Johnson's thirty-one-page brief to Menoher was printed as "Air Service Information Circular No. 181: Legal Questions Affecting Federal Control of the Air" (Washington, DC: Government Printing Office, 1921).

24. Handbook of the National Conference of Commissioners on Uniform State Laws and Proceedings of the Thirtieth Annual Conference, St. Louis, Missouri, Aug. 19–24, 1920, 7, 120; William P. MacCracken Jr., interview by Charles E. Planck, Mar. 31, 1962, transcript, CM-008000-01, NASM; Michael Osborn and Joseph Riggs, *Mr. Mac: William P. MacCracken, Jr. on Aviation, Law, and Optometry* (Memphis, TN: Southern College of Optometry, 1970), 35; Boston, Preliminary Report of the American Bar Association Committee on the law of Aviation, Jan. 3, 1921, American Bar Association 1920–26, Aviation Law Committee, box 36, MacCracken Papers, HHPL. The NCCUSL's Special Committee on Uniform Aviation Law included Baltimore lawyer John Hinkley as chair, George G. Bogert of Cornell University Law School, Charles V. Imlay (Washington, DC), Thomas J. Brooks (IN), N. E. Corthell (WY), A. T. Stovall (MS), and George B. Young (VT) among its members. For a history of the NCCUSL, see Allison Dunham, "A History of the National Conference of Commissioners on Uniform State Laws," *Law and Contemporary Problems* 30 (Spring 1965): 233–249.

25. "Report of the Special Committee on the Law of Aviation," *Report of the Forty-Fourth Annual Meeting of the American Bar Association Held at Cincinnati, Ohio, August 31, September 1 and 2, 1921* (Baltimore: Lord Baltimore Press, 1921), 502, 505–506.

26. *Report of the Forty-Fourth Annual Meeting of the American Bar Association*, 83–90, 94–95.

27. "Report of the Special Committee on the Law of Aviation," 502, 505–506.

28. Amendments 16 through 19, the four "Progressive Era amendments," were ratified between February 3, 1913, and August 18, 1920.

29. *Aeronautics: Seventh Annual Report of the National Advisory Committee for Aeronautics, 1921* (Washington, DC: Government Publishing Office, 1922), 12; Memorandum Regarding the Suggestion That a Constitutional Amendment is Necessary to Establish Federal Control of Civil Aviation, Aug. 26, 1921, folder 8, box 416, MAA Papers, AHC.

30. W. Jefferson Davis, "Laws of the Air," *US Air Service* 4, no. 5 (Dec. 1920): 20.

31. Rooker to MAA, Jan. 3, 1921, folder 7, box 134, MAA Papers, AHC.

32. *Report of the Forty-Fourth Annual Meeting of the American Bar Association*, 83–90, 94–95; Osborn and Riggs, *Mr. Mac*, 39.

33. Osborn and Riggs, *Mr. Mac*, 39.

34. The NCCUSL formed its own aviation committee in 1920 to prevent the creation of conflicting state laws in the absence of clear federal jurisdiction. Bogert became chair of the committee in 1921. *Handbook of the National Conference of Commissioners on Uniform State Laws and Proceedings of the Thirtieth Annual Conference, St. Louis, Missouri, August 19–24, 1920*, 7, 120.

35. Woolsey to Crisp and Bradley, Feb. 2, 1920, folder 5, box 133, MAA Papers, AHC.

36. Testimony of Frank Russell, Dec. 19, 1921, *Civil Aviation in the Department of Commerce, Hearings Before a Subcommittee of the Committee on Commerce, United States Senate, Sixty-Seventh Congress, Second Session, on S. 2815*, 67th Cong. (1921), 24.

37. Testimony of Frank Russell, Dec. 19, 1921, *Civil Aviation in the Department of Commerce, Hearings Before a Subcommittee of the Committee on Commerce, United States Senate, Sixty-Seventh Congress, Second Session, on S. 2815*, 67th Cong. (1921), 24.

38. A Bill to Create a Bureau of Civil Aeronautics in the Department of Commerce, to Encourage and Regulate the Operation of Civil Aircraft in Interstate and Foreign Commerce, and for Other Purposes, S. 2448, 67th Cong. (1921).

39. Memorandum Regarding the Suggestion That a Constitutional Amendment is Necessary to Establish Federal Control of Civil Aviation, Aug. 26, 1921, folder 8, box 416, MAA Papers, AHC.

40. Coffin to Patrick, Nov. 2, 1921, box 3, Chief of the Air Service and the Air Corps, 1922–1927, Records of the Office of the Chief of the Air Corps, 1917–44, Records of the Army Air Forces, RG 18, NARA II; Minutes of Regular Meeting of Executive Committee, Sept. 15, 1921, folder 12, box 96, Office of the Secretary Records, 1903–1924, RU 45, Smithsonian.

41. Moffett to Bell, Nov. 7, 1921, and Coffin to Moffett, Nov. 16, 1921, both in folder 11, box 227, MMA Papers, AHC; Coffin to Patrick, Nov. 17, 1921, box 3, Records of the Office of the Chief of the Air Corps, 1917–44, Records of the Army Air Forces, RG 18, NARA II.

42. Coffin and Bradley met with Patrick and Moffett on December 5. Three days later, Curtiss vice president and MAA secretary Frank H. Russell joined them for further discussions with the NACA's Ames, Marvin, and George Lewis. Coffin to Patrick, Nov. 2, 1921, box 3, Chief of the Air Service and the Air Corps, 1922–1927, Records of the Office of the Chief of the Air Corps, 1917–44, Records of the Army Air Forces, RG 18, NARA II; Memorandum Regarding the Wadsworth-Hicks Bill, Referring to the Draft of December 6, folder 8, box 416, MAA Papers, AHC; Walcott to Shaughnessy, Dec. 12, 1921, folder 15.02, box 44, Office of the Second Assistant Postmaster General, Records of the Post Office Department, RG 28, NARA I.

43. In accordance with the New York Aviation Commission's earlier charge that the state's national representative work with the executive branch to pass legislation, this revised industry bill was quickly introduced into both Houses of Congress by Republican representative Frederick C. Hicks as H.R. 9407 and Senator Wadsworth as S. 2815 on December 9 and 15, respectively.

44. *Civil Aviation in the Department of Commerce, Hearings Before a Subcommittee of the Committee on Commerce, United States Senate, Sixty-Seventh Congress, Second Session, on S. 2815*, 67th Cong. (1921).

45. Organizations in support of S. 2815 included the American Legion's California Department, the Aero Club of America, the *Aviation and Aircraft Journal*, the Boston Chamber of Commerce, and the Chicago City Council. J. Parker Kirlin and John M. Wooley to the NACA, "In RE S. 2815 Being a Bill to Create a Bureau of Civil Aviation, etc.," Jan. 16, 1922, Aviation, Air Law, Federal, 1020-22, Proposed Bills, box 50, MacCracken Papers, HHPL; Allen to Hoover, Dec. 24, 1921, Commerce Department, Aeronautics, Bureau of, Legislation, 1921, box 121, Commerce Papers, HHPL; Aero Club Resolution, received at Commerce Department Jan. 21, 1922, Commerce Department, Aeronautics, Bureau of, Legislation, 1922, box 121, Commerce Papers, HHPL; "The Wadsworth-Hicks Civil Aviation Bill," *Aviation and Aircraft Journal* 11 (Dec. 26, 1921): 730–731; "Boston C. of C. on Air Laws," *Aviation* 12 (Feb. 13, 1922): 194; Small to Ely, Jan. 27, 1922, folder 2, box 122, MAA Papers, AHC.

46. Wadsworth submitted this revised bill, S. 3076, on January 16, 1922. A Bill to Create a Bureau of Aeronautics in the Department of Commerce, to Encourage and Regulate the Operation of Civil Aircraft in Interstate and Foreign Commerce, and for Other Purposes, S. 3076, 67th Cong. (1922); Committee on Commerce, Bureau of Aeronautics in Department of Commerce, S. Rep. No. 67-460 (1922); 67th Cong. Rec. 2531-32 and 2545-48 (1922). S. 3076 included an additional section that limited the issuance of federal licenses only to US citizens and corporations owned and controlled by citizens on Hoover's personal recommendation. Hoover did not take any issue with the original bill's extensive grant of federal authority. Draft of a Bill, enclosed in Hoover to Wadsworth, Dec. 20, 1921, and Wadsworth to Hoover, Dec. 22, 1921, both in Commerce Department, Aeronautics, Bureau of, Legislation 1921, box 121, Commerce Papers, HHPL.

47. Bradley to Patrick, Feb. 2, 1922, and Patrick to Bradley, Feb. 3, 1922, both cited in Walterman, "Airpower and Private Enterprise," 396.

48. Memorandum Regarding Air Law, undated (sometime between the passage of S. 3076 on Feb. 14 and the ABA/NCCUSL meeting on Feb. 25, 1922), folder 8, box, 416, MAA Papers, AHC.

49. *Minutes of the Joint Meeting of Committee of the Conference of Commissioners on Uniform State Laws and Committee on the American Bar Association on the Law of Aeronautics*, Feb. 25, 1922, box 36, MacCracken Papers, HHPL.

50. *Minutes of the Joint Meeting of Committee of the Conference of Commissioners on Uniform State Laws and Committee on the American Bar Association on the Law of Aeronautics*, Feb. 25, 1922, box 36, MacCracken Papers, HHPL (emphasis added).

51. *Minutes of the Joint Meeting of Committee of the Conference of Commissioners on Uniform State Laws and Committee on the American Bar Association on the Law of Aeronautics*, Feb. 25, 1922, box 36, MacCracken Papers, HHPL; "Report of the Committee on the Law of Aeronautics," *Report of the Forty-Fifth Annual Meeting of the American Bar Association Held at San Francisco, California, August 9–11, 1922* (Baltimore: Lord Baltimore Press, 1922), 415.

52. *Railroad Commission of Wisconsin v. Chicago, Burlington and Quincy Railroad Co.* 257 US 563 (1922).

53. Bogert to MacCracken, Mar. 15, 1922, cited in Komons, *Bonfires to Beacons*, 52–53.

54. Klein to Emmet, May 25, 1922, Commerce Department, Aeronautics, Bureau of, Legislation, 1922, box 121, Commerce Papers, HHPL; Winslow, Samuel Ellsworth, Biographical Directory of the United States Congress, accessed Sept. 25, 2019, http://bioguide.congress.gov/scripts/biodisplay.pl?index=W000639.

55. William P. MacCracken Jr., interview by Charles E. Planck, Mar. 31, 1962, transcript, CM-008000-01, NASM; Osborn and Riggs, *Mr. Mac*, 40–41.

56. Roosevelt to Hoover, Apr. 8, 1922; Weeks to Hoover, Apr. 11, 1922; telegram, Bradley to Hoover, Apr. 11, 1922; Coffin to Hoover, Apr. 18, 1922, all in Commerce Department, Aeronautics, Bureau of, Legislation, 1922, box 121, Commerce Papers, HHPL.

57. Klein to Emmet, May 25, 1922, Commerce Department, Aeronautics, Bureau of, Legislation, 1922, box 121, Commerce Papers, HHPL.

58. Lamb to Hoover, May 26, 1922, Commerce Department, Aeronautics, Bureau of, Legislation, 1922, box 121, Commerce Papers, HHPL.

59. Hoover to Winslow, June 13, 1922, Commerce Department, Aeronautics, Bureau of, Legislation, 1922, box 121, Commerce Papers, HHPL. According to aviation historian Nick A. Komons, Winslow—under the influence of Representative Curry of California—informed MacCracken and Lamb that he "wanted a bill establishing a department of aviation," and the two had "little choice" but to do so. MacCracken had come to accept the need for uniform federal control but had been an outsider to the administrative consensus emerging among the interested executive departments and industry for the past two years. According to the NACA's Samuel Stratton, MacCracken's initial draft was "not . . . approved by the Secretary of Commerce," and he quickly dropped the unified department framework. Winslow secured the remaining copies of this initial Commerce Department legislation in his office "under lock and key." Komons, *Bonfires to Beacons*, 56; Minutes of Regular Meeting of Executive Committee, June 29, 1922, box 94, folder 13, Office of the

Secretary Records, 1903–1924, RU 45, Smithsonian; Klein to Emmet, June 14, 1922, Commerce Department, Aeronautics, Bureau of, Legislation, 1922, box 121, Commerce Papers, HHPL.

60. A Bill to Authorize the Creation and Establishment in the Department of Commerce of a Bureau of Aviation for the Purpose of Regulating the Use of the Air in the Operation of Aircraft in Commerce, for the Uniform Regulation of Commerce Carried on by Means of Aircraft in Navigating the Air, for the Prevention of Interference with Such Commerce or Subjecting the Same to Unjust Discrimination, and to Encourage, Foster, and Develop Civil Aviation, and for Other Purposes," enclosed in Hoover to Winslow, June 19, 1922, House of Representatives, Winslow, Samuel E. 1921–1922, box 277, Commerce Papers, HHPL; Minutes of Regular Meeting of Executive Committee, Aug. 31, 1922, box 94, folder 13, Office of the Secretary Records, 1903–1924, RU 45, Smithsonian.

61. James J. O'Hara to Emmet, Aug. 7, 1922, Commerce Department, Aeronautics, Bureau of, Legislation, 1922, box 121, Commerce Papers, HHPL.

62. "Report of the Committee on the Law of Aeronautics," *Report of the Forty-Sixth Annual Meeting of the American Bar Association Held at Minneapolis, Minnesota, August 29–31, 1923* (Baltimore: Lord Baltimore Press, 1923), 384–385. Davis joined the ABA's Special Committee on the Law of Aviation once MacCracken took over as chair in its second year. Like Davis, Bogert called for the ABA to endorse ratification of the air convention. See George Bogert, "Problems in Aviation Law," *Cornell Law Quarterly* 6, no. 3 (Mar. 1921), 279.

63. "Proposes Control of Civil Aviation," *New York Times*, Jan. 8, 1923.

64. Memorandum upon Federal Civil Aeronautics Legislation, enclosed in Lee to MacCracken, 6 No. 1925, Aviation, Air Law, Federal, 1925–26, Acts and Reports, S. 41, box 51, MacCracken Papers, HHPL. MacCracken's fellow ABA aviation committee member W. Jefferson Davis also noted that the Civil Air Navigation Act of 1923 "follows closely the regulations prescribed by the International Air Navigation Convention of 1919"; Jefferson Davis, "The Civil Aeronautics Act of 1923," *American Bar Association Journal* 9 (July 1923): 420–422.

65. Civil Aeronautics Act of 1923, H.R. 13715, 67th Cong. (1923). The initial version of the bill, introduced in January 1923, explicitly stated that states retained sovereignty over their airspace "except in so far as granted to the United States as an incident to the powers delegated by the States to the United States in the Constitution." This clear attempt to assuage states' rights advocates was dropped from subsequent versions of the Winslow bill, although the declaration of national air sovereignty remained.

66. As mentioned in chapter 2, over three years earlier US delegates Major General Patrick and Admiral Knapp had expressed constitutional concerns about whether the federal government possessed the authority to force state, municipal, and private airports to comply with the convention's equal access provision.

67. William P. MacCracken Jr., "Aeronautical Legal Problems," an address delivered before the First National Air Institute under Auspices of the Detroit Aviation Society, Oct. 11, 1922, Commerce Department, Aeronautics, Bureau of, Legislation, 1922, box 121, Commerce Papers, HHPL. MacCracken's belief that the United States would pass regula-

tions aligned with the convention can be seen in the inclusion of a clause that required US aircraft and pilots to adhere to US regulations when over the high seas or countries without their own regulations. With his stated commitment to uniformity, it is difficult to believe that MacCracken would include such a provision if it would lead to US pilots and aircraft adhering to regulations in international flight that differed from those of Canada and the convention. Civil Aeronautics Act of 1923, H.R. 13715, 67th Cong. (1923).

68. The NAA's membership list reads like a who's who of aviation at the time and included Orville Wright, Edward P. Warner of MIT, NACA Executive Committee chair Joseph Ames, America's top wartime ace Eddie Rickenbacker, aircraft manufacturer William Boeing, aeronautical engineers Archibald Black and Glenn L. Martin, Edmund Ely of the National Aircraft Underwriters Association, *Aviation* editor Lester Gardner, and chairman of the Society of Automotive Engineers' Aeronautic Division Henry M. Crane, to name but a few. I refer to Archibald Black and Glenn L. Martin as "aeronautical engineers" based not on their receipt of an aeronautical engineering diploma from an accredited engineering program but on contemporary recognition of their professional expertise. Documentary evidence shows that members of the engineering profession clearly thought of the two as aeronautical engineers. The phrase "Consulting Aeronautical Engineer" graced the top of Black's letterhead, and Lester Gardner's *Who's Who in American Aeronautics* explicitly listed his profession as "Aeronautical Engineer." Although self-taught, Glenn Martin served as vice president of the Society of Automotive Engineers (SAE) and on both the Aeronautical Division of the SAE's Standards Committee and the Aero Club's Technical Committee alongside formally trained engineers such as Henry Crane, Grover C. Loening, and Jerome C. Hunsaker. In the first decades of the twentieth century, when formal university programs for aeronautical engineering remained rare—MIT's first course was offered under Hunsaker's guidance in 1914, and its undergraduate program officially began twelve years later—professional legitimacy was conferred on the first generation of aeronautical engineers via recognition from peers and organizations such as the SAE, the MAA, and the Air Service's Engineering Division in Dayton, Ohio. While Black and Martin's professional colleagues clearly recognized them as practicing aeronautical engineering, individuals such as James V. Martin and W. W. Christmas fought in vain for similar recognition and framed their exclusion from the nascent profession as part of a conspiracy to uphold an "aircraft trust"; Black to Hoover, July 21, 1921, Commerce Department, Aeronautics, Bureau of, Legislation, 1921, box 121, Herbert Hoover Papers; Lester Gardner, *Who's Who in American Aeronautics* (New York: Garner Moffat, 1922), 26; MIT AeroAstro History, accessed Feb. 25, 2019, https://aeroastro.mit .edu/about/history.

69. The Advanced Committee on Organization, "Advance Information on the Second National Aero Congress to Form a National Aeronautic Association, Oct. 12, 13, and 14, 1922," folder 2, box 432, MAA Papers, AHC.

70. Articles of Association of the National Aeronautic Association of USA (Incorporated), approved Oct. 14, 1922, Aviation, National Aeronautic Association, Bulletins, 1922, box 61, MacCracken Papers, HHPL; Report of the Legislative Committee of the Second National Convention of the National Aeronautic Association, USA, held Oct. 1, 1923, Aviation, National Aeronautic Association, Correspondence, 1922–26, box 61, MacCracken

Papers, HHPL. W. Jefferson Davis served alongside MacCracken on both the ABA's aviation committee and the NAA's Legislative Committee.

71. National Aeronautic Association Press Release, Jan. 10, 1923, folder 4, box 432, MAA Papers, AHC.

72. National Aeronautic Association Press Release, Nov. 10, 1922, folder 3, box 432, MAA Papers, AHC; Bill Robie, *For the Greatest Achievement: A History of the Aero Club of America and the National Aeronautic Association* (Washington, DC: Smithsonian Institution Press, 1993), 111–113; Komons, *Bonfires to Beacons*, 58–61.

73. "Report of the Committee on the Law of Aeronautics," in *Report of the Forty-Sixth Annual Meeting of the American Bar Association held at Minneapolis, Minnesota, August 29–31, 1923*, 384; "A.S.M.E. for Civil Air Bureau," *Aviation* 14 (Mar. 5, 1923): 273; "Report of American Legion's Air Committee," *Aviation* 15 (Nov. 26, 1923): 656; *Journal of the Proceedings of the Board of Commissioners of Cook County*, Jan. 22, 1923; "ACCA Legislative Bulletin," no. 13, 1923, folder 9, box 420, MAA Papers, AHC. "ACCA Information Bulletin," no. 9, Jan. 25, 1923, and "ACCA Information Bulletin," no. 11, Mar. 22, 1925, both in folder 2, box 248, MAA Papers, AHC.

74. Whereas the Senate bill derived from Woolsey's limited legislation defined only three terms (aircraft, civil aircraft, and commerce) and offered a single penalty for infraction, MacCracken's bill defined ten times as many terms and divided infractions into civil and criminal with varying penalties.

75. Minutes of Special Meeting of Executive Committee, Jan. 19, 1923, box 94, folder 13, Office of the Secretary Records, 1903–1924, RU 45, Smithsonian. Under MacCracken's bill, the NACA would now include representatives from the new Bureau of Civil Aeronautics, the Assistant Postmaster General's Office, and the Coast Guard.

76. Weeks to Winslow, Jan. 31, 1923, and Ames to Winslow, Feb. 2, 1923, both in folder HR67A-D15 HR13715-13715, box 124, RG 233, Records of the US House of Representatives, 67th Congress, NARA I. In addition to the bill's airfield provision, the NACA also expressed concern over a planned increase in pay for military officers assigned to the new bureau, the potential for industry's undue influence on the Commerce Department through the creation of a new Civil Aeronautics Consulting Board, and potential duplication of weather reporting and aeronautical research. Secretary of War John Weeks's department also questioned the need for the Civil Aeronautics Consulting Board and took serious umbrage to placing aviation in the Canal Zone under the jurisdiction of a civilian in the Commerce Department.

77. Minutes of Special Meeting of Executive Committee, Jan. 19, 1923, box 94, folder 13, Office of the Secretary Records, 1903–1924, RU 45, Smithsonian.

78. Conference on Winslow Bill, Feb. 5, 1923, folder 13, box 94, Office of the Secretary Records 1903–1924, RU 45, Smithsonian. At this meeting, Lt. John Parker Van Zandt of the Air Service, Cdr. Harry B. Cecil of the Navy's Bureau of Aeronautics, John Victory of the NACA, NAA general manager Harold E. Hartney, and Bradley of the MAA agreed that "no opposition would be made" to the grant of federal authority to create airfields "if it would in any way impede the passage of the bill" and not to defend the proposed Civil Aeronautics Consulting Board if any objection arose during congressional hearings.

79. Initially submitted in the fourth session of the 67th Congress as H.R. 13715 on

January 18, 1923, the Winslow bill was resubmitted twice during the 68th Congress, in the first session as H.R. 3243 on December 13, 1923, and in the second session as H.R. 10522 on December 9, 1924. This latter version incorporated Secretary of War Weeks's demand to exclude Canal Zone aviation from Commerce Department regulation.

80. Calvin Coolidge, Message of the President of the United States to Congress, Dec. 6, 1923; Winslow to Coolidge, Dec. 10, 1923, and Coolidge to Winslow, Dec. 20, 1923, both in reel 26, Coolidge Papers, LOC.

81. To placate the NACA's concern over runaway appropriations, H.R. 10522—the final version of the Winslow bill—explicitly mentioned the power to establish federally funded "aerial lighthouses" and radio aids. It still, however, included airfields within the definition of air navigation facilities at MacCracken's insistence. To create a Bureau of Civil Air Navigation in the Department of Commerce, encourage and regulate the navigation of civil aircraft, and for other purposes, H.R. 10522, 68th Cong. (1924).

82. In addition to Hoover, testimony in favor of the Winslow bill came from solicitor Stephen B. Davis and communications expert P. E. D. Nagle of the Commerce Department, Lt. John Parker Van Zandt of the Air Service, Cdr. Marc A. Mitscher of the Navy Bureau of Aeronautics, Lt. Cdr. Stephen S. Yeandle of the Coast Guard, Charles Marvin of the NACA, Commissioner General of Immigration William W. Husband, and Second Assistant Postmaster General Paul Henderson. In the first weeks of the 68th Congress, the Senate once again passed a bill restricting federal control to interstate and foreign flights, S. 76.

83. *Bureau of Civil Air Navigation in the Department of Commerce: Hearings Before the Committee on Interstate and Foreign Commerce*, 68th Cong. (1924), 65–66.

84. The section titled "States Rights" meticulously cataloged how the bill did "not interfere with the political sovereignty of a State over its air space" and left issues of "private property rights, . . . criminal jurisdiction," and the establishment of "air space reservations" with the states. But this assurance, buried within a committee report, paled in comparison to the complete assertion of federal air sovereignty within the text of the bill. House Committee on Interstate and Foreign Commerce, Civil Air Navigation Bill, H.R. Rep. No. 68-1262 (1925).

85. Russell to Keys, June 23, 1925, Russell, F.H. (1925), box 11, Clement Melville Keys Papers, NASM.

86. Komons, *Bonfires to Beacons*, 58.

87. An undated handwritten note by MacCracken—more than likely from some time in 1924 due to references concerning a meeting with Judge Stephen B. Davis, the new Commerce Department solicitor, and the Lampert Committee established in March of that year—shows that both the international air convention and the "Canadian situation" were in the forefront of MacCracken's thoughts during this period; Handwritten note, undated, Aviation, National Aeronautic Association, Legislative Committee, 1924–25, box 62, MacCracken Papers, HHPL.

88. Office of the Legislative Drafting Counsel, "Preliminary Note: The Relation of the Convention to the Proposed Civil Air Navigation Act," in *International Air Navigation Convention, Text of Articles and Annexes Together with Amendments Issued Thereunder up to June, 1924* (Washington, DC: Government Publishing Office, Dec. 1, 1924), 1; H.R. Rep.

No. 68-1262 (1925), 15. The preliminary note cited previously connected specific articles throughout the Winslow bill to nearly thirty of the convention's articles and seven out of eight of its technical annexes.

89. Lester Gardner, *Who's Who in American Aeronautics* (New York: Gardner Publishing, 1925), 28; Leon Harris, *Only to God: The Extraordinary Life of Godfrey Lowell Cabot* (New York: Atheneum, 1967), 202–204. At the time of Cabot's testimony, Coffin and MacCracken both served as governors at large for the NAA and chaired the Program Committee and Legislative Committee, respectively. Aeronautical Chamber of Commerce of America, *Aircraft Yearbook 1925*, 259.

90. Statement of Godfrey L. Cabot in *Hearings Before the President's Aircraft Board*, vol. 4 (Washington, DC: Government Publishing Office, 1925), 1482, 1485.

91. Maurer Maurer, *Aviation in the US Army, 1919–1939* (Washington, DC: Government Printing Office, 1987), 45; *Annual Report of the Chief of the Air Service for Fiscal Year Ending June 30, 1923*, 78. The Lassiter Board, named after its chair Maj. Gen. William Lassiter, advised Secretary of War Weeks in October 1923 that the woeful situation in the Air Service required annual appropriation of $25 million over the next ten years, with $15 million set aside exclusively for the procurement of new aircraft. Report of Lassiter Committee Appointed to Consider Needs of Air Service, dated Oct. 11, 1923, in Aeronautical Chamber of Commerce of America, *Aircraft Yearbook 1924* (New York: Aeronautical Chamber of Commerce of America, 1924), 298.

92. Despite the report of Attorney General Thomas Watt Gregory at the time of the MAA's creation that it did not violate the Sherman Antitrust Act, Bradley and his organization remained under nearly constant investigation for the next decade. The most well known of these include those led by Charles Evans Hughes in 1918 and Republican representative James Frear in 1919. Gregory to Baker, Oct. 6, 1917, folder 2, box 10 and Aircraft Investigations, 1917–1925, folder 6, box 46, both in MAA papers, AHC; Komons, *Bonfires to Beacons*, 39–41.

93. H.R. 192, 68 *Cong. Rec.* H 4816 (daily ed. Mar. 24, 1924); Komons, *Bonfires to Beacons*, 72–73. Thomas W. Walterman argues that Progressive House Republicans fused their indignation at the scandals of the Harding presidency, warnings from Air Service officials, and the promised efficiency of the unified department model into a "point of attack on the Administration." As such, he portrays the Lampert Committee as the result of a larger struggle between the socially conscious "purest strain" of Progressivism associated with Theodore Roosevelt and the business-minded and efficiency-focused version of the 1920s best embodied in Commerce Secretary Hoover. Thomas W. Walterman, "Airpower and Private Enterprise: Federal-Industrial Relations in the Aeronautics Field, 1918–1926" (PhD diss., Washington University, 1970), vi, 412. For more on the debate over the shape of the aeronautical industry and its proper relationship with government, see Jacob A. Vander Meulen, *The Politics of Aircraft: Building an American Military Industry* (Lawrence: University Press of Kansas, 1991).

94. In addition to Lampert, the committee included Republicans Albert H. Vestal of Indiana, Randolph Perkins of New Jersey, Charles L. Faust of Missouri, and Frank R. Reid of Illinois as well as Democrats Anning S. Prall of New York, Patrick B. O'Sullivan of Connecticut, William N. Rogers of New Hampshire, and Clarence F. Lea of California,

who, as a member of Winslow's Committee on Foreign and Interstate Commerce, was well versed in the particulars of civil aviation regulation and the provisions of the 1919 convention.

95. *Report of the Select Committee into Operations of the United States Air Services*, H.R. Rep. no. 68-1653, at 1 (1925) (hereafter cited as the *Lampert Committee Report*). Members of the committee heard witness testimony in "Washington, New York, Pasadena, and San Diego," with California Representative Clarence Lea serving as a committee of one at the two latter locations prior to the return of the 68th Congress in December. Locations visited by the Lampert Committee included Bolling Field, the Anacostia Naval Air Station, the Bureau of Standards, Langley Field, the NACA's Langley Laboratory, McCook Field, Wilbur Wright Field, and the Dayton International Air Races.

96. Mitchell found support among Representative Fiorello La Guardia, the NAA's Harold E. Hartney, *Aviation* editor Lester D. Gardner, and several Air Service aviators, while Secretary of War Weeks, Commerce Secretary Hoover, NACA members Major General Patrick and Rear Adm. William A. Moffett, Brig. Gen. Hugh A. Drum, Connecticut's freshman senator and wartime aviator Hiram S. Bingham, Lt. John Parker Van Zandt of the Air Service, the NACA's Joseph Ames, and NAA president Godfrey Cabot all supported the creation of a Commerce Department bureau.

97. Secretary of the Navy Curtis Wilbur shared President Coolidge's feelings on the matter at hearings before the House Committee on Military Affairs on a bill to create a unified department, H.R. 10147, underway at the same time. *Hearings Before the Committee on Military Affairs on H.R. 10147*, 68th Cong. 2, 134 (1925).

98. *Lampert Committee Hearings*, 704, 706, 1202, 1204. With the United States outside of the international convention, the State Department had to obtain advance permission from each country flown over during the 1924 World Flight. For more on the World Flight, see Carroll V. Glines, *Around the World in 175 Days: The First Round-the-World Flight* (Washington, DC: Smithsonian Institution Press, 2001), and Ernest A. McKay, *A World to Conquer: The Epic Story of the First Around-the-World Flight* (New York: Arco Publishing, 1981).

99. "The Origins and Purpose of the Survey of Civil Aviation," July 15, 1925, 1925 Commercial—General, box 1, Records of the Office of the Chief of Air Service, 1917–21, Records of the Training and War Plans Division, Records of the Army Air Forces, RG 18, NARA II.

100. Drake to Coffin (telegram), May 28, 1925, Drake to Coffin, June 6, 1925, and MacPherson to Wallace, July 25, 1925, all in 83272 [1 of 3], box 568, General Records of the Department of Commerce, RG 40, NARA II.

101. Coffin to Hoover, Apr. 18, 1922, Commerce Department, Aeronautics, Bureau of, Legislation, 1922, box 121, Commerce Papers, HHPL.

102. Also known as the Kelly Air Mail Act after its sponsor, Republican Clyde Kelly of Pennsylvania, the Air Mail Act of 1925 marked the culmination of a multiyear effort to transfer air mail carriage to private industry, one that the MAA, ACCA, and NAA strongly supported. See William M. Leary, *Aerial Pioneers: The US Air Mail Service, 1918–1927* (Washington, DC: Smithsonian Institution Press, 1985).

103. Air Mail Act, Pub. L. No. 359, 43 Stat. 805 (1925).

104. Komons argues that the Air Mail Act made "federal regulation . . . a foregone

conclusion." Nothing is inevitable, but even if we accept this premise the Air Mail Act did not stipulate the administrative framework or the extent of federal control within subsequent federal legislation; Komons, *Bonfires to Beacons*, 66.

105. Canada's director of civil aviation John A. Wilson hoped the Post Office would assume full "responsibility for inspection of aircraft and pilots," while Lester Gardner suggested in his magazine *Aviation* that the Post Office should simply adopt and extend the convention's registration and identification scheme to all US aircraft. Wilson, "Civil Aviation in the United States," undated memorandum, John A. Wilson Fonds, MG 30, microfilm reel 10781, LAC; "The Contract Air Mail," *Aviation* 18, no. 17 (Apr. 27, 1925): 457.

106. Robert F. van der Linden, *Airlines and Air Mail: The Post Office and the Birth of the Commercial Aviation Industry* (Lexington: University Press of Kentucky, 2002), 22.

107. In addition to Drake as chair, the Joint Committee also included New York College of Engineering professor Joseph Roe as vice-chair, NAA Governor for Pennsylvania and NAT board of directors member Charles T. Ludington, former MAA/ACCA employee and current NAT traffic manager Luther K. Bell, NACA member William F. Durand, and MIT professor of aeronautical engineering Edward Warner. They were aided by aeronautical engineer Alexander Klemin, Lt. John Parker Van Zandt of the War Department, and the Commerce Department's Transportation Division; Gregg to Hoover, Aug. 4, 1925, Aviation—General 1925, box 2497, Records of the Bureau of Foreign and Domestic Commerce, General Records, RG 151, NARA II.

108. Coolidge to Morrow, Mar. 11, 1925, reel 25, series I, subseries 2, correspondence, Dwight W. Morrow Papers, Amherst College Archives and Special Collections, Robert Frost Library, Amherst, MA.

109. "Navy Plane Missing in a Stormy Sea, Its Fuel Exhausted," *New York Times*, Sept. 2, 1925. The crew was ultimately found alive a week later, adrift on their floating aircraft.

110. "Squall Splits Big Airship," *Special to The New York Times*, Sept. 4, 1925.

111. "President Expects a New Shenandoah" and "Secretary Denies Politics in Flight," both in *New York Times*, Sept. 5, 1925. For more on these two incidences, see Thomas Wildenberg, *Billy Mitchell's War with the Navy: The Interwar Rivalry over Air Power* (Annapolis, MD: Naval Institute Press, 2014), 132–136, and William F. Trimble, *Admiral William A. Moffett: Architect of Naval Aviation* (Annapolis, MD: Naval Institute Press, 1994), 155–160.

112. After his fiery testimony before the Lampert Committee, Secretary of War Weeks had "reverted [Mitchell] to his permanent grade of colonel" and transferred him to San Antonio, Texas. Alfred F. Hurley, *Billy Mitchell: Crusader for Air Power* (Bloomington: Indiana University Press, 1975), 98.

113. "Mitchell Charges 'Gross Negligence' in Shenandoah Loss," *New York Times*, Sept. 6, 1925 (emphasis added).

114. Wildenberg, *Billy Mitchell's War with the Navy*, 142.

115. Bingham testified before the Lampert Committee on Feb. 19, 1925, that the House should immediately pass the Winslow bill to allow a reconciliation committee to merge the House and Senate bills as soon as possible; *Lampert Committee Hearings*, 2747–2748. He would go on to become president of the NAA from 1928 to 1934.

116. The other four members were retired Maj. Gen. James G. Harbor, retired Rear

Adm. Frank F. Fletcher, Sixth Circuit Court of Appeals judge Arthur C. Denison, and lone Democrat Carl Vinson of the House Committee on Naval Affairs. *Report of the President's Aircraft Board, November 30, 1925* (Washington, DC: Government Publishing Office, 1925), 1; Komons, *Bonfires to Beacons*, 75–76.

117. Drake to Warner, Oct. 1, 1925, 83272 [1 of 3], box 568, General Records of the Department of Commerce, RG 40, NARA II. Joint Committee chair Drake transmitted its "findings and recommendations" to Morrow and Coolidge on November 6; Drake to Morrow, Nov. 6, 1925, and Drake to Coolidge, Nov. 6, 1925, both in 83272 [1 of 3], box 568, General Records of the Department of Commerce, RG 40, NARA II. According to McCracken, the Morrow Board did not possess the funds to officially bring him on as counsel, so Coffin retained MacCracken's services at his own expense; Osborn and Riggs, *Mr. Mac*, 50; William P. MacCracken Jr., interview by Charles E. Planck, Mar. 31, 1962, transcript, CM-008000-01, NASM.

118. Between September 21 and October 16, 1925, the Morrow Board heard testimony from ninety-nine representatives from the military, government agencies, and aeronautics industry on the current state of the military air services, the sensitive issue of promotion and pay for military aviators, and how best to foster the growth of commercial aviation. To counter perceptions of bias, the board "gave the greater portion of the time to hearing those men with actual air experience." The majority of military aviators refused to support a unified service and instead favored Major General Patrick's compromise proposal to elevate the Air Service to a distinct corps within the War Department analogous to that of the Marine Corps within the Navy. Acting secretary of war Dwight F. Davis, Brigadier General Drum, Major General Patrick, Rear Admiral Moffett, Postmaster General New, Paul Henderson, Secretary Hoover, and Cabot reaffirmed their support for placing commercial aviation within the Commerce Department, "where it belongs." To disarm Mitchell, Morrow allowed him to "read chapters from [his book] *Winged Defense* for hours in a dull monotone" that constitutes eighty-nine pages in the hearing's official transcripts. Disengagement proved the best weapon against a man who—with disciplinary action pending and no new arguments to support his position—had "reached a point where only theatrics were left, and they did not materialize." *Report of the President's Aircraft Board, November 30, 1925* (Washington, DC: Government Publishing Office, 1925), 2; *Hearings Before the President's Aircraft Board* (Washington, DC: Government Publishing Office, 1925), 5–6, 54–55, 72–73, 308, 319, 333–334, 1189, 1482–1486; James J. Cooke, *Billy Mitchell* (Boulder, CO: Lynne Rienner, 2002), 182.

119. The Joint Committee submitted its initial report to Coolidge and the Morrow Board on November 6, 1925, and published its final report on January 23, 1926. The Morrow Board released its report on November 30, and the Lampert Committee followed with its report two weeks later.

120. *Report of the President's Aircraft Board, November 30, 1925*, 6, 10, 19, 29.

121. Select Comm. of Inquiry into Operations of the United States Air Services, Report, H.R. Rep. No. 1653, at 8–9 (1925).

122. "Recommendation and Summarized Findings of the Committee on Civil Aviation of the US Department of Commerce and American Engineering Council," Aviation, Commerce Dept., Committee on Civil Aviation, 1925, Recommendations and Summarized

Findings, box 56, MacCracken Papers, HHPL; *Civil Aviation: A Report by the Joint Committee on Civil Aviation of the US Department of Commerce and the American Engineering Council* (New York: McGraw-Hill, 1926), 8, 107 (emphasis added). The italicized phrase did not exist in the Joint Committee's draft recommendations sent to the Morrow Board, but was added into its final report on January 23, 1926.

123. *Civil Aviation*, 8, 107.

124. Michigan Republican Clarence J. McLeod's H.R. 196 proposed a $100,000,000 fund to provide direct federal loans for commercial operations. H.R. 6516, submitted by Illinois Republican John J. Jerome, proposed the creation of a five-member air board along the lines of the Interdepartmental Board's original draft bill from seven years earlier. True to form, California representative Curry submitted another rendition of his bill for a Department of Air. Based on discussions with Ames, Representative Parker—now chair of the House Committee on Interstate and Foreign Commerce—made sure bills that went contrary to the Morrow Board did not make it out of his committee; A Bill to provide governmental aid in commercializing aviation in the United States, its Territories, and possessions, H.R. 196, 69th Cong. (1926); A Bill to provide a bureau of civil, commercial, and strategic aeronautics within the United States government, H.R. 6516, 69th Cong. (1926); Ames to Parker, Jan. 20, 1926, folder 15-6, box 135, National Advisory Committee for Aeronautics, General Correspondence, 1915–1942, Records of the National Aeronautics and Space Administration, RG 255, NARA II.

125. Hoover to Bingham, Sept. 23, 1925, Bureau of Aeronautics, Legislation, 1925–1926, box 121, Commerce Papers, HHPL.

126. Bingham to Hoover, Oct. 13, 1925, Bureau of Aeronautics, Legislation, 1925–1926, box 121, Commerce Papers, HHPL.

127. Calvin Coolidge, Message of the President of the United States to Congress, Dec. 8, 1925; A Bill to encourage and regulate the use of aircraft in commerce, and other purposes, S. 41, 69th Cong. (1925); A Bill to encourage and regulate the use of aircraft in commerce, and other purposes, H.R. 4772, 69th Cong. (1925).

128. A Bill to encourage and regulate the use of aircraft in commerce, and other purposes, S. 41, 69th Cong. (1925). The original draft also called for the absorption of the NACA into the Department of Commerce, a long-held desire of Hoover, but it was removed in a revised version of the bill introduced six days later.

129. Komons, *Bonfires to Beacons*, 81.

130. Hoover to Jones, Dec. 9, 1925, Bureau of Aeronautics, Legislation, 1925–1926, box 121, Commerce Papers, HHPL.

131. 69 *Cong. Rec.* S917-926 (daily ed. Dec. 15, 1925).

132. Cabot to Hoover, Apr. 17, 1925, NAA, Legislative Committee, 1924–25, box 62, MacCracken Papers, HHPL; Cabot to Parker, Dec. 7, 1925, folder 72965, box 369, General Records of the Department of Commerce, General Correspondence, RG 40, NARA II. According to Komons, Democrat Clarence Lea and Republican Schuyler Merritt of Connecticut strongly opposed Parker's attempt to pass his complement to Bingham's bill that had received Hoover's blessing. It was to these two that Coffin and MacCracken turned in their attempt to secure the full extent of federal authority previously allowed for in the Winslow bill. See Komons, *Bonfires to Beacons*, 82–83.

133. "First Tentative Draft, November 16, 1925," Aviation, Air Law, Federal, 1925 Drafts and Notes, box 51, MacCracken Papers, HHPL; An Act to encourage and regulate the use of aircraft in commerce, and for other purposes, S. 41, H. Rep. No. 69-572 (1926).

134. Ames to Parker, Feb. 5, 1926, folder 15-6, box 135, National Advisory Committee for Aeronautics, General Correspondence, 1915–1942, Records of the National Aeronautics and Space Administration, RG 255, NARA II.

135. History would ultimately prove Ames correct, and Congress placed intrastate flights under federal control in the 1938 Civil Aeronautics Act.

136. These included member of the House Committee on Foreign and Interstate Commerce George Huddleston of Alabama, Luther Johnson of Texas, Henry Tucker of Virginia, and William Bankhead of Alabama.

137. *Cong. Rec.* H 7312-7330 (daily ed. Apr. 12, 1926). Huddleston and Kansas Republican Homer Hoch both submitted amendments to limit the bill to a strict interpretation of the commerce clause.

138. NACA members present at this April 22, 1926, special meeting also included Bureau of Standards director George K. Burgess, Orville Wright, Maj. John F. Curry, Capt. E. S. Land, and David W. Taylor. Minutes of Meeting of Special Subcommittee on the Encouragement and Regulation of Aircraft in Commerce, Apr. 22, 1926, folder 13-10, box 53, National Advisory Committee for Aeronautics, General Correspondence, 1915–1942, Records of the National Aeronautics and Space Administration, RG 255, NARA II; Komons, *Bonfires to Beacons*, 84–86.

139. Victory to Walcott, May 3, 1926, folder 13-10: Subcommittee on Encouragement and Regulation of Aircraft in Commerce, 1926, box 53, Records of the National Aeronautics and Space Administration, RG 255, NARA II.

140. *Cong. Rec.* S9351-9356 (daily ed. May 13, 1926).

141. Latchford to Harrison, May 15, 1926, box 7695, Records of the State Department, RG 59, NARA II.

142. The Air Commerce Act represented one of several changes prompted by the aerial crisis of 1925. Coolidge's signature on the Naval Aircraft Expansion Act and the Air Corps Act in the weeks following the promulgation of the Air Commerce Act created additional assistant secretary positions within the Navy and War Departments as per the Morrow Board's recommendations. MIT aeronautical engineering instructor and NACA researcher Edward Warner accepted the position in the Navy Department, and Guggenheim Fund for the Promotion of Aeronautics trustee Frederick Trubee Davison accepted the complementary post in the War Department. They were to cooperate with the new Assistant Secretary of Commerce for Aeronautics on matters of air policy that affected both civil and military aviation. Naval Aircraft Expansion Act, Pub. L. No. 69-422, 44 Stat. 668 (1926); Air Corps Act, Pub. L. No. 69-446, 44 Stat. 721 (1926).

143. Osborn and Riggs, *Mr. Mac*, 63. MacCracken was not Hoover's first or even second choice to fill the new position in his department, but former second assistant postmaster Paul Henderson of National Air Transport and Philadelphia businessman Hollinshead N. Taylor both elected to remain in private industry. Both Walcott and Senator Bingham expressed concern over MacCracken's nomination, but Hoover assured them that he would maintain a watchful eye on his new assistant secretary; Komons,

Bonfires to Beacons, 87; Walcott to Coolidge, reel 109, Calvin Coolidge Papers, 1915–1932, LOC; Bingham to Hoover (telegram) enclosed in McIntyre to Hoover, June 29, 1926, and Hoover to Bingham (telegram), July 30, 1926, both in Bureau of Aeronautics, MacCracken, William P., 1926–1928, box 120 Commerce Papers, HHPL; undated confidential memo.

Chapter 5. The Need for Regulatory Compatibility

1. Manufacturers Aircraft Association, *Aircraft Yearbook* (New York: Doubleday, Page, 1920), 43; "Mail Delivered by Airplane at Seattle," *Vancouver Daily World*, Mar. 3, 1919.

2. Hubbard's route allowed US mail from Seattle to more rapidly connect with the Asia-bound *Africa Maru* before its regular departures from Victoria. R. E. G. Davies, *Airlines of the United States since 1914* (Washington, DC: Smithsonian Institution Press, 1972), 10–12.

3. Cornelius Vanderbilt Jr., "An Air Mail Record," *New York Times*, Oct. 16, 1921; Peter M. Bowers, *Boeing Aircraft since 1916* (Annapolis, MD: Naval Institute Press, 1966, 1968), 49; Jim Brown, *Hubbard: The Forgotten Boeing Aviator* (Seattle, WA: Peanut Butter Publishing, 1996), 96–102.

4. F. Robert van der Linden, *Airlines and Air Mail: The Post Office and the Birth of the Commercial Aviation Industry* (Lexington: University Press of Kentucky, 2002), 29–30.

5. Minutes of the Canadian Air Board, May 17, 1920, "Air Board" series, vol. 3510, Department of National Defence Fonds, RG 24, LAC; Wilson to Christie, May 23, 1923, folder 32-6, International Air Navigation, 1927, box 177, National Advisory Committee for Aeronautics, General Correspondence, 1915–1942, Records of the National Aeronautics and Space Administration, RG 255, NARA II. In November 1923, the Navy Department agreed to a Canadian request to provide airworthiness inspections for US aircraft types approved for military service prior to their entrance into Canada. Wilson to Christie, May 23, 1923; Christie to McNamee, May 30, 1923; McNamee to Naval Operations, General Board, Naval War College, Bureau of Aeronautics, and the NACA; all in folder 32-6, International Air Navigation, 1927, box 177, National Advisory Committee for Aeronautics, General Correspondence, 1915–1942, Records of the National Aeronautics and Space Administration, RG 255, NARA II; Chief of Bureau of Aeronautics, "Examination for Air Worthiness for American Civil Aircraft Desiring to Enter Canada," Nov. 12, 1923, enclosed in Denby to Hughes, Nov. 19, 1923, box 7699, Records of the State Department, RG 59, NARA II.

6. *Report of the Air Board for the Year 1920* (Ottawa: Thomas Mulvey, 1921), 7–8; J. A. Wilson to the Undersecretary of State, May 18, 1920, G-1 vol. 1256, Records of the Department of External Affairs, RG 25, LAC; H. W. Brown, Acting Deputy Minister, Department of Militia and Defence, "Inter State Flying between Canada and the United States," Oct. 2, 1922, enclosed in Brown to Pope, Oct. 2, 1922, vol. 1308, RG 25, G-1, Records of the Department of External Affairs, LAC.

7. Hubbard's Boeing B-1 was not the only US-owned aircraft operating in Canada, and at least four others received Air Board–issued registrations between 1920 and 1923. These included a Junkers-Larsen JL-6 owned by J. M. Larsen of New York City, a Dayton-Wright FP2 (N-CAED) contracted for experimental work by the Spanish River Pulp and Paper Company of Ontario, a Standard J-1 (N-CACH) rented by the Curtiss Aeroplane and

Motor Corporation to the Fairchild Aerial Surveys, and a Curtiss HS-2L owned by Pacific Airways of Seattle. But the vast majority of American aircraft that crossed into Canada did so "without complying with the Canadian Air Regulations," much to the displeasure of the Canadian government. On February 26, 1922, an Army airplane landed at Camp Borden, Ontario, without prior clearance or identifying documentation, and a US Fishery and Forestry Department flying boat entered Canadian territory without the necessary advance permission and landed at Prince Rupert, British Columbia, on a flight from Seattle to Ketchikan, Alaska, that July. Three years later, American J. Kalec took un-authorized aerial photographs of property on the Canadian side of Lake Erie outside Port Talbot, Ontario, from an aircraft owned by the Detroit firm C. A. Pfeffer and Associates "with the object of converting it into a summer resort." National Aviation Museum, "Register of Canadian Civil Aircraft, 1920–29," Ottawa, Canada, 1964, in box 353, Fairchild Industries Collection, NASM; Sanford to Sec. of State, Dec. 7, 1920, Records of the Department of State, RG 59 (National Archives Microfilm Publication M51, roll 1435); Wilson to Pope, Mar. 8, 1922, Wilson to Pope, July 20, 1922, both in vol. 1308, G-1, Records of the Department of External Affairs, RG 25, LAC; Chilton to Kellogg, Sept. 25, 1925, box 7699, Records of the State Department, RG 59, NARA II.

8. At the request of the US State Department, the Canadian Government continued to extend this temporary courtesy until the creation of a bilateral air agreement in 1929. With the Air Commerce Act now in effect, the Canadian Government in April 1927 in-sisted on a "more formal" reciprocal agreement that would allow Canadian aircraft to enter the United States. A temporary agreement crafted the next month morphed into a permanent bilateral agreement effective October 1929. The criteria for US aircraft to enter Canadian airspace remained "substantially the same" as the earlier special courtesies, them-selves based on the provisions of the 1919 convention. The complete absence of specific regulations in the 1929 US–Canadian air agreement shows the extent that operational and registrational practices had converged between the two nations by that time, and the agreement served as "a model [for] aircraft agreements" with Italy, Belgium, the Nether-lands, Germany, South Africa, and Great Britain prior to World War II; Beaudry to Kel-logg, Apr. 2, 1927, box 6635, Records of the State Department, RG 59, NARA II; Stephen Latchford, "Aviation Relations between the United States and Canada Prior to Negotia-tion of the Air Navigation Arrangement of 1929," *Journal of Air Law* 2, no. 3 (July 1931): 341; Auchincloss to Latchford, Oct. 11, 1929, box 6635, Records of the State Department, RG 59, NARA II; Correspondence regarding these bilateral agreements can be found in *Papers Relating to the Foreign Relations of the United States*, while the agreements them-selves can be found within the *Executive Agreement Series* (Washington, DC: Government Publishing Office, 1929–1946), http://hdl.handle.net/2027/osu.32435056937998.

9. Stephen Krasner, "Structural Causes and Regime Consequences: Regimes as Intervening Variables," in *International Regimes*, ed. Stephen Krasner (Ithaca, NY: Cornell University Press, 1983), 9.

10. S. H. Philben, "The Need for Federal Control in Commercial Aviation," *Journal for the Society of Automotive Engineers* 9, no. 2 (Aug. 1921): 154.

11. The Los Angeles County Board incorporated the convention's rules of the air almost verbatim in June 1920, as did the City of Los Angeles five months later. Tasked

with creating legislation in the Wilson administration, Interdepartmental Board chair
Col. John F. Curry drafted a comprehensive set of air regulations that directly adopted
the convention's provisions for lights and rules of the air. In March 1920, the Army Air
Service's *Air Service News Letter* provided its readers with six rules of the air for the
"voluntary adoption by all operators" that came directly from the convention. In his
Landing Field Guide and Pilot's Log Book, World War I pilot Bruce Eytinge presented the
1919 convention's rules of the air, qualifications for pilots, and medical requirements as
the preferred standard operating procedures for US flyers. And as they anticipated the
passage of a federal law under the new Harding administration in the summer of 1921,
the NACA and the War Department agreed that "air regulations for the United States
should be in accord, as far as practicable, with the International Convention" to allow for
unencumbered "freedom of intercommunication" between the United States and Canada;
Los Angeles County, CA, Ordinance No. 620 (passed June 16, 1920); City of Los Angeles,
CA, Ordinance No. 41,008 (passed Nov. 11, 1920); Note from Carpenter attached to
"General Regulations for Flying," Nov. 26, 1919, box 7695, Records of the State Department,
RG 59, NARA II; US Air Service, *Air Service News Letter* 4, no. 13 (Mar. 22, 1920):
10–11 (four verbatim and two with slight modification); Bruce Eytinge, ed, *Landing Field
Guide and Pilot's Log Book* (New York: Eytinge and Uden, 1920), 19–23; "Book Reviews,"
Aviation and Aircraft Journal 10, no. 21 (May 16, 1921): 639; Minutes of First Meeting of
Special Subcommittee to Prepare Regulations for Air Navigation, June 7, 1921, and War
Department Memorandum for Major Van Nostrand, June 17, 1921, enclosed in Van
Nostrand to Victory, July 2, 1921, both in folder 54-2, box 276, National Advisory Committee
for Aeronautics, General Correspondence, 1915–1942, RG 255, NARA II.

    12. As the airplane became a tool to circumvent terrestrial Prohibition Era checkpoints,
the existence of compatible cross-border regulations would also help to deter the
use of aircraft in smuggling operations. "Planes Used to Smuggle Rum from Canada to
US," *New York Tribune*, Dec. 2, 1920. For a discussion of aerial smuggling in the 1920s, see
David Courtwright, *Sky as Frontier: Adventure, Aviation, and Empire* (College Station: Texas
A&M Press, 2005), 202–203; Stephen T. Moore, *Bootleggers and Borders: The Paradox of
Prohibition on a Canada-US Borderland* (Lincoln: University Press of Nebraska, 2014),
66–67; and Roger D. Connor, "Boardwalk Empire of the Air: Aerial Bootlegging in Prohibition
Era America," paper presented at the 2014 International Association for the
History of Transport, Traffic and Mobility, http://www.t2m.org/wp-content/uploads
/2014/09/Roger%20D%20Connor_Boardwalk%20Empire%20of%20the%20Air%20Aerial
%20Bootlegging%20in%20Prohibition%20Era%20America.pdf (accessed May 8, 2018).

    13. Ellis Hawley famously coined the term "associationalism" to describe a government/industry
relationship wherein government facilitates the work of manufacturers,
engineers, and interested civil organizations to coordinate and standardize industrial
activity. As Hawley and others have shown, Herbert Hoover actively supported attempts
to rationalize, standardize, and codify practices in various industries throughout his
tenure as secretary of commerce and president. *But associationalism is a method of action
and not a reason to act.* While the insurance industry and engineering profession worked
within an associationalist framework in their adoption and diffusion of the convention's
operational and registrational provisions, they did so in response to a larger impetus: the

perceived need to maintain compatibility with Canadian regulations and international standard procedures. For more on associationalism, see Ellis Hawley, "Three Facets of Hooverian Associationalism: Lumber, Aviation, and Movies, 1921–1930," in *Regulation in Perspective: Historical Essays*, ed. Thomas K. McCraw (Cambridge, MA: Harvard University Press, 1981), 99; Brian Balogh, *The Associational State: American Governance in the Twentieth Century* (Philadelphia: University of Pennsylvania Press, 2015); Kathryn Steen, *The American Synthetic Organic Chemicals Industry: War and Politics, 1910–1930* (Chapel Hill: University of North Carolina Press, 2014).

14. A substantial body of literature on policy convergence exists, most notably in the areas of environmental policy and immigration. See David Vogel, *Trading Up: Consumer and Environmental Regulation in a Global Economy* (Cambridge, MA: Harvard University Press, 1995); Daniel W. Drezner, "Globalization and Policy Convergence," *International Studies Review* 3, no. 1 (Spring 2001): 53–78; Per-Olof Busch and Helge Jörgens, "The International Sources of Policy Convergence: Explaining the Spread of Environmental Policy Innovations," *Journal of European Public Policy* 12, no. 5 (Oct. 2005): 860–884; Katharina Holzinger and Christoph Knill, "Causes and Conditions of Cross-National Policy Convergence," *Journal of European Public Policy* 12, no. 5 (Oct. 2005): 775–796; and Sebastian Strunz et al., "Policy Convergence as a Multifaceted Concept: The Case of Renewable Energy Policies in the European Union," *Journal of Public Policy* 17, no. 3 (July 2018): 361–387.

15. "Longest Air Trip Begins Tomorrow," *New York Times*, June 5, 1922. Rickenbacker's charge to deliver "a letter from President Harding to the convention of Shriners in San Francisco" and plan to provide Chief of Air Service Maj. Gen. Mason M. Patrick with a comprehensive survey of the nation's airfields provided a veneer of official sanction, and the media closely followed the flight.

16. W. David Lewis, *Eddie Rickenbacker: An American Hero in the Twentieth Century* (Baltimore: Johns Hopkins University Press, 2005), 256; "Lightning Delays Rickenbacker Flight," *New York Times*, June 9, 1922; "Rickenbacker's Bad Day," *Los Angeles Times*, June 9, 1922; "Flight Is Once More Cut Short," *Los Angeles Times*, June 10, 1922.

17. "Rickenbacker Plane is Wrecked at Omaha," *Washington Post*, June 12, 1922. Rickenbacker continued to San Francisco by train to deliver Harding's letter.

18. Among Schroeder's resident aircraft engineers were former Boeing test pilot Hunter A. Munter in Seattle, Washington, and future founder of Beech Aircraft Walter H. Beech in Wichita, Kansas. Announcing Certificates of Airworthiness for Individual Aircraft, May 1, 1922, in Schroeder to Chief of Air Service, May 20, 1922, Underwriters' Laboratories, box 4, Records of the Office of the Chief of Air Service, 1917–21, Records of the Training and War Plans Division, Airways Section, Records of the Army Air Forces, RG 18, NARA II; Approved Applications of Hunter A. Munter and Walter H. Beech, both in folder 25-1, collection 2015–8, UL Archive; Schroeder to Meissner, Jan. 12, 1923, folder 25-1, collection 2015–8, UL Archives.

19. "The Story of an Unairworthy Airplane," *Aviation* 13, no. 3 (July 17, 1922): 71.

20. According to aviation historian W. David Lewis, Rickenbacker was well aware of the difficulties with the JL-6 due to a mishap-laden long-distance flight in 1920; Lewis, *Rickenbacker*, 246–248.

21. Flint to Walcott, Mar. 26, 1917, folder 3, box 159, MAA Papers, AHC.

22. Williams to Walcott, May 8, 1917, folder 3, box 159, MAA Papers, AHC.

23. National Advisory Committee for Aeronautics, "Insurance for Aviators," received May 23, 1917, folder 1, box 409, MAA Papers, AHC.

24. Minutes of Meeting of Executive Committee, Nov. 15, 1917, folder 8, box 94, RU 45, Smithsonian.

25. Howard Mingos, "Is Flying Safe?," June 1, 1919, folder 8, box 453, MAA Papers, AHC.

26. Curtiss Current Comment, Jan. 23, 1920, folder 8, box 453 MAA Papers, AHC.

27. Hedley C. Wright, "Aviation Insurance Outlined," *The Spectator: A Weekly Review of Insurance*, Feb. 17, 1921. For more on the rise of the American insurance industry and the quantification of human activities, see Dan Bouk, *How Our Days Became Numbered: Risk and the Rise of the Statistical Individual* (Chicago: University of Chicago Press, 2015). For a detailed study of the quantification of technological risk and attempts to minimize it, see Arwen P. Mohan, *Risk: Negotiating Safety in American Society* (Baltimore: Johns Hopkins University Press, 2013).

28. Kailin Tuan, "Aviation Insurance in America," *Journal of Risk and Insurance* 32 (Mar. 1965): 2–4; Stephen Binnington Sweeney, *The Nature and Development of Aviation Insurance* (published PhD thesis, University of Pennsylvania, 1927), 10, 13. Horatio Barber—a British "early bird" aviator and wartime head of the Royal Air Force's Central Flying School—became the association's manager and chief underwriter, served on Lloyd's Aviation Technical Committee, and "was instrumental in establishing Lloyd's register of aircraft and pilots" along the lines of its shipping registry. *Who's Who in American Aeronautics*, compiled by Lester D. Gardner (New York: Aviation Publishing Corporation, 1928), 6; "Aerial Insurance," *Flight*, Mar. 13, 1919, 339.

29. Proceedings of the New York State Senate, Apr. 14, 1919, *Journal of the Senate of the State of New York at their One Hundred and Forty-Second Session*, 2 (Albany, NY: J. B. Lyon Company, 1919), 1239; Laws of New York, Chap. 391, 392, and 393, approved May 5, 1919.

30. Walter G. Cowles, "Aircraft Insurance," paper delivered Nov. 21, 1919, in Proceedings of the Casualty Actuary and Statistical Society of America, 1919–1920 Vol. VI, Numbers 13, 14 (Lancaster, PA: New Era Printing, 1921), 40–41, 45–46.

31. Hugh Lewis and Gwilym Hugh Lewis, *Aviation and Insurance: Notes for Underwriters* (London: Stone and Cox, 1920), 43, 47. Workmen's compensation requirements varied by state. According to Cowles, annual premiums ranged from $195 to $410 among the thirty-three states that required it in 1919. Cowles, "Aircraft Insurance."

32. The Travelers, *Airplanes and Safety* (Hartford, CT: The Travelers, 1921), 101; Walter G. Cowles Obituary, *Proceedings of the Casualty Actuarial Society* 30, no. 60 (Nov. 19, 1943): 103.

33. Cowles to Bradley, Apr. 6, 1920, folder 1, box 122, MAA Papers, AHC. In this letter, Cowles recounts an incident that occurred "something less than a year ago."

34. Air Service fatality rates sat at 10.8 percent for the period between April 1917 and the beginning of 1920, and this increased to 11.58 percent two years after the armistice. The total loss rate from Air Service crashes increased even more during these respective

periods, from 12 percent to 46.95 percent. Bradley, Coffin, and others believed that commercial operations would be far safer but could not back up that assumption with data. Air Service Information Circular No. 340, "Statistics Compiled from Reports on Crashes in the US Army Air Service during the Calendar Years 1918–1921 Inclusive, and Results of Physical Examinations for Flying during the Calendar Years 1920 and 1921," 6 (May 1, 1922). According to the NACA, military aviation averaged one fatality for every 989 flying hours in 1921; National Advisory Committee for Aeronautics, "Air Travel Safety Records: Miles per Fatality," Sept. 1925, W-425020-01, NASM. For more on how the military services played a central role in the elimination of pilot error to reduce the uncertainty of the "human element" in flight, see Timothy P. Schultz, *The Problem with Pilots: How Physicians, Engineers, and Airpower Enthusiasts Redefined Flight* (Baltimore: Johns Hopkins University Press, 2018).

35. Kailin Tuan, "Aviation Insurance in America," *Journal of Risk and Insurance* 32, no. 1 (Mar. 1965), 3; the Travelers Insurance Company, *Aviation and Safety*, 96–97. According to Tuan, the Traveler's aviation program included life insurance, accident insurance, daily tickets for individual passenger insurance, workmen's compensation, and public liability and property damage.

36. Cowles, "Aircraft Insurance," 33, 41–45. Frederick L. Hoffman of the Prudential Insurance Company of America shared Cowles's assessment. Hoffman to Bradley, June 27, 1919, folder 6, box 121, MAA Papers, AHC.

37. "The Aircraft Field," *Spectator*, Feb. 5, 1920. *The Spectator* noted that the cost of insurance for a $3,000 aircraft came to roughly $600 per month.

38. Fairchild to MAA, Feb. 14, 1921, folder 6, box 92, MAA Papers, AHC.

39. "Why Europe Leads in Aerial Insurance," *Weekly Underwriter*, Feb. 7, 1920.

40. "Association of Aircraft Writers," *Standard*, July 31, 1920. Members of the NAUA included Aetna Life Insurance Company, Aetna Casualty and Surety Company, Automobile Insurance Company of Hartford, National Liberty Insurance Company, Fireman's Fund Insurance Company, Home Insurance Company, and Globe and Rutgers Insurance Company; Manufacturers Aircraft Association, *Aircraft Yearbook* (Boston: Small, Maynard, 1921), 274.

41. Edmund Ely, "Aircraft Insurance," Mar. 15, 1920, folder 1, box 409, MAA Papers, AHC.

42. The British passed their Air Navigations Regulations of 1919 on April 30, 1919. Based on the British draft convention that served as the core of the 1919 convention, these domestic regulations nicely aligned with those of the international air agreement.

43. "Association of Aircraft Writers," *Standard*, July 31, 1920.

44. Harry Chase Brearley, *A Symbol of Safety: An Interpretive Study of a Notable Institution Organized for Service—Not Profit* (New York: Doubleday, Page, 1923), 17–19.

45. Minutes of Special Meeting of the Board of Directors of the Manufacturers Aircraft Association, Oct. 15, 1920, folder 10, box 2, MAA Papers, AHC. The members of the MAA's Special Insurance Committee were Bradley, Aeromarine's Inglis Uppercu, and Frank Russell. Minutes of Special Meeting of the Board of Directors of the Manufacturers Aircraft Association, Aug. 16, 1920, folder 10, box 2, MAA Papers, AHC.

46. Klemin taught aeronautics at MIT under Jerome Hunsaker before the war,

headed up the Air Service's research department at McCook Field during the hostilities, and went on to become a successful consultant to several aeronautical firms before becoming the head of NYU's Guggenheim School of Aeronautics. Klemin's 1928 *Who's Who* entry lists his occupation as "aeronautical instruction and engineering" and refers to him specifically as an "aeronautical engineer." Alexander Klemin, *Who's Who in American Aeronautics*, 3rd ed. (New York: Aviation Publishing, 1928), 65; Small to Klemin, Jan. 8, 1921, folder 6, box 78, MAA Papers, AHC; Klemin to Small, Jan. 6, 1921, and Russell to Klemin, Jan. 8, 1921, both in folder 2, box 122, MAA Papers, AHC.

47. Small's initial plan also included an aircraft inspection for a type certificate as called for in the 1919 convention that would use the War Department's minimum engineering standards, but manufacturers proved reluctant to participate due to costs and a reliance on military specifications ill-suited to the civilian market. Small dropped the aircraft type certificate component from the UL's system at Bradley's instance, and the costs of inspection under the UL system now fell solely on the individual owner/operator seeking insurance coverage. Small to Bradley, Mar. 4, 1921, folder 2, box 122, MAA Papers, AHC; Bradley to Small, Mar. 28, 1921, folder 2, box 122, MAA Papers, AHC.

48. R. H. Lucas, "Summary of Underwriters' Laboratories Aircraft Activity, 1921–1925," June 1958, folder 5, collection 2014, UL Archives; Underwriters' Laboratories, "Aircraft Register" and "Register of Aircraft Pilots," July 5, 1921, folder 4, box 410, MAA Papers, AHC.

49. To register a private or commercial aircraft, the UL required the personal data of the owner and operator, the number of crew and potential passengers, detailed specifications of the aircraft and engines, and a $5 fee. *Register of Aircraft Pilots, Rules of the Air, Aircraft Register, Identification Marking, Aircraft Nomenclature*, Underwriters' Laboratories, box 4, Records of the Office of the Chief of Air Service, 1917–21, Records of the Training and War Plans Division, Airways Section, Records of the Army Air Forces, RG 18, NARA II; "Laboratories Registry Recognized," *Insurance Press*, Sept. 28, 1921, 38.

50. Small to Bradley, June 9, 1921, and Bradley to Small, June 13, 1921, both in folder 2, box 122, MAA Papers, AHC; Lucas, "Summary."

51. Pilots needed to submit their general information, training and flight experience, and a full account of any and all previous accidents. Although not a license in the normal sense, membership on the UL's pilot register was meant to assure potential employers and passengers of a pilot's competence and allowed the pilot to purchase certain lines of insurance through a NAUA member company. Pilot registrations remained valid for twelve months, expired if the pilot went ninety days without flying, could "be cancelled or suspended at any time for cause," and were subject to automatic suspension following a crash for the duration of the insurance inquiry. Criteria for the UL's required medical examination came from Annex E of the convention. Underwriters' Laboratories, *Register of Aircraft Pilots, Rules of the Air, Aircraft Register, Identification Marking, Aircraft Nomenclature*, Aug. 1921.

52. Schroeder to Meissner, Jan. 12, 1923, folder 25-1, collection 2015-8, UL Archives.

53. Application for Insurance to the Hartford Fire Insurance Company and/or Hartford Accident and Indemnity Company, ca. 1921, folder 1, box 409, MAA Papers, AHC.

54. Uppercu purchased the very flying boats he had produced for the Navy as war

surplus, modified them for civilian use, and began a scheduled air service from Key West to Havana in 1920 that catered to Americans looking for relief from Prohibition. Under the presidency of Charles F. Redden, Aeromarine added scheduled service between Miami and Nassau and seasonal scheduled service between Cleveland and Detroit as well as New York and Atlantic City over the next three years. See William M. Leary Jr., "At the Dawn of Commercial Aviation: Inglis M. Uppercu and Aeromarine Airways," *Business History Review* 53 (Summer 1979): 180–193.

55. Leary, "At the Dawn of Commercial Aviation," 192; Aeronautical Chamber of Commerce of America, *Aircraft Year Book, 1923* (New York: Aeronautical Chamber of Commerce of America, 1923), 21.

56. In 1923, four passengers died when the *Columbia* sank after a forced water landing, and the *Ponce de Leon* succumbed to a storm shortly thereafter. The fact that "neither aircraft was replaced" illustrates Aeromarine's "shaky economic foundations." Leary, "At the Dawn of Commercial Aviation," 190.

57. Edmund Ely, "The Status of Aircraft Insurance," an address delivered before the First National Air Institute under Auspices of the Detroit Aviation Society, Oct. 11, 1922.

58. Ely, "The Status of Aircraft Insurance." Ely and Small became two important allies of Bradley and Coffin in their push for federal regulation. As discussed in chapter 3, they both signed on to industry's request that newly inaugurated President Harding convene a board under Hoover to recommend aviation policy and helped secure Hoover's support for a Commerce Bureau–only bill alongside Bradley, Coffin, and Keys at their July 18, 1921, meeting. Minutes of a Special Meeting of the Board of Directors of the Manufacturers Aircraft Association, May 26, 1921, folder 1, box 3, MAA Papers, AHC; A Brief for the Advancement of Civilian Aviation in the United States, folder 4, box 406, MAA Papers, AHC; Small to Delegates of Aircraft Industry and Associated Interests Attending Washington Conference July 18, 1921, and July 28, 1921, folder 2, box 122, MAA Papers, AHC.

59. Aeronautical Chamber of Commerce of America, *Aircraft Year Book, 1921* (New York: Aeronautical Chamber of Commerce of America, 1921), 274.

60. *Annual Report of the Commissioner of Insurance for the Year Ending Dec. 31, 1921, Part 1* (Boston: Wright and Potter, 1922), x.

61. "Miscellaneous Business Records for the Last Four Years," *Standard* 90, no. 22 (June 3, 1922), 825–845.

62. "Discontinue Aircraft Business," *Insurance Press* 54, no. 1,386 (May 10, 1922), 9.

63. Tuan, "Aviation Insurance in America," 4.

64. Baldwin to Timberlake, Aug. 6, 1925, Correspondence 1924, box 1, Sanford Aviation Insurance Collection, NASM.

65. "Major Schroeder Resigns to Become Affiliated with the Ford Projects," *Laboratories Data* 6, no. 4 (April 1925): 78.

66. Totals based on List #3: Aircraft Register, List #4: Certificates for Registration and Classification as Aircraft Pilots, and List #2: Airworthiness Certificates Granted, all in collection 2014, UL Archives.

67. Aeronautical Chamber of Commerce of America, *Aircraft Year Book, 1924* (New York: Aeronautical Chamber of Commerce of America, 1924), 105.

68. "Registering Aircraft," *Aviation* 12, no. 6 (Feb. 6, 1922), 159.

69. Henry Middlebrook Crane, *Who's Who in American Aeronautics*, 2nd ed. (New York: Gardner Publishing, 1925), 33–34; "Officers of the Society," *Journal of the Society of Automotive Engineers* 12, no. 2 (Feb. 1925): 160.

70. "Aeronautic Banquet a Success," *Journal of the Society of Automotive Engineers* 17, no. 5 (Nov. 1925): 421.

71. "Meetings of the Society," *Journal of the Society of Automotive Engineers* 17, no. 5 (Nov. 1925): 419.

72. According to Layton, the American Society of Civil Engineers (ASCE), the American Institute of Mining Engineers (AIME), the American Society of Mechanical Engineers (ASME), and the American Institute of Electrical Engineers (AIEE) constitute the four "founder societies" established in the 1870s and 1880s. The professionalization of engineering and the role of standardization in that process remains beyond the scope of this study. For details, see Edwin Layton, *The Revolt of the Engineers: Social Responsibility and the American Engineering Profession* (Baltimore: Johns Hopkins University Press, 1986), and JoAnne Yates and Craig N. Murphy, *Engineering Rules: Global Standard Setting since 1880* (Baltimore: Johns Hopkins University Press, 2019). For a detailed analysis of the standardization work of the AESC, see Andrew L. Russell, *Open Standards and the Digital Age: History, Ideology, and Networks* (Cambridge: Cambridge University Press, 2014), 41–57.

73. Tom D. Crouch, *Rocketeers and Gentlemen Engineers: A History of the American Institute of Aeronautics and Astronautics . . . and What Came Before* (Reston, VA: American Institute of Aeronautics and Astronautics, 2006), 14.

74. Hanks to Williams, "Aeronautic Division—Progress of Standardization," July 7, 1917, folder 7, box 199, MAA Papers, AHC.

75. "Aeronautical Engineers and Aircraft Manufacturers Meet," *Air Service Journal* 1, no. 13 (Oct. 4, 1917): 406; Charles Manly, *Who's Who in American Aeronautics*, 2nd ed. (New York: Gardner Publishing, 1925), 79; List of Standards Completed and in Progress, enclosed in Kettering to Foss, Apr. 2, 1918, folder 3, box 198, MAA Papers, AHC; "Presidential Address of George W. Dunham," *Journal of the Society of Automotive Engineers* 2, no. 1 (Jan. 1918): 3. After their creation within the SAE's Aeronautic Division and approval by Coffin, the International Aircraft Standards Board Council—a division of the Advisory Commission of the Council of National Defense under the direction of F. G. Diffin—would then "correlate" standards with various Allied governments.

76. Yates and Murphy, *Engineering Rules*, 82; Conference of the International Aircraft Standards Commission, Report of Meeting of Delegates, Oct. 14, 15, and 16, 1918 (London: Waterlow and Sons, 1918), 32. The IASC arose out of a March 1918 meeting in London where William F. Durand of the NACA, Manly, SAE general manager Coker L. Clarkson, British Engineering Standards Committee secretary Charles le Maistre, and Allied representatives agreed on the need to create a "permanent aircraft standards organization"; "Inter-Allied Conference on Standards in London," *Journal of the Society of Automotive Engineers* 2, no. 3 (Mar. 1918): 257.

77. Russell, *Open Standards and the Digital Age*, 64; A. A. Stevenson, "American Engineering Standards Committee: Its Organization and Work," paper presented before the National Federation of Construction Industries, Chicago, Mar. 25, 1920. The five

professional engineering societies at the AESC's founding included the American Society of Civil Engineers (ASCE), the American Institute of Mining Engineers (AIME), the American Society of Mechanical Engineers (ASME), the American Institute of Electric Engineers (AIEE), and the American Society of Testing Materials (ASTM).

78. As a result of two Bureau of Standards–sponsored meetings in 1919, the AESC's sectional committee process was extended to the development of national safety codes, best practices that went beyond strictly technical matters such as rules of the air for aviation. Morton G. Lloyd, "Preparation of Safety Codes under the Auspices of the American Engineering Standards Committee," *Monthly Labor Review* 15 (Sept. 1922): 483.

79. Russell, *Open Standards and the Digital Age*, 66–67; Paul G. Agnew, "Work of the American Engineering Standards Committee," *Annals of the American Academy of Political and Social Science* 137 (May 1928), 15.

80. New York Senator James W. Wadsworth's War Department–sponsored Senate Joint Resolution 56 to provide official representation to the IASC faced "strenuous" objections in the House, Bradley worked in vain with the SAE to secure presidential authority for Director of Air Service Charles T. Menoher to appoint IASC delegates, and archival evidence even suggests that Assistant Secretary of War Benedict Crowell's American Aviation Mission discussed in chapter 2 initially formed as a potential IASC delegation. Devendorf to Stratton, June 28, 1919, folder American Engineering Standards Committee Jan. 1918–July 1919, box 35, Jan. 1918–July 1919, Records of the National Aeronautics and Space Administration, RG 255, NARA II; Bradley to le Maistre (telegram), Apr. 18, 1919, folder 5, box 122, MAA Papers, AHC; Crowell to Baker (telegram), Apr. 22, 1919, folder 72, box 4, Records of the Office of the Chief of Air Service, 1917–21, Records of the Air Service Advisory Board, RG 18, NARA II.

81. Ames to Agnew, Sept. 6, 1920, folder 2-12, American Engineering Standards Committee July–Dec. 1920, box 35, and Ames to Clarkson, Feb. 28, 1921, folder 63-5 Society of Automotive Engineers July 1916–Dec. 1923, box 285, both in RG 255, Records of the National Aeronautics and Space Administration, NARA II; Ames to le Maistre, Mar. 9, 1921, both in folder 6, box 122, MAA Papers, AHC. Agnew, Briggs, Bradley, and representatives of the interested executive departments and engineering societies gathered in the NACA's conference room on August 11, 1920, to discuss US participation in the IASC. Those present at this meeting included Cdr. Patrick N. L. Bellinger, Lt. Comdr. Sydney M. Kraus, and Lt. Comdr. Garland Fulton of the Navy; Capt. Henry W. Harms of the War Department; Morton G. Lloyd of the Bureau of Standards; Charles I. Stanton of the Post Office; W. J. Humphreys of the Department of Agriculture; George Lewis from the NACA; L. C. Hill of the SAE; George Burgess, John Mathews, and C. Warwick from the American Society of Testing Materials; John Cautley and C. B. LePage of the American Society of Mechanical Engineers; and AESC chair A. A. Stevenson. Minutes of Conference on American Representation on the International Aircraft Standards Commission, Aug. 11, 1920, folder 2-12, box 35, National Advisory Committee for Aeronautics, General Correspondence, 1915–1942, RG 255, Records of the National Aeronautics and Space Administration, NARA II.

82. Under Stratton's leadership, the Bureau of Standards expanded from an agency focused "solely [on] testing weights, measures, and scales" to one "more akin to testing

laboratories, such as the Underwriters Laboratory," as it became the primary verification body for the integrity of federally purchased materials in the first decades of the twentieth century. During World War I, engineers within the Bureau of Standards played a central role in the development of the Liberty engine and tested a plethora of aeronautical instruments. Lee Vinsel, "Virtue via Association: The National Bureau of Standards, Automobiles, and Political Economy, 1919–1940," *Enterprise and Society* 17, no. 4 (Dec. 2016): 7; Rexmond C. Cochrane, *Measure for Progress: A History of the National Bureau of Standards* (Washington, DC: US Department of Commerce, 1966), 90–91, 179–186.

83. Peter Briggs Myers and Johanna M. H. Levelt Sengers, *Lyman James Briggs, 1874–1963* (Washington, DC: National Academy Press, 1999). At the behest of Secretary of State Robert Lansing, mid-level representatives of federal agencies began meeting as the Economic Liaison Committee in March 1919 to better coordinate governmental activities in support of US overseas commerce. Agencies represented included the Departments of State, Treasury, War, Navy, Interior, Agriculture, Commerce, and Labor; the Federal Reserve Board; the Federal Trade Commission; the US Shipping Board; the US Railroad Administration; the War Finance Corp.; the US Tariff Commission; and the International High Commission. Subsequent discussions on the increased activity of European aeronautical interests in Latin America among the State Department, the Commerce Department's Bureau of Foreign and Domestic Commerce, and the War and Navy Departments resulted in the creation of an aviation subcommittee under the chairmanship of acting director of the Bureau of Foreign and Domestic Commerce Ralph S. MacElwee. Foreign Trade Promotion Work, H. Doc. No. 66-650, at 34; MacElwee to Redfield, 14 July 1919, Aeroplanes, General, 1919–1929, box 2511, Records of the Bureau of Foreign and Domestic Commerce, General Records, RG 151, NARA II; Joint Army and Navy Board on Aeronautics to Baker and Daniels, Aug. 7, 1919, Aviation, General, 1918–1922, box 2497, Records of the Bureau of Foreign and Domestic Commerce, General Records, RG 151, NARA II.

84. Minutes of the First Meeting of the Sub-Committee on Commercial Aviation, Jan. 12, 1920, folder 4, box 159, MAA Papers, AHC; Minutes of the Fifth Meeting of the Sub-Committee on Commercial Aviation, Feb. 12, 1920, folder 13-3, box 41, National Advisory Committee for Aeronautics, General Correspondence, 1915–1942, Records of the National Aeronautics and Space Administration, RG 255, NARA II. Those attending the subcommittee's first meeting on January 12, 1920, were MacElwee (chairman), Col. Sevier (Engineering and Standardization Branch), Hickam (Air Service), Cmd. Childs (Navy), George Lewis (NACA), William R. Manning (State Department, Office of the Foreign Trade Advisor), Eggerton (Post Office), Mr. Stuart (Forest Service), Charles A. McQueen (Commerce Department Latin American Division), F. H. Eldridge (Commerce Department Far Eastern Division), and Lyman Briggs (Bureau of Standards). Bradley joined the subcommittee beginning with its February 5 meeting. By August 1920, the subcommittee's meeting place had changed from MacElwee's office to the NACA's conference room and minutes were taken on NACA letterhead.

85. Clarkson to Bradley, May 27, 1920, folder 6, box 198, MAA Papers, AHC.

86. Lewis to King, Aug. 17, 1922, folder number unknown (damaged), box 281 and Minutes of the Twenty-First Meeting of the Subcommittee on Commercial Aviation, Sept. 16, 1920, folder 13-3, box 41, both in National Advisory Committee for Aeronautics,

General Correspondence, 1915–1942, Records of the National Aeronautics and Space Administration, RG 255, NARA II.

87. Manly to Jones, Nov. 17, 1919, folder 2-12 American Engineering Standards Committee Aug.–Dec. 1919, box 35, National Advisory Committee for Aeronautics, General Correspondence, 1915–1942, RG 255, Records of the National Aeronautics and Space Administration, NARA II; Crane to Clarkson, May 17, 1920, folder 6, box 198, MAA Papers, AHC.

88. Report Adopted by the SAE Aeronautic Subdivision on Performance and Testing, Nov. 26, 1920, folder 6, box 198, MAA Papers, AHC; "Standards Committee Meeting, Aeronautical Division Report," *Journal of the Society of Automotive Engineers* 8 (Feb. 1921): 171–172.

89. In addition to Crane as chairman, the Aeronautic Division of the SAE's Standards Committee in 1921 included prominent figures in aeronautical engineering such as former chief aeronautical engineer of the Army Virginius Clark, the Navy's Jerome C. Hunsaker, Grover C. Loening, Glenn L. Martin, and E. C. Zoll of the Post Office. "The 1921 Standards Committee," *Journal of the Society of Automotive Engineers* 8 (Dec. 1921): 364.

90. Minutes of a Special Meeting of the Board of Directors of the Manufacturers Aircraft Association, May 26, 1921, folder 1, box 3, MAA Papers, AHC.

91. Coffin to Agnew, Mar. 2, 1921, folder 6, box 122, MAA Papers, AHC.

92. Sectional Committee for Aeronautical Safety Code, Minutes of Meeting Held Sept. 2, 1921, folder 10, box 222, MAA Papers, AHC.

93. The convention left matters of technical standardization to the proposed International Commission for Air Navigation (ICAN), which held its first meeting from July 11 to July 28, 1922, and would continue to meet in the interwar period.

94. Lewis to King, Aug. 17, 1922, folder number unknown (damaged), box 281, National Advisory Committee for Aeronautics, General Correspondence, 1915–1942, Records of the National Aeronautics and Space Administration, RG 255, NARA II.

95. Halstead to Lewis, Sept. 7, 1922, folder number unknown (damaged), box 281, National Advisory Committee for Aeronautics, General Correspondence, 1915–1942, Records of the National Aeronautics and Space Administration, RG 255, NARA II.

96. Among the seventy-six criteria within Part 4: Signals and Signaling Equipment, thirty-nine aligned with the convention, twenty-nine went beyond the convention's stipulations but did not contradict them, and only eight differed. The inclusion of a detailed rating system for air fields and services offered at airdromes derived from a 1919 Air Service proposal within Part 5: Airdromes and Airways meant that only six of that section's thirty-one rules paralleled the convention, but the other twenty-two did not directly conflict with it. Of the fifty-seven rules of the air in Part 6: Traffic and Pilotage Rules, thirty-three corresponded with the convention, twenty-three additional rules did not clash with it, and only one—the required safe distance after takeoff before an aircraft could make a turn—was "not strictly in conformity." Halstead believed that the subcommittee could divert from the convention on those rare occasions "where reasons are sufficiently good," such as when it rejected the convention's stipulation that a navigator be present during all flights as an unnecessary restriction on commercial aviation. Comparison of Part 4 with ICAN, enclosed in Halstead to Subcommittee Members, Aug. 8, 1923, folder 58-4, Safety in Aviation, Jun.–Dec, 1923, box 281, National Advisory Committee for Aeronautics, General Correspondence, 1915–1942, Records of the National Aero-

nautics and Space Administration, RG 255, NARA II; Comparison of Part 6: with the International Convention, enclosed in Halstead to Subcommittee Members, Apr. 7, 1923, and Comparison of Part 5: Airdromes with the International Convention, enclosed in Halstead to Subcommittee Members, Apr. 11, 1923, both in American Aeronautical Safety Code, box 4, Training and War Plans Division, Records of the Army Air Forces, RG 18, NARA II; Halstead to Members of the Subcommittee, June 23, 1922, American Aeronautical Safety Code, box 4, Training and War Plans Division, Records of the Army Air Forces, RG 18, NARA II.

97. At its February 20, 1923, meeting, the Sectional Committee declared "that wherever Subcommittees make departures from the requirements of the International Convention, the reasons for such choice shall be submitted by the Subcommittee to the Sectional Committee." Minutes of the Sectional Committee Meeting for the Aeronautical Safety Code, Feb. 20, 1923, folder 58-4, Safety in Aviation, Jan.–May, 1923, box 281, National Advisory Committee for Aeronautics, General Correspondence, 1915–1942, Records of the National Aeronautics and Space Administration, RG 255, NARA II.

98. When ICAN secretary Albert Roper contacted Halstead in May to stress the need for uniformity with the convention, the latter confidently replied that "the requirements of the American Aeronautical Safety Code will parallel closely those existing in the annexes." Minutes of Meeting of the Subcommittee on Airdromes, Traffic Rules, Etc., Jan. 16, 1923, folder 58-4, Safety in Aviation, Jan.–May, 1923, box 281, and Christie to Lewis, Feb. 7, 1923, folder 32-6, International Air Navigation, 1922–1923, box 177, both in National Advisory Committee for Aeronautics, General Correspondence, 1915–1942, Records of the National Aeronautics and Space Administration, RG 255, NARA II. Roper to Halstead, Mar. 24, 1923, and Halstead to Roper, Apr. 16, 1923, both in folder 58-4, Safety in Aviation, Jan.–May, 1923, box 281, National Advisory Committee for Aeronautics, General Correspondence, 1915–1942, Records of the National Aeronautics and Space Administration, RG 255, NARA II.

99. Wilson personally attended the Sectional Committee's May and October 1923 meetings to "influence them a little in keeping as nearly along the same lines as the Canadian regulations as possible." At the Sectional Committee's final meeting on April 23, 1925, its members approved "several changes [to bring] the Code more into line with Canadian Air Regulations," such as the inclusion of a requirement that pilots have experience with an aircraft type prior to operating it in commercial service. Wilson to Scott, Mar. 13, 1923, John A. Wilson Fonds, MG 30, microfilm reel 10779, LAC; Minutes of the Sectional Committee for Aeronautical Safety Code, Apr. 23, 1925, folder 58-4, Safety in Aviation, 1924–1925, box 281, National Advisory Committee for Aeronautics, General Correspondence, 1915–1942, Records of the National Aeronautics and Space Administration, RG 255, NARA II; "Aeronautical Safety Code Meeting," undated memorandum, John A. Wilson Fonds, MG 30, microfilm reel 10781, LAC.

100. "American Aeronautical Safety Code," *Journal of the Society of Automotive Engineers* 14, no. 1 (Jan. 1925): 1.

101. Society of Automotive Engineers, *American Aeronautical Safety Code Approved by the American Engineering Standards Committee Oct., 1925* (New York: Society of Automotive Engineers, 1925), 23, 26, 28.

102. The Sectional Committee experienced some turnover in individual members during its four years of existence, most notably the replacement of Aero Club of America representatives with those of the National Aeronautic Association. SAE members approved the AASC by a vote of 228 for, 1 against, and 77 abstentions. Lloyd to Burnett and Burgess, June 10, 1925, folder 58-4, Safety in Aviation, 1924–1925, box 281, National Advisory Committee for Aeronautics, General Correspondence, 1915–1942, Records of the National Aeronautics and Space Administration, RG 255, NARA II; Burgess to AESC, June 11, 1925, folder 58-4, Safety in Aviation, 1924–1925, box 281, National Advisory Committee for Aeronautics, General Correspondence, 1915–1942, Records of the National Aeronautics and Space Administration, RG 255, NARA II; "Standards Committee Meeting," *Journal of the Society of Automotive Engineers* 17, no. 1 (July 1925): 30; "Standards Approved by Letter-Ballot," *Journal of the Society of Automotive Engineers* 17, no. 3 (Sept. 1925): 226.

103. McCracken later referred to Grimshaw as "the draftsman of our Air Commerce Regulations." MacCracken to Borges, Jan. 8, 1927, box 436, Civil Aeronautics Administration Central Files, RG 237, NARA II. Klemin to Lewis, Sept. 14, 1926, and Lewis to Langley Memorial Aeronautical Laboratory, Sept. 16, 1926, both in folder 15-6, box 135, National Advisory Committee for Aeronautics, General Correspondence, 1915–1942, Records of the National Aeronautics and Space Administration, RG 255, NARA II.

104. The other three designations in MacCracken's draft regulations, commercial ("C"), state ("S"), and private ("P"), corresponded with those in Annex A of the convention. McCracken obtained copies of the convention and the bulletins of the International Commission for Air Navigation from the State Department while serving on the Morrow Board as Coffin's assistant. Latchford to Harrison, Nov. 6, 1925, box 5618, and MacCracken to Harrison, Dec. 28, 1925, box 5619, both in Records of the State Department, RG 59, NARA II.

105. MacCracken's draft regulations included the AASC's requirement for a second glide landing and a demonstration of emergency maneuvers not found in the convention, and his reduction of the required safe distance between planes down to three hundred feet noticeably differed from both the convention and the AASC. "Regulations for the Licensing of Pilots and Aircraft of the United States as Proposed by the Department of Commerce in Collaboration with the Army Air Corps and the Naval Air Service," enclosed in Klemin to Lewis, Sept. 14, 1926, folder 15-6, box 135, National Advisory Committee for Aeronautics, General Correspondence, 1915–1942, Records of the National Aeronautics and Space Administration, RG 255, NARA II.

106. The existence of an additional chapter titled "Foreign Aircraft" that did not make it into either the initial draft or the 1926 Air Commerce Regulations shows that this draft committee clearly understood the need to maintain compatibility with the convention's rules of the air. This chapter would have allowed foreign aircraft access to US airspace based on reciprocity but only "if the Secretary of Commerce finds that the law[s] of such country . . . are adequate for the navigation of such aircraft in the United States," and those nations that had "ratified or adhered" to the convention were automatically considered to possess such compatible regulations. In order for this to be the case, US regulations would therefore have to be compatible with the convention. The lack of US ratification of the convention probably best explains why this chapter—with its direct

mention of an unratified treaty—did not make it into the draft that MacCracken distributed for comment, but its existence provides powerful evidence that MacCracken and his
compatriots drafted the Air Commerce Regulations with a clear understanding of the
need to maintain compatibility with established practices under the convention. "Foreign
Aircraft" (unpublished draft chapter), in Regulations for the Licensing of Pilots and
Aircraft of the United States as Proposed by the Department of Commerce in Collaboration with the Army Air Corps and the Naval Air Service, conference copy, Oct. 4, 1926,
Aeronautics Branch, Air Commerce Regulations, Printed Material, 1926–1931, box 12,
MacCracken Papers, HHPL.

107. Committee on Interstate and Foreign Commerce, Civil Air Navigation Bill, H.R.
Rep. No. 68-1262 at 8, 15-16 (1925); Bell to Drake, Oct. 9, 1925, enclosed in Bell to Keys,
undated, Lampert and Hoover Committee, box 18, Clement Melville Keys Papers, NASM.

108. Bradley to Cuthell, Feb. 11, 1926, folder 6, box 130, MAA Papers, AHC; Lloyd to
Agnew, June 23, 1925, folder 58-4, Safety in Aviation, 1924–1925, box 281, National Ad-
visory Committee for Aeronautics, General Correspondence, 1915–1942, Records of the
National Aeronautics and Space Administration, RG 255, NARA II.

109. Aircraft designer Grover Loening feared that restrictive regulations would
deliver "a death blow to our little struggling industry." Testimony of Grover Loening,
Hearings Before the President's Aircraft Board, vol. 4 (Washington, DC: Government Publishing Office, 1925), 1450.

110. "Publisher's News Letter," *Aviation* 21 (Sept. 27, 1926): 570; "The New Air Regulations," *Aviation* 21 (Oct. 11, 1926): 621.

111. MacCracken forwarded his draft regulations to Bradley in early September. This
special committee of the ACCA included general secretary Bradley, president of Wright
Aeronautical Corporation Charles L. Lawrence, E. A. Johnson of Johnson Airplane and
Supply Company, J. G. Ray of Pitcairn Aviation, Mr. Ireland of Ireland Aircraft, C. S. Jones
of Curtiss Flying Service, and R. H. Depew and F. R. Weymouth of Fairchild Aviation
Corporation. MacCracken to Bradley, Sept. 11, 1926, all in 83272 [1 of 3], box 568, General
Records of the Department of Commerce, RG 40, NARA II; Report of the First Meeting
of the Special Committee for Studying the Proposed Regulations for the Licensing of
Pilots and Aircraft of the United States, Sept. 27, 1926, folder 3, box 475, MAA Papers,
AHC.

112. One of the individuals MacCracken invited to the insurance conference was
Horatio Barber, who had continued to underwrite aircraft through Lloyd's after the
collapse of the NAUA. MacCracken to Bradley, Sept. 11, 1926, and MacCracken to Barber,
Sept. 11, 1926, both in folder 83272 [3 of 3], box 568, General Records of the Department
of Commerce, RG 40, NARA II; Aeronautical Chamber of Commerce of America, *Aircraft
Year Book, 1927* (New York: Aeronautical Chamber of Commerce of America, 1927), 10.

113. McCracken held two conferences with aircraft manufacturers, the first on October
4 and the second four days later, where he accepted the industry's recommendation that
aircraft did not have to undergo physical airworthiness testing if they adhered to the following formulae:

$$\left(\frac{Gross\ Weight}{Horsepower} \times \frac{Gross\ Weight}{Wing\ Area} = 200\ or\ less\ \text{AND}\ \frac{Gross\ Weight}{Horsepower} + \frac{Gross\ Weight}{Wing\ Area} = 30\ or\ less \right).$$

Mandatory flight tests were reinserted into the June 1, 1928, revision of the Air Commerce Regulations. MacCracken to Pitcairn, Oct. 8, 1926, folder 83272 [1 of 3], box 568, General Records of the Department of Commerce, RG 40, NARA II. Commerce Department, Aeronautics Branch, "Final Draft of Air Commerce Regulations," Nov. 15, 1926, folder 2, box 475, MAA Papers, AHC. See Department of Commerce, Aeronautics Branch, Information Bulletin No. 7, *Air Commerce Regulations Effective as Amended June 1, 1928* (Washington, DC: Government Publishing Office, 1928).

114. Department of Commerce, Aeronautics Branch, *Air Commerce Regulations Effective December 31, 1926* (Washington, DC: Government Publishing Office, 1926).

115. R. S. Moore, "Comparison of ICAN Regulations and US Proposed Regulations for Airworthiness of Aircraft," Nov. 18, 1926, box 435, Records of the Civil Aeronautics Administration, RG 237, NARA II.

116. Department of Commerce, New Amendments Added to Air Regulations, not for release earlier than Monday afternoon, Apr. 11, 1927, folder 6, box 475, MAA Papers, AHC; Department of Commerce, Aeronautics Branch, Information Bulletin No. 7, *Air Commerce Regulations Effective as Amended June 1, 1928* (Washington, DC: Government Publishing Office, 1928). Unlike the convention and the AASC, the Air Commerce Regulations applied · only to aircraft. Airworthiness and licensing requirements for airships and balloons remained subject to "special orders of the Secretary of Commerce," and this regulatory uncertainty around lighter-than-air craft may have contributed to their reduced commercial development within the United States.

117. Note by International Commission for Air Navigation Secretary General Albert Roper, Feb. 14, 1929, International Conference on Air Navigation (Paris), 1929. Director of the Königsberg Institute of Aerial Law Hans Oppikofer echoed Roper's assessment a year later when he noted in a League of Nations report that the general rules for aerial traffic had become "uniform and homogenous . . . under the pressure of like technical conditions." Hans Oppikofer, "International Commercial Aviation and National Administration," in *Enquires into the Economic, Administrative and Legal Situation of International Air Navigation* (Geneva: League of Nations, 1930), 157.

Chapter 6. Shattered Expectations

1. Charles L. Ponce de Leon, "The Man Nobody Knows: Charles A. Lindbergh and the Culture of Celebrity," in *The Airplane in American Culture*, ed. Dominick A. Pisano (Ann Arbor: University of Michigan Press, 2003), 84.

2. Thomas Kessner, *The Flight of the Century: Charles Lindbergh and the Rise of American Aviation* (Oxford: Oxford University Press, 2010), 134, 140, 147. Lindbergh's Latin American goodwill tour arose out of a request from recently appointed US ambassador to Mexico Dwight Morrow—the very same individual who chaired the President's Aircraft Board in 1925—that Lindbergh visit Mexico City to cement closer ties with that government after tense negotiations to prevent the nationalization of US oil interests. Morrow had come to know Lindbergh personally while serving as the famous aviator's financial advisor at J. P. Morgan prior to his appointment as US ambassador to Mexico. Lindbergh met Anne Morrow during his 1927 goodwill tour, and when the two wed in 1929 Dwight became the famous aviator's father-in-law. For more on Morrow's efforts to come to an

understanding over Mexico's post-Revolution oil laws, see Lorenzo Meyer, *Mexico and the United States in the Oil Controversy, 1917–1942*, trans. Muriel Vasconcellos (Mexico City: El Colegio de México, 1972), 125–138.

3. For more on the responses to Lindbergh's goodwill tour in the United States and Latin America, see Jenifer van Vleck, *Empire of the Air: Aviation and the American Ascendency* (Cambridge, MA: Harvard University Press, 2013), 58–64.

4. "Lindbergh Reaches Goal at Havana," *New York Times*, Feb. 9, 1928.

5. Remarks of the President of Cuba, General Gerardo Machado, in greeting to Colonel Charles A. Lindbergh, at the Presidential Palace, Havana, Cuba, Feb. 8, 1928, folder 031.4—Visit of Lindbergh, box No. 1, entry 152, RG 43, NARA II.

6. "Full Text of President Coolidge's Address," *New York Times*, Jan. 17, 1928.

7. Calvin Coolidge, *Annual Message to Congress*, Dec. 6, 1927.

8. R. E. G. Davies, *Airlines of Latin America since 1919* (Washington, DC: Smithsonian Institution Press, 1984), 86.

9. "Lindbergh Air Host to Cuban Notables," *New York Times*, Feb. 12, 1928; Form letter to delegates of the Sixth International Conference of American States inviting them to join Colonel Lindbergh, Feb. 11, 1928, folder 031.4—Visit of Lindbergh, box 1, entry 152, RG 43, NARA II. Lindbergh proved the viability of Caribbean air mail service when he arrived in Havana with a sack of mail from Santo Domingo and two from Port au Prince. "Lindbergh Looks on His Cuban Hop as Forerunner of Air Mail Service," *New York Times*, Feb. 9, 1928. Lindbergh became a consultant for Pan Am soon after his goodwill tour and undertook the company's inaugural flight from Miami to Panama in February 1929.

10. "Cuban Air Line Planned," *New York Times*, Feb. 13, 1928; "Cuba is Negotiating for Air Mail Line," *New York Times*, Feb. 25, 1928; 1928 Foreign Air Mail Act, Pub. L. No. 107-49, 45 Stat. 248 (1928). Numerous studies exist on the creation of the Pan American System. See Matthew Josephson, *Empire of the Air: Juan Trippe and the Struggle for World Airways* (New York: Harcourt, Brace and Company, 1944), Marylin Bender and Selig Altschul, *The Chosen Instrument: Pan Am, Juan Trippe, and the Rise and Fall of an American Entrepreneur* (New York: Simon and Schuster, 1982), and van Vleck, *Empire of the Air*.

11. In their efforts to negotiate agreements, Pan Am vice president John MacGregor and P. E. D. Nagle received direct support from the US State Department. Secretary of State Frank Kellogg informed US diplomatic missions to render "all possible assistance" to the efforts of Pan Am's representatives but admonished them that State Department personnel could not "initiate or conduct private business negotiations with foreign governments on behalf of private American citizens." Even with this assistance, MacGregor found negotiations with Guatemala and Honduras particularly troublesome. Kellogg to Geissler, Mar. 16, 1928, White to Hoover, Mar. 3, 1928, Caffery to Kellogg, May 5, 1928, and Kellogg to Summerlin, July 11, 1928, all in *Papers Relating to the Foreign Relations of the United States 1928*, vol. 1 (Washington, DC: Government Publishing Office, 1942), 778, 780, 786, 827. The most detailed study of the US government's support for Pan Am's early expansion into Latin America remains Wesley P. Newton, *The Perilous Sky: US Aviation Diplomacy and Latin America, 1919–1931* (Coral Gables, FL: University of Miami Press, 1978).

12. In May 1928, Kellogg informed Ambassador Morrow of Trippe's belief that in light of the "convention signed at Havana by three Mexican delegates the Mexican Government

would raise no objection to free passage of the company's planes." For Trippe, signature of the Havana Convention, not ratification, was sufficient to obtain access to Mexican airspace. In response, the Mexican government granted "provisional" access, but "definite permission" would have to await full ratification by both parties. Kellogg to Morrow, May 11, 1928, and Schoenfeld to Kellogg, June 23, 1928, both in *Papers Relating to the Foreign Relations of the United States 1928*, vol. 1 (Washington, DC: Government Publishing Office, 1942), 781; Trippe to Butterworth, Jan. 2, 1931, Correspondence: Senate Investigation, Legislation and Regulation, other, Box 26 of 28, PAA.

13. Payer to Lawton, Aug. 25, 1933, Copy 7, Chapter VII and Topic 28b, box 3, entry 203, RG 43, National Archives at College Park, College Park, MD. The governments in question included Mexico, Guatemala, British Honduras (modern-day Belize), Salvador, Honduras, Nicaragua, Costa Rica, Panama, Colombia, Venezuela, Puerto Rico, the Dominican Republic, Haiti, and Cuba.

14. By August 1922, Bolivia, Brazil, Cuba, Ecuador, Guatemala, Panama, the United States, and Uruguay had signed the 1919 convention, and Nicaragua and Peru had ratified it. International Commission for Air Navigation, "Official Bulletin No. 1" (Aug. 1922), 26.

15. Edward P. Warner, "The International Convention for Air Navigation and the Pan American Convention for Air Navigation: A Comparative and Critical Analysis," *Air Law Review* 3, no. 3 (July 1932): 221–308. In 1947, the International Civil Aviation Organization (ICAO) began overseeing the implementation of a new postwar aviation regime. See David MacKenzie, *ICAO: A History of the International Civil Aviation Organization* (Toronto: University of Toronto Press, 2010). If discussed at all, scholars relegate the Havana Convention to a brief summary in the context of a related issue. See van Vleck, *Empire of the Air*, and Dawna L. Rhoades, *Evolution of International Aviation: Phoenix Rising*, 3rd ed. (London: Routledge, 2016). In *The Perilous Sky*, Newton discusses the Havana Convention in various places but does not provide an analysis of its creation, interpretation, or effectiveness.

16. Studies of technological failures have shown that we can discover just as much, if not more, from a technology's inability to meet expectations as from its successful development and widespread adoption, and this holds true for policy as well. Examples include Shelley McKellar, "Negotiating Risk: The Failed Development of Atomic Hearts in America, 1967–1977," *Technology and Culture* 54, no. 1 (Jan. 2013): 1–39, and Edward Jones-Imhotep, *The Unreliable Nation: Hostile Nature and Technological Failure in the Cold War* (Cambridge, MA: MIT Press, 2017).

17. Joan H. Wilson, *American Business and Foreign Policy, 1920–1933* (Lexington: University Press of Kentucky, 1971), xvi.

18. Cabot to Kellogg, Mar. 10, 1925; Davis to Cabot, Mar. 27, 1925, both in box 5618, Records of the State Department, RG 59, NARA II.

19. Arthur Bliss Lane, Memorandum, Mar. 23, 1925, box 5618, Records of the State Department, RG 59, NARA II.

20. Harrison to Grew, Mar. 26, 1925, box 5618, Records of the State Department, RG 59, NARA II.

21. Cabot to Kellogg, Mar. 26, 1925, box 5618, Records of the State Department, RG 59, NARA II.

22. Arthur Bliss Lane to Harrison, Apr. 6, 1925, and Kellogg to Cabot, Apr. 6, 1925,

both in box 5618, Records of the State Department, RG 59, NARA II. An undated draft letter in response to Cabot's offer, written by Stephen Latchford of the State Department's
Solicitor's Office, pointed to the "impropriety" of Cabot's offer that NAA representatives
serve as diplomatic middlemen. Draft of letter to Cabot, undated, box 5618, Records of
the State Department, RG 59, NARA II.

23. Cabot to Kellogg, Apr. 14, 1925, and Harrison to Cabot, May 1, 1925, both in box
5616, Records of the State Department, RG 59, NARA II.

24. Harrison to Kellogg, June 11, 1925, box 5616, Records of the State Department,
RG 59, NARA II; Halstead to Lewis, June 13, 1925, box 177, National Advisory Committee
for Aeronautics, General Correspondence, 1915–1942, Records of the National Aeronautics
and Space Administration, RG 255, NARA II.

25. Harrison served as US embassy secretary in Tokyo in 1908, Peking from 1909 to
1910, London from 1910 to 1912, and the Legation to Bogota from 1912 to 1915. Leland
Harrison, *Register of the Department of State* (Washington, DC: Government Publishing
Office, 1922), 24, 129, 234.

26. Joan H. Wilson, *American Business and Foreign Policy, 1920–1933* (Lexington: Uni-
versity Press of Kentucky, 1971), xvi. Although still a prominent idea among the public, a
wealth of scholarship refutes the notion of the 1920s as an isolationist decade. See Frank
Costigliola, *Awkward Dominion: American Political, Economic, and Cultural Relations with
Europe, 1919–1933* (Ithaca, NY: Cornell University Press, 1984), and Benjamin Allen Coates,
*Legalist Empire: International Law and American Foreign Relations in the Early Twentieth
Century* (Oxford: Oxford University Press, 2016).

27. Harrison to Fletcher, Sept. 26, 1921, box 5614, Records of the State Department,
RG 59, NARA II; "Convention Relating to the Regulation of Aerial Navigation, dated 13
October 1919," International Commission for Air Navigation, *Bulletin Official No. 6*, June
1924.

28. Brancker to Skinner, Nov. 23, 1922, enclosed in Linnell to Hughes, Dec. 4, 1922,
box 5614, Records of the State Department, RG 59, NARA II.

29. James Monroe, Seventh Annual Message, Dec. 2, 1823.

30. Jay Sexton, *The Monroe Doctrine: Empire and Nation in Nineteenth-Century America*
(New York: Hill and Wang, 2011), 4; Mark T. Gilderhus, "The Monroe Doctrine: Meanings
and Implications," *Presidential Studies Quarterly* 36, no. 1 (Mar. 2006): 6.

31. For a detailed discussion of the interpretation and application of the Monroe Doctrine during this period, see Juan Pablo Scarfi, "In the Name of the Americas: The Pan-
American Redefinition of the Monroe Doctrine and the Emerging Language of American
International Law in the Western Hemisphere, 1898–1933," *Diplomatic History* 40, no. 2
(2016): 189–218.

32. John Barrett, "Practical Pan-Americanism," *North American Review* 202, no. 718
(Sept. 1915), 414.

33. Ricardo D. Salvatore, "The Making of a Hemispheric Intellectual-Statesman: Leo S.
Rowe in Argentina (1906–1919)," *Journal of Transnational American Studies* 2, no. 1 (2010):
5, http://escholarship.org/uc/item/92m7b409. For more on the US use of Pan Americanism as a tool of foreign policy, see David Sheinin, "Rethinking Pan Americanism: An
Introduction," in *Beyond the Ideal: Pan Americanism in Inter-American Affairs*, ed. David

Sheinin (Westport, CT: Praeger, 2000), and Ricardo D. Salvatore, *Disciplinary Conquest: US Scholars in South America, 1900–1945* (Durham, NC: Duke University Press, 2016). For a study of the relatively peaceful shift from British to US global dominance, see Kori Schake, *Safe Passage: The Transition from British to American Hegemony* (Cambridge, MA: Harvard University Press, 2017).

34. One could interpret the historical consolidation and projection of US economic, military, and cultural power—particularly from 1865 to 1933—as an attempt to develop a new global core that would, in the context of Immanuel Wallerstein's World System Analysis framework, necessitate the establishment of an interconnected periphery. For a distillation of Wallerstein's thought on the core/periphery dynamic, see "Africa in a Capitalist World," in *The Essential Wallerstein* (New York: New Press, 2000), 39–70.

35. Delegates met at the First International Conference of American States in Washington, DC, from October 21, 1889, to April 19, 1890; for the Second Conference in Mexico City from October 22, 1901, to January 31, 1902; at Rio de Janerio for the Third Conference from July 23 to August 27, 1906; and for the Fourth Conference in Buenos Aires from July 12 to August 30, 1910. For an in-depth discussion of the areas covered by these inter-American conferences, see James Brown Scott, ed., *The International Conferences of American States, 1889–1928* (New York: Oxford University Press, 1931).

36. The First Conference of American States created the Commercial Bureau of American Republics to collect and distribute "information as to the productions . . . customs, laws, and regulations" among members of the newly established International Union of American Republics. When the Fourth Conference of American States adopted the moniker of Pan American Union (PAU) in 1910, it also expanded the bureau's duties to include report preparation for the various conferences, the creation of a conference agenda, the execution of conference resolutions, and the compilation and classification of treaties and conventions. "Third Report of the Committee on Customs Regulations as adopted by the Conference, April 14, 1890," *International American Conference, Reports of Committees and Discussions Thereon*, vol. 1 (Washington, DC: Government Printing Office, 1890), 404; Reorganization of the "Union of American Republics," Article I, in Fourth International Conference of American States, S. Doc. No. 64-744, 156.

37. *The Bulletin of the Pan American Union*, the organization's official publication, provided US commercial interests with detailed information on the geography, finances, commercial interactions, industries, and transportation infrastructure in the various Latin American republics. For more on the importance of the *Bulletin* to US industry, see James F. Vivian, "The Commercial Bureau of American Republics, 1894–1902: Advertising Policy, the State Department, and the Governance of the International Union," *Proceedings of the American Philosophical Society* 118, no. 6 (Dec. 1974), and Jeffrey Sommers, "Haiti and the Hemispheric Imperative to Invest: The Bulletin of the Pan American Union," *Journal of Haitian Studies* 9, no. 1 (Spring 2003): 68–94.

38. Ricardo D. Salvatore, "Imperial Mechanics: South America's Hemispheric Integration in the Machine Age," *American Quarterly* 58, no. 3 (Sept. 2006); Blaine to Harrison, Proposed Intercontinental Railway, May 12, 1890, in "Message from the President of the United States transmitting a letter of the Secretary of State relative to proposed reciprocal commercial treaties between the United States and other American Republics," Ex.

Doc. No. 51-158, 1890. Secretary of State James G. Blaine, the orchestrator of the first Inter-American Conference, prioritized transportation networks as a means to ensure "the development and prosperity of our sister Republic[s] and at the same time . . . the expansion of our commerce" and secured a resolution providing for a commission of engineers to conduct a survey for a Pan-American railway that would connect New York City and Buenos Aires. In 1898, a US-sponsored survey under the direction of future Pennsylvania Railroad president Alexander J. Cassett provided a detailed plan for the erection of the 5,456 miles of track still required in the various countries along the route, but five years later only 171 miles had been completed. Despite Pan-Americanism's cooperative spirit, the American republics had very different priorities, resources, and needs. Whereas the United States pushed for a north to south line that would better serve its economic desires, states like Colombia chose to invest their limited resources in short east to west lines that complimented existing waterway networks. S. Doc. No. 61-744, *Fourth International Conference of American States*, Jan. 16, 1911, 182.

39. "Aeronautics Federation of the Western Hemisphere," *Bulletin of the Pan American Union* 42, no. 2 (Feb. 1916): 208, 210. Santos-Dumont presented two papers on the power of aviation to unify the American Republics at the Second Pan American Scientific Conference, "How the Airplane May Effect Closer Alliance of the South American Countries with the United States" and "Airships to Assist in Joining North and South America." James Brown Scott, *Second Pan American Scientific Conference: The Final Act and Interpretive Commentary Thereon* (Washington, DC: Government Printing Office, 1916), 465.

40. This shift from continental imperial expansion under the ideology of Manifest Destiny to one of overseas expansion prompted a fierce debate over the very definition of the American ideal and the United States' place in the world. For more on this transformation, see Heather Cox Richardson, *West from Appomattox: The Reconstruction of America after the Civil War* (New Haven, CT: Yale University Press, 2007), and Stephen Kinzer, *The True Flag: Theodore Roosevelt, Mark Twain, and the Birth of American Empire* (New York: Henry Holt, 2017).

41. The Panama Canal continues to capture the popular imagination, and many studies about its creation exist. See David McCullough, *The Path between the Seas: The Creation of the Panama Canal, 1870–1914* (New York: Simon and Schuster, 1977); Walter LaFeber, *The Panama Canal: The Crisis in Historical Perspective* (Oxford: Oxford University Press, 1990); Matthew Parker, *Panama Fever: The Epic Story of One of the Greatest Human Achievements of All Time—The Building of the Panama Canal* (New York: Doubleday, 2007); Julie Greene, *The Canal Builders: Making America's Empire at the Panama Canal* (New York: Penguin Press, 2009); and Noel Maurer and Carlos Yu, *The Big Dig: How America Took, Built, Ran, and Ultimately Gave Away the Panama Canal* (Princeton, NJ: Princeton University Press, 2011).

42. "TS 431: Panama; Panama Canal," 33 Main Section Stat. 2234, 2242.

43. As historian Benjamin Allen Coates noted, the Roosevelt Corollary proclaimed "a de facto US protectorate over the hemisphere." Theodore Roosevelt, Fourth Annual Message to Congress, Dec. 6, 1904; Coates, *Legalist Empire*, 112. For more on the creation of a customs receivership in the Dominican Republic, see Emily S. Rosenberg, *Financial Missionaries to the World: The Politics and Culture of Dollar Diplomacy, 1900–1930* (Cam-

bridge, MA: Harvard University Press, 1999), and Cyrus Veeser, "Inventing Dollar Diplomacy: The Gilded-Age Origins of the Roosevelt Corollary to the Monroe Doctrine," *Diplomatic History* 27, no. 3 (June 2003): 301–326.

44. Quoted in Walter LaFeber, *Inevitable Revolutions: The United States in Central America* (New York: W. W. Norton, 1984), 37.

45. Examples include Taft's decision to send Marines to Nicaragua and his support of the United Fruit Company's imperialistic expansion in Honduras; the occupation of Haiti and the Dominican Republic in the Wilson administration; and the continued presence of US troops in Panama, Haiti, and the Dominican Republic under Harding. For more on US interventions in Latin America, see LaFeber, *Inevitable Revolutions*, and Cyrus Veeser, *A World Safe for Capitalism: Dollar Diplomacy and America's Rise to Global Power* (New York: Columbia University Press, 2002).

46. Taft Agreement, Orders of the Secretary of War, Dec. 3, 1904, in *Executive Orders Relating to the Panama Canal* (Mount Hope, Canal Zone: Panama Canal Press, 1922), 29–31.

47. Dan Hagedorn, *Conquistadors of the Sky: A History of Aviation in Latin America* (Washington, DC: Smithsonian Institution Press, 2008), 130–134.

48. George C. Day to Commandant, 15th Naval District, July 26, 1920, folder 426-1, 1918–1922, box 146, RG 80, General Records of the Department of the Navy, General Board Subject File, 1900–1949, NARA I; Telegram from Geissler to Hughes, Dec. 16, 1922, in Newton, *The Perilous Sky*, 43.

49. Nancy Mitchell details how the fear of German economic and military intervention in the Western Hemisphere compelled the United States to take a more imperialistic stance as a preventative measure. In this light, US expansionism could be viewed as a series of preemptive and defensive actions similar to the rationale that the Romans used as they expanded outward some two millennia earlier. See Nancy Mitchell, *The Danger of Dreams*.

50. In 1925 alone, SCADTA carried 1,134 passengers and 26,343 pounds of mail out of 225,300 total pounds of goods in 527 flights totaling 183,206 miles without a single accident. Aeronautical Chamber of Commerce of America, *Aircraft Yearbook 1926* (New York: Aeronautical Chamber of Commerce of America, 1926), 150. For more on SCADTA's operations in Colombia, see Davies, *Airlines of Latin America since 1919*, 207–220.

51. Morrow to Washington, DC, May 16, 1923, folder 426-1, 1923–1935, box 146, RG 80, General Records of the Department of the Navy, General Board Subject File, 1900–1949, NARA I.

52. Major C. S. Ridley of the office of the Governor of the Canal Zone, Major Follett Bradley of the Panama Canal Zone Department, and Lieut. Comm. Ralph F. Wood of the US Navy compiled these initial policy recommendations prior to meeting with officials of the Panamanian government. Ridley, Bradley, and Wood to American Minister to Panama John G. South, Feb. 1, 1924, folder 426-1, 1923–1935, box 146, RG 80, General Records of the Department of the Navy, General Board Subject File, 1900–1949, NARA I.

53. Secretary of State Charles Evan Hughes to Secretary of the Navy Edwin Denby, Mar. 1, 1924, folder 426-1, 1923–1935, box 146, RG 80, General Records of the Department of the Navy, General Board Subject File, 1900–1949, NARA I.

54. By this time, both the United States and Panama had signed but not yet ratified

the 1919 convention. Secretary of the Navy Curtis D. Wilbur to Secretary of State Hughes, Apr. 11, 1924, folder 426-1, 1923–1935, box 146, RG 80, General Records of the Department of the Navy, General Board Subject File, 1900–1949, NARA I. Wilbur was appointed secretary of the navy after Denby was forced to resign as a result of the Teapot Dome scandal on March 10, 1924.

55. Imperfected Treaty between the United States and Panama for Settlement of Points of Difference, signed July 28, 1926, in *Papers Relating to the Foreign Relations of the United States 1926*, vol. 2 (Washington, DC: Government Publishing Office, 1941), 828–870.

56. Arosemens to Muse, May 16, 1929, in *Papers Relating to the Foreign Relations of the United States 1929*, vol. 3 (Washington, DC: Government Publishing Office, 1944), 722–723.

57. Hoover to Hughes, Sept. 13, 1923, Aviation-General, 1923–1924, box 2497, RG 151, Records of the Bureau of Foreign and Domestic Commerce, NARA II; Hughes to Denby, Aug. 25, 1923, and Rear Admiral W. L. Rodgers to Denby, Sept. 21, 1923, both in folder 438, Apr. 1–Dec. 31, 1923, box 168, RG 80, General Records of the Department of the Navy, General Board Subject File, 1900–1949, NARA I.

58. Acting Secretary of War Dwight F. Davis to Kellogg, Aug. 14, 1925, box 436, RG 237, Civil Aeronautics Administration Central Files, NARA II. SCADTA received permission to undertake the trial flight under strict overflight and operational restrictions with the understanding that such a grant in no way implied future permissions. Von Bauer's efforts in Washington to secure an air mail contract in 1925 came to naught, but his tenacity only stimulated a desire to establish a US carrier to preempt foreign activity in Central America and the Caribbean. For more, see Newton, *The Perilous Sky*, 64–70.

59. In addition to Harrison, those present included Assistant Postmaster General Warren Irving Glover, Adm. Hilary P. Jones, Glenn H. Griffith of the Treasury Department's Customs Division, Maj. George Strong of the War Department, and Assistant Secretary J. Walter Drake and Transportation Division head Eugene S. Gregg from the Commerce Department. For a step-by-step discussion of the work of the Interdepartmental Aviation Conference, see Newton, *The Perilous Sky*, 70–75.

60. Air Navigation Convention, Message from the President of the United States, 69th Cong., 1st sess., June 16, 1926.

61. Latchford to Carter, Dec. 20, 1928, box 5617, Records of the State Department, RG 59, NARA II.

62. Harrison, Memorandum regarding telephone conversation with Mr. MacCracken, Dec. 7, 1926, box 5619, Records of the State Department, RG 59, NARA II.

63. White to Latchford, May 9, 1928, box 5617, Records of the State Department, RG 59, NARA II.

64. A multilateral air convention perfectly complemented the ideology of legalism—an approach to foreign policy that sought to prevent interstate conflict and promote prosperity through "the codification of important areas of customary international law into positive treaty form" —which had become firmly entrenched within the US legal profession and the State Department by the early twentieth century. Predominantly through the efforts of lawyer-diplomats such as James Brown Scott, James Bassett Moore, Elihu Root, and other members of the American Society of International Law, legalism became the dominant foreign policy ideology in the Progressive Era because it

provided a morally unassailable rationale to increase the United States' global standing, allowed for an expansionist agenda "without the political headaches of overt colonialism," and justified intervention based on a deep-rooted "racial condescension" that presupposed the superiority of white Anglo-Saxon Protestant morality and "civilized" norms. Francis Anthony Boyle, *Foundations of World Order: The Legalist Approach to International Relations, 1898–1922* (Durham, NC: Duke University Press, 1999), 22; Coates, *Legalist Empire*, 81; Gilderhus, "The Monroe Doctrine," 11.

65. Victory to Ames, Mar. 26, 1923, folder 13, box 94, Office of the Secretary Records, 1903–1924, RU 45, Smithsonian. This message was conveyed to NACA secretary John Victory by a "proper subordinate official of the State Department."

66. Pan American Union, *Fifth International Conference of American States, Special Handbook for the Use of the Delegates* (Washington, DC: Government Publishing Office, 1922), 26; Fifth International Conference on American States, Revised Program Submitted to the Governing Board by the Secretary of State of the United States, dated May 22, 1922, box 11, Leo Stanton Rowe Papers, OAS.

67. After a brief stint in Havana, Fletcher's career spanned posts in Peking (1903–1905 and 1907–1909), Lisbon (1905–1907), Chile (1910–1916), Mexico (1916–1919), Belgium and Luxembourg (1922–1924), and Italy (1924–1929). Henry Prather Fletcher, *Register of the Department of State* (Washington, DC: Government Publishing Office, 1925), 128; "Henry Prather Fletcher (1873–1959)," Department of State, Office of the Historian, accessed Dec. 9, 2019, https://history.state.gov/departmenthistory/people/fletcher-henry-prather.

68. Harding to Fletcher, Jan. 9, 1923, box 1, Henry P. Fletcher Papers, LOC. Harding "was pleased to comply with [his] ambition," although he regretted the Fifth Conference would preclude Fletcher and his wife Beatrice from joining the Hardings on a planned couples vacation to Florida.

69. Fletcher was joined on the delegation by future secretary of state Frank Kellogg, PAU director Leo S. Rowe, former Ohio Democratic senator Atlee Pomerene, former Delaware Democratic senator and member of the Du Pont family Willard Saulsbury, Rockefeller Foundation president George R. Vincent, and Washington DC banker William Eric Fowler. Phillips to Fletcher, Feb. 14, 1923, box 10, Fletcher Papers, LOC; Samuel Guy Inman, *Inter-American Conferences 1826–1954: History and Problems* (Washington, DC: University Press of Washington, DC, 1965), 102.

70. Hughes personally blocked an attempt to invite a representative of the League of Nations. "Acts for Santiago Conference Amity," *Washington Post*, Dec. 3, 1922; "Hughes Bars League at Chile Conference," *New York Times*, Feb. 8, 1923; Inman, *Inter-American Conferences*, 92–93.

71. US Marines remained in multiple Central American and Caribbean nations, and the Harding administration continued to refuse to recognize the postrevolutionary Mexican government, a fact that prevented its representatives from attending the Santiago conference.

72. "Two Santiago Plans Opposed," *Los Angeles Times*, Apr. 21, 1923; "America Dodges "Monroe" Treaty," *Christian Science Monitor*, May 1, 1923; "Assails Fletcher for Report on Santiago Parley," *Chicago Daily Tribune*, June 3, 1923.

73. Fletcher to Kellogg, July 8, 1927, box 13, Henry P. Fletcher Papers, LOC. Although

the stalemate over political issues left the impression that the Fifth Conference accomplished nothing from March 25 to May 3, this was not the case. Delegates signed treaties designed to prevent inter-American conflicts, facilitate communication of customs provisions, and—what Fletcher deemed one of the conference's most significant achievements—to protect trademarks. Conventions between the United States and Other American Republics Signed at the Fifth International Conference of American States, *Papers Relating to the Foreign Relations of the United States 1923*, vol. 1 (Washington, DC: Government Publishing Office, 1938), 297–320; "Sees Improved South American Relations," *Wall Street Journal*, May 29, 1923.

74. Notes on the Meeting of the Delegation, Mar. 28, 1923, box 1, entry 133, RG 43, NARA II; Committee No. 4: Communications, Motions Made by the Delegates of the United States of America, box 4, entry 133, RG 43, NARA II; Minutes of the Thirteenth Plenary Session, May 2, 1923, in *Fifth International American Conference, Manuscript Minutes in English*, OAS.

75. PAU members Argentina, Bolivia, Brazil, Colombia, Costa Rica, Cuba, Chile, the Dominican Republic, Ecuador, El Salvador, Guatemala, Honduras, Mexico, Nicaragua, Panama, Paraguay, Peru, Uruguay, and Venezuela signed the Madrid Convention on November 1, 1926, and it was "ratified by the King of Spain on April 13, 1927." Hammond to Kellogg, May 2, 1927, box 5621, Records of the State Department, RG 59, NARA II. In September 1926, the Spanish government of Primo de Rivera withdrew from the League of Nations in protest against its lack of a permanent seat on the Council. With Spain a nonsignatory to the 1919 convention, Rivera saw an aviation agreement as a way to draw Latin American nations closer to its sphere of influence and elevate its world power status. Spain would rejoin the League of Nations two years later. For Spain's relationship with the League of Nations during this period, see Ismael Saz, "Foreign Policy under the Dictatorship of Primo de Rivera," in *Spain and the Great Powers in the Twentieth Century*, ed. Sebastian Balfour and Paul Preston (London: Routledge, 1999): 53–72.

76. John Jay Ide, "International Aeronautical Organizations," *US Air Services* 15, no. 9 (Sept. 1930), 31.

77. Latchford to Harrison, Dec. 6, 1926, box 5617, Records of the State Department, RG 59, NARA II.

78. Cunningham memorandum, Nov. 9, 1926, box 5621, Records of the State Department, RG 59, NARA II; Rogers to Klein, Dec. 22, 1926, box 436, Civil Aeronautics Administration Central Files, RG 237, NARA II.

79. The PAU's Governing Board authorized its new chairman and Hughes's successor as Secretary of State Frank B. Kellogg to invite member states to appoint delegates to the Inter-American Commercial Aviation Committee (IACAC) on May 5, 1926, two weeks before Coolidge signed the Air Commerce Act.

80. "A Tribute to Esteban Gil Borges," *Bulletin of the Pan American Union* 77, no. 11 (Nov. 1943): 618–621. At the suggestion of Assistant Secretary of Commerce J. Walter Drake, Borges consulted with Morton Lloyd and Arthur Halstead of the Bureau of Standards, who had both recently worked on the American Aeronautical Safety Code. Halstead stressed to Borges the need to incorporate Wallace's original reservations, eliminate any reference to the League of Nations and the Permanent Court of International Justice, and

replace the annexes with minimum standards that did not infringe on the operational requirements of "other international agreements." "Suggestions for the Modification of the International Air Convention," enclosed in Halstead to Borges, June 19, 1926, folder 32-6, International Air Navigation, 1927, box 177, National Advisory Committee for Aeronautics, General Correspondence, 1915–1942, Records of the National Aeronautics and Space Administration, RG 255, NARA II.

81. Rogers to Klein, Dec. 22, 1926, box 436, Civil Aeronautics Administration Central Files, RG 237, NARA II.

82. This committee also included Borges, Aeronautics Branch liaison James D. Summers, Assistant Commerce Department Solicitor Ira Grimshaw, and a representative from the State Department (most likely Stephen Latchford of the Solicitor's Office), the Bureau of Foreign Commerce (most likely Rogers), and the Bureau of Standards (most likely Halstead). Evidence exists that these revisions to the IACAC's agenda occurred in consultation with the War and Navy Departments, but neither were officially represented on this drafting committee. Borges to MacCracken, Jan. 4, 1927, MacCracken to Borges, Jan. 8, 1927, Borges to MacCracken, Mar. 15, 1927, Harrison to MacCracken, Mar. 14, 1927, all in box 436, Civil Aeronautics Administration Central Files, RG 237, NARA II.

83. Additional elements within this draft agenda would further promote US economic interests. The subjection of foreign aircraft to the laws of member states on landing would allow the United States to confiscate machines that violated its patent laws, and an explicit right to engage in cabotage (the embarking and disembarking of passengers and cargo between two points within a single state) would permit US airlines to offer intrastate service within member states. In addition, the draft charged the PAU to work toward uniform regulations among member states with the assumption that it would promote the United States' recently passed Air Commerce Regulations. Inter-American Commission on Commercial Aviation, "Data Prepared by the Pan American Union as a Contribution to the Work of the Commission," Inter-American Commission on Commercial Aviation 1927, box 8, John Lansing Callan Papers, LOC.

84. In addition to Assistant Secretary of Commerce for Aeronautics MacCracken, the US delegation also included Harry P. Guggenheim of the Daniel Guggenheim Fund for the Promotion of Aeronautics and John Lansing Callan, who eight years earlier had played a central role in the creation of the 1919 convention. The five PAU member states not represented at the IACAC were Ecuador, Haiti, Honduras, Nicaragua, and Paraguay. Castle to MacCracken, Apr. 27, 1927, Inter-American Commercial Aviation Commission, 1927, box 14, William P. MacCracken Jr. Papers, HHPL.

85. While the Colombian proposal left the issue of munitions transport up to each member state and allowed for cabotage, as in the US proposal, it exempted aircraft "from seizure or embargo" due to claims of patent infringement. Contribution to the Labors of the Inter-American Commission of Commercial Aviation, Prepared by the Colombia Delegation, Project of a Pan American Convention of Aerial Navigation, in Inter-American Commercial Aviation Commission, Minutes of Meetings (hereafter referred to as IACAC), May 2–19, 1927, 19–25, box 8, John Lansing Callan Papers, LOC.

86. Although Lt. Courtney Whitney of the Air Corps informed MacCracken that the War Department considered nationality requirements for aircraft registration vital for

"military reasons," MacCracken refused to directly oppose the majority, possibly due to a desire to not upset potential purchasers of US aircraft. The All-American Aircraft Display, a government-sponsored industrial exposition at Bolling Field outside of Washington, DC, from May 2 to May 6 was timed to explicitly coincide with the IACAC conference to promote such sales. IACAC, 39, 66; Whitney to MacCracken, May 10, 1927, box 436, Civil Aeronautics Administration Central Files, RG 237, NARA II; Aeronautical Chamber of Commerce of America, *Aircraft Yearbook 1928* (New York: Aeronautical Chamber of Commerce of America, 1928), 317–321.

87. IACAC, 71, 76–78. Only Cuba, Venezuela, and the United States voted to retain the ability to transport munitions. MacCracken and his fellow delegates did obtain agreement on the ability to refuse to recognize certificates and permits of other member states due to safety concerns, as well as the right to engage in cabotage. Final Act of the Inter-American Commercial Aviation Commission, May 2–19, 1927, Washington, DC.

88. IACAC, 105. The IACAC left several issues unresolved. Delegates deferred to the PAU on whether legal jurisdiction within an aircraft during flight rested with the nation flown over or the aircraft's nation of registration and whether Canada and European possessions in the Western Hemisphere could adhere to the convention. In addition, delegates agreed that they lacked the expertise to draft a comprehensive set of laws and regulations for convention adherents. US delegate Harry P. Guggenheim did recommend, however, that the recently promulgated Air Commerce Regulations "be sent to the delegates in order that they may study them and use them as a basis for regulations in their respective countries." IACAC, 73, 84, 99.

89. State Department Solicitor Stephen Latchford agreed on the need for full access to a state's own prohibited zones, and Warner also recommended that Canada and Newfoundland be permitted to adhere to any hemispheric convention. Warner to Wilbur, July 5, 1927, folder 438, 1918–1922, box 169, General Records of the Department of the Navy, General Board Subject File, 1900–1949, RG 80, NARA I; Latchford to Morgan, Aug. 10, 1927, box 437, Civil Aeronautics Administration Central Files, RG 237, NARA II.

90. Davison to Kellogg, July 27, 1927, box 437, Civil Aeronautics Administration Central Files, RG 237, NARA II.

91. Stephen Latchford of the Solicitor's Office and Stokeley W. Morgan, assistant chief of the Division of Latin American Affairs, represented the State Department at this meeting. Memorandum of a Conference with Regard to the Instructions to be Issued to the American Delegates, Oct. 26, 1927, and Morgan to MacCracken, Oct. 27, 1927, both in box 437, Civil Aeronautics Administration Central Files, RG 237, NARA II.

92. Memorandum of a Conference with Regard to the Instructions to be Issued to the American Delegates, Oct. 26, 1927, and Morgan to MacCracken, Oct. 27, 1927, both in box 437, Civil Aeronautics Administration Central Files, RG 237, NARA II. Those present also recommended opening the convention to non-PAU states in the Western Hemisphere and incorporating the 1919 convention's signals for distress.

93. Kellogg to Fletcher, July 26, 1927, box 13, Henry P. Fletcher Papers, LOC.

94. Fletcher to Kellogg, July 8, 1927, and Fletcher to Kellogg, Aug. 26, 1927, both in box 13, Henry P. Fletcher Papers, LOC; Fletcher to Kellogg, telegram, Oct. 15, 1927, box 14, Henry P. Fletcher Papers, LOC.

95. Fletcher to Hughes, Dec. 29, 1927, reel 4, Charles Evans Hughes Papers, LOC.

96. Instructions to Delegates, Special Political Memorandum, in *Papers Relating to the Foreign Relations of the United States 1928*, vol. 1 (Washington, DC: Government Publishing Office, 1942), 549.

97. Instructions to Delegates and Instructions to Delegates, Special Political Memorandum, both in *Papers Relating to the Foreign Relations of the United States 1928*, vol. 1 (Washington, DC: Government Publishing Office, 1942), 551, 573, 580. In addition to Hughes and Fletcher, the US delegation also included PAU director Leo S. Rowe, US ambassador to Mexico and former head of the President's Aircraft Board Dwight D. Morrow, secretary for the Carnegie Endowment for International Peace and AIIL president James Brown Scott, US ambassador to Cuba Noble B. Judah, former Democratic senator from Alabama Oscar W. Underwood, former New York Supreme Court justice Morgan Joseph O'Brien, and Stanford University president Ray Lyman Wilbur. Kellogg to Hughes, Oct. 1, 1927, reel 4, Charles Evans Hughes Papers, LOC.

98. Sixth International Conference of American States, Biographical Sketches of Delegates, box 1, entry 160, RG 43, NARA II; IACAC, 39, 66. Although not an official member of the Colombian delegation, Peter Paul von Bauer also traveled to Havana to convince Latin American representatives of the need to allow non-US carriers access rights to the Canal Zone. "German Stirs Latins against US at Havana," *Chicago Daily Tribune*, Jan. 13, 1928.

99. Espil pointed out that any air agreement that did not facilitate airmail service "would not be worthwhile." Discussions on the definition of state aircraft occurred within a drafting subcommittee composed of Olaya Herrera, Espil, Fletcher, Communications Committee chairman Sampaio Correa of Brazil, and Aquiles Elorduy of Mexico. Sixth International Conference of American States, Minutes of the Third Session of the Communication Committee, Jan. 24, 1928, Committee No. 4 Minutes, box 1, entry 157, RG 43, NARA II; Newton, *The Perilous Sky*, 153.

100. Sixth International Conference of American States, Minutes of the Third Session of the Communication Committee, Jan. 24, 1928, Committee No. 4 Minutes, box 1, entry 157, RG 43, NARA II; Hughes to State (telegram), Jan. 24, 1928, box 436, Civil Aeronautics Administration Central Files, RG 237, NARA II.

101. Hughes to State, telegram, Feb. 1, 1928, box 6, entry 152, RG 43, NARA II.

102. Memorandum, matters to be covered in meeting of subcommittee on communications, Feb. 2, 1928, box 6, entry 152, RG 43, NARA II. Although nowhere near the clear statement in support of US action as the original Fletcher amendment, this compromise wording provided a level of interpretive flexibility, and Kellogg conceded that it constituted the best possible compromise. Kellogg to American Delegation in Havana, telegram, Feb. 2, 1928, box 6, entry 152, RG 43, NARA II; Hughes to State, Feb. 4, 1928, box 1, entry 155, RG 43, NARA II.

103. Hughes to State, Feb. 7, 1928, box 6, entry 152, RG 43, NARA II; Report of the Delegates of the United States of America to the Sixth International Conference of American States, Held at Havana, Cuba, January 16 to February 20, 1928, with Appendixes (Washington, DC: Government Publishing Office, 1928), 35.

104. This article, the subject of the Fletcher amendment, began as Article 31 and became Article 30 in the final version of the Havana Convention.

105. Other examples of the Communication Committee's work to align the Havana Convention with the two existing international air agreements include the incorporation of the 1919 convention's provision that adherents exchange meteorological information, publish uniform maps, establish a "uniform system of signals," and collaborate on the development of radio as well as the Madrid Convention's requirement for third-party arbitration. Hughes to State, telegram, Jan. 27, 1928, and Hughes to State, telegram, Jan. 27, 1928, both in box 436, Civil Aeronautics Administration Central Files, RG 237, NARA II.

106. Hoover to Kellogg, June 8, 1928, box 437, Civil Aeronautics Administration Central Files, RG 237, NARA II.

107. Memorandum, "Question of Withdrawing from the Senate the Convention Relating to Air Navigation Concluded at Paris in 1919," Jan. 4, 1929, box 5617, Records of the State Department, RG 59, NARA II; Thurston to White, Feb. 9, 1929, box 6479, Records of the State Department, RG 59, NARA II.

108. Kellogg to Coolidge, Feb. 14, 1929, box 5617, Records of the State Department, RG 59, NARA II.

109. Kellogg's subsequent official request that Senate Committee on Foreign Relations chair William Borah of Idaho take no action on the 1919 convention simply formalized the earlier gentleman's agreement between Harrison and MacCracken. Kellogg to Borah, Feb. 20, 1929, box 5617, Records of the State Department, RG 59, NARA II.

110. Presidential Proclamation No. 4971, Sept. 28, 1928.

111. Davies, *Airlines of Latin America since 1919*, 86; "Lindbergh Delivers Canal Mail on Time," *New York Times*, Feb. 7, 1929.

112. PANAGRA, began as a joint venture between Pan Am and the shipping giant W. R. Grace and Company in 1929, had secured permission to operate in Peru and Ecuador. Davies, *Airlines of Latin America since 1919*, 224.

113. Like previous US Foreign Airmail Contracts that Pan Am relied on, potential carriers had to show that they possessed access rights to the airspace of every country along the proposed route before they could even place a bid. Secretary of State Kellogg charged Jefferson Caffery, the US envoy in Bogotá, with securing "an agreement to operate in Colombia" prior to February 28, or "all bids will . . . have to be rejected." Advertisement for Foreign Air Mail Service, Jan. 31, 1929, United States Post Office Advertisements, folder 2 of 2, Legislation and Regulation, Airmail box 2 of 6, PAA; Kellogg to Caffery, Feb. 9, 1929, in *Papers Relating to the Foreign Relations of the United States 1929*, vol. 2 (Washington, DC: Government Publishing Office, 1943), 879–880.

114. US-registered commercial aircraft could "fly along the Atlantic and Pacific Coasts of Colombia" and land, repair, refuel, and take on and disembark passengers at recognized coastal ports while Colombian-registered aircraft obtained access to "Atlantic and Pacific ports" of the contiguous United States as well as the Panama Canal Zone "provided they adhered to specified routes." Noticeable for its simplicity, the agreement contained no technical annexes and no mention of the three multilateral agreements in existence. Kellogg to Olaya Herrera, Feb. 23, 1929, and Olaya Herrera to Kellogg, Feb. 23, 1929, both

in *Papers Relating to the Foreign Relations of the United States 1929*, vol. 2 (Washington, DC: Government Publishing Office, 1943), 881–884.

115. Like the Havana Convention, E.O. 5047 incorporated the definition of private aircraft from the 1919 convention, based aircraft nationality on the country of registration rather than the nationality of the owner, and left the requisite certificates of registration and competency up to the laws of the state of registration. Pilots needed to pass "a satisfactory examination" on Canal Zone regulations prior to their arrival in the Canal Zone. Presidential Proclamation No. 5047, Feb. 18, 1929.

116. An element of the unpopular 1926 US-Panama Treaty, the Aviation Board allowed the United States to institute strict regulations under the cover of Panama's right to unrestricted aerial sovereignty. Under the US drafted air regulations for Panama, nonregularly scheduled flights needed to provide a minimum two-hour notice, aircraft were restricted to specified coastal routes on the Atlantic and Pacific sides, and aircraft landing in Panama City and Colon near the Canal Zone remained under full quarantine prior to inspection. In addition, pilots needed to provide the requisite customs documentation, a personal bill of health approved by "an American Consular Office or Medical Official," a complete passenger list detailing the race of all on board, any requisite sanitary certificates, and proof of "clearance from last port." "Decree by Which Commercial Aviation in the Republic of Panama is Regulated," enclosed in White to Wilbur, Apr. 24, 1929, folder 426-1, 1923–1935, RG 80, General Records of the Department of the Navy, General Board Subject File, 1900–1949, NARA II.

117. Davies, *Airlines of Latin America since 1919*, 226.

118. Alfred Wegerdt, "Germany and the Aerial Navigation Convention at Paris, October 13, 1919," translated by William Bigler, *Journal of Air Law* 1, no. 1 (Jan. 1930): 15, 27. Wegerdt wrote this article to encourage revisions to the 1919 convention so that Germany, a wartime belligerent, could fully participate in it. The ICAN was "particularly anxious to have the participation of the United States Government," and the interested executive departments and the NACA in Washington recognized the potential benefits of participation. Report of Barton Kyle Yount, Army air attaché in Paris, Feb. 18, 1929, Aviation, Commerce Department, International Commission for Air Navigation, Correspondence, folder 1927-Apr., 1929, box 56, William P. MacCracken Jr. Papers, HHPL.

119. Assistant State Department solicitor Joseph R. Baker joined MacCracken. In addition, three individuals already stationed in Paris—the NACA's technical assistant in Europe John Jay Ide, Assistant Military Attaché Major Barton K. Yount, and Assistant Naval Attaché Lt. Comm. William D. Thomas—served as technical advisors. Stimson to Hoover, Apr. 25, 1929, box 5617, Records of the State Department, RG 59, NARA II. Assistant Treasury Secretary Seymour Lowman to Stimson, Mar. 29, 1929, Secretary of War Good to Stimson, Apr. 1, 1929, Acting Secretary of the Navy to Stimson, Apr. 11, 1929, Postmaster Brown to Stimson, Apr. 3, 1929, Secretary of Commerce Lamont to Stimson, Apr. 7, 1929, Ames to Stimson, Apr. 18, 1929, all in box 5617, Records of the State Department, RG 59, NARA II.

120. This would require the full incorporation of the US reservations to the 1919 convention as embodied within the Havana Convention: eliminating its customs provisions, allowing a state's own aircraft to access its own prohibited zones for all reasons

except commercial interstate flight, permitting more flexible agreements between members and nonmembers, and clearly separating the 1919 convention from the League of Nations. Stimson to MacCracken, May 20, 1929, box 5617, Records of the State Department, RG 59, NARA II.

121. Convention adherents present included Belgium, Great Britain, Canada, the Irish Free States, the Union of South Africa, India, Bulgaria, Chile, Denmark, France, Greece, Italy, Japan, the Netherlands, Poland, Portugal, the Saar Territory, the Kingdom of the Serbs, Croats, and Slovenes, Siam, Sweden, Czechoslovakia, and Uruguay, with Persia and Romania absent. Nonmember states in attendance included Germany, the United States, Austria, Brazil, China, Colombia, Cuba, Spain, Estonia, Finland, Haiti, Hungary, Luxembourg, Norway, Switzerland, and Venezuela. Although Panama had officially accepted an invitation to participate, "it was unable to send a Representative." M. Metternich and Commander Brivonesi represented the League of Nations in an "advisory capacity." International Commission for Air Navigation, Extraordinary Session of June 1929, Minutes, Sittings of June 10–15, 1929 (hereafter referred to as Extraordinary Session), box 15, William P. MacCracken Jr. Papers, HHPL.

122. After much debate, the Final Protocol of the Extraordinary Session did not fully address US reservations to the 1919 convention. A state's own aircraft could only access its prohibited zones as "an exceptional measure and in the interest of public safety," special agreements between members and nonmembers could "not be contradictory to the general principles" of the 1919 convention but those principles remained undefined, and the ICAN's connection to the League of Nations and the convention's customs provisions endured. Armour to White, telegram of message from MacCracken and Baker, June 14, 1929, and White to Amour, June 14, 1929, both in box 5617, Records of the State Department, RG 59, NARA II.

123. MacCracken and Baker signed the Final Resolution of the Extraordinary Session on behalf of the United States with reservations to Articles 3, 5, 34, and Annex H, exactly the same as before minus the reservation to the Permanent International Court of Justice. Despite his limited success, MacCracken believed that the modified convention "on the whole . . . represent[s] substantial improvements . . . worth the efforts," and the NAA called on the State Department to use "all proper methods . . . to procure the adoption" of the modified 1919 convention. MacCracken to Gardner, June 17, 1929, box 14, William P. MacCracken Jr. Papers, HHPL; MacCracken and Baker to Stimson, June 28, 1929, box 5617, Records of the State Department, RG 59, NARA II; Bingham to Stimson, Sept. 12, 1929, box 5618, Records of the State Department, RG 59, NARA II.

124. When Assistant Secretary Francis White requested feedback on ratification in the wake of the Extraordinary Session, only the Post Office and the Treasury Department continued to express support for ratification with existing reservations. The Department of the Navy called for three new reservations that completely "exempted [the Canal Zone] from the operation of this Convention," while the Labor Department wanted an additional reservation to protect its freedom of action on immigration matters. NACA Executive Committee chair Joseph S. Ames informed White that members now believed that the United States should "withhold action on the question of ratification unless some situation arises in connection with our international aerial operations which makes

ratification desirable." Clarence Young—appointed assistant secretary of commerce for aeronautics upon MacCracken's resignation in October 1929—requested that the 1919 convention "remain in abeyance" until either all ICAN member states had ratified the Extraordinary Session's amendments, a major European nonmember state (i.e., Germany) ratified the convention, or a lack of US adherence negatively affected Pan Am's ability to obtain overflight rights in Latin America. Brown to Stimson, Nov. 26, 1929, Adams to Stimson, Dec. 23, 1929, Ames to White, Dec. 17, 1929, all in box 5618, Records of the State Department, RG 59, NARA II; Young to Stimson, Dec. 11, 1929, Mellon to Stimson, May 24, 1930, and Reitzel to Assistant Secretary of Labor, Feb. 14, 1930, both in box 435, Records of the Civil Aeronautics Administration, RG 237, NARA II.

125. The 1919 convention sat in the Senate until President Roosevelt requested its withdraw from consideration on January 15, 1934. *Papers Relating to the Foreign Relations of the United States 1926*, vol. 1 (Washington, DC: Government Publishing Office, 1941), 145.

126. White and Young, Memorandum on the Havana Convention on Commercial Aviation, Feb. 9, 1931, box 436, Records of the Civil Aeronautics Administration, RG 237, NARA II.

127. "TS 840: Multilateral; Commercial Aviation (Inter-American)," US Statutes at Large 47 (1922–1933): 1901–1911. By the end of 1934, the Dominican Republic (1932), Haiti (1933), Costa Rica (1933), Honduras (1933), and Chile (1934) had also ratified the Havana Convention. See *Papers Relating to the Foreign Relations of the United States 1933*, vol. 4 (Washington, DC: Government Publishing Office, 1950), and *Papers Relating to the Foreign Relations of the United States 1934*, vol. 4 (Washington, DC: Government Publishing Office, 1951).

128. The Havana Convention charged the PAU to work to achieve uniform regulations among member states, with the assumption that US dominance would result in a "California Effect" that would encourage the adoption of its regulations. When the State and Commerce Departments worked to compile the IACAC's draft agenda in 1926, Borges had suggested that the PAU distribute and promote a US-drafted model air law. As a member of the US delegation to the IACAC in 1927, Harry Guggenheim recommended sending the recently approved Air Commerce Regulations "to the delegates in order that they may study them and use them as a basis for regulations in their respective countries." Seven months after the Havana Convention came into effect in the United States, Assistant Secretary of State Francis White recommended that the United States export the Air Commerce Regulations to Latin American states through the PAU. Inter-American Commission on Commercial Aviation, "Data Prepared by the Pan American Union as a Contribution to the Work of the Commission," Inter-American Commission on Commercial Aviation 1927, box 8, John Lansing Callan Papers, LOC; IACAC, 99; Memorandum on the Havana Convention, Sept. 5, 1931, enclosed in White to Lamont, Sept. 14, 1931, box 437, Records of the Civil Aeronautics Administration, RG 237, NARA II. For more on the "California Effect," see David Vogel, *Trading Up: Consumer and Environmental Regulation in a Global Economy* (Cambridge, MA: Harvard University Press, 1995).

129. "TS 840: Multilateral; Commercial Aviation (Inter-American)," US Statutes at Large 47 (1922–1933): 1901–1911. At two joint meetings in late September 1931, members

of the State Department and Aeronautics Branch found this requirement to apply foreign provisions utterly "impractical" and called for the adoption of US regulations through the PAU. Because the laws of PAU member states remained generally less extensive than those of the United States, Pan Am could utilize its own standard operating procedures based on the more detailed Air Commerce Regulations and still remain "in conformity with . . . the laws and regulations" of the various American republics. Memorandum, Conference Pertaining to the Interpretation of Provisions of Havana Convention on Commercial Aviation, Oct. 2, 1931, box 437, Records of the Civil Aeronautics Administration, RG 237, NARA II; Operating Contract between Pan Am and the Government of Haiti, Feb. 22, 1929, Haiti, Operating Contract and Air Agreement, Corporate and General, Geographic Locations, Caribbean, box 2 of 5, PAA.

130. The Guatemalan General Bureau of Civil Aviation agreed to allow US private aircraft to enter its airspace provided that pilots obtained "clearance papers from the Guatemalan Consulate before departure." Discussions with the Mexican government hit an "impasse" after it requested the creation of a separate bilateral agreement on private aircraft with the United States, and such an agreement remained elusive during the first years of the Roosevelt administration. Whitehouse to Hull, Mar. 29, 1933, and Norweb to Hull, June 1, 1934, *Papers Relating to the Foreign Relations of the United States 1933*, vol. 4 (Washington, DC: Government Publishing Office, 1950), 506–507, 614–615; Norweb to Hull, Apr. 18, 1935, *Papers Relating to the Foreign Relations of the United States 1935*, vol. 4 (Washington, DC: Government Publishing Office, 1953), 229.

131. Only after ratification did the ABA's Committee on Aeronautical Law study the Havana Convention at the request of Assistant Secretary of Commerce for Aeronautics Clarence Young. Its chairman John C. Cooper Jr., a Florida lawyer who would become vice president of Pan Am in 1934 and a staunch advocate for a liberal international air regime after World War II, concluded that the Constitution's Supremacy Clause would allow for the extension of federal power to fully apply the provisions of the Havana Convention that remained outside the scope of federal authority under the Air Commerce Act. Young to Cooper, Nov. 24, 1931, box 437, Records of the Civil Aeronautics Administration, RG 237, NARA II; John C. Cooper, *The Right to Fly* (New York: Henry Holt and Company, 1947); "Report of the Standing Committee on Aeronautical Law," *Report of the Fifty-Fifth Annual Meeting of the American Bar Association Held at Washington, DC, October 12–15, 1932* (Baltimore: Lord Baltimore Press, 1932); John C. Cooper, "The Pan American Convention on Commercial Aviation and the Treaty-Making Power," *American Bar Association Journal* 19, no. 1 (Jan. 1933).

132. At the Seventh International Conference of American States in Montevideo, Uruguay, in 1933, the Roosevelt administration chose to refrain from initiating any revisions to the Havana Convention out of concern that such discussions could lead to the inclusion of new provisions that might negatively affect Pan Am's current operations. Latchford, Memorandum regarding the Havana Convention on Commercial Aviation, June 24, 1933, Entry #P 184: Records Relating to the Havana Convention on Commercial Aviation, 1926–1944, RG 59, NARA II; Cordell Hull, Instructions to Delegates, Nov. 10, 1933, in *Papers Relating to the Foreign Relations of the United States 1933*, vol. 4 (Washington, DC: Government Publishing Office, 1950), 43, 122–123.

133. Kenneth Colegrove, "The International Aviation Policy of the United States," *Journal of Air Law* 2, no. 4 (Oct. 1931): 454.

134. Van Vleck, *Empire of the Air*, 171. The Roosevelt administration's postwar aeronautical vision was based on FDR's five aerial freedoms: the right of overflight, the right to land, the right to transport passengers and goods *to* and *from* the home country and foreign states, and the right to cabotage (the ability to embark and disembark passengers and goods at intermediate points). The Havana Convention had provided for all five of these, but it did not apply worldwide and, as we have seen, its application in the Western Hemisphere remained elusive. For more on FDR's approach to postwar aviation and the tension it created in the "special relationship" between the United States and Britain, see Alan Dobson, *FDR and Civil Aviation: Flying Strong, Flying Free* (New York: Palgrave Macmillan, 2011), and Alan Dobson, *Peaceful Air Warfare: The United States, Britain, and the Politics of International Aviation* (Oxford: Clarendon Press, 1991).

Conclusion

1. Trump's position represented merely another episode within a larger debate that has dominated American political discourse throughout the twentieth century: the extent to which the United States should engage with the rest of the world. Remarks by President Trump to the 73rd Session of the United Nations General Assembly, New York, Sept. 25, 2018, https://www.whitehouse.gov/briefings-statements/remarks-president-trump -73rd-session-united-nations-general-assembly-new-york-ny/. For more on the development of the globalist ideal and the importance of World War II in that process, see Or Rosenboim, *The Emergence of Globalism: Visions of World Order in Britain and the United States, 1939–1950* (Princeton, NJ: Princeton University Press, 2017).

2. "Indonesia Plane Crash Adds to Country's Troubling Safety Record," *New York Times*, Oct. 28, 2018.

3. "Ethiopian Airlines Plane is the 2nd Boeing Max 8 to Crash in Months," *New York Times*, Mar. 10, 2019.

4. "Two-Thirds of the 737 Max 8 Jets in the World Have Been Pulled From the Skies," *New York Times*, Mar. 12, 2019.

5. "E.U. Suspends Boeing 737 Max 8 Operations, but F.A.A. Stands Firm," *New York Times*, Mar. 12, 2019; Boeing, "Boeing in Brief," accessed Jan. 8, 2019, https://www.boeing .com/company/general-info/.

6. "US Joins Other Nations in Grounding Boeing Plane," *New York Times*, Mar. 13, 2019; FAA Emergency Order of Prohibition, Mar. 13, 2019, https://www.faa.gov/news /updates/media/Emergency_Order.pdf.

7. The 1938 Civil Aeronautics Act effectively extended federal control to intrastate flight alongside the inclusion of direct economic regulation over air carriers, while the Federal Aviation Act of 1958 established what became known as the Federal Aviation Administration to regulate civilian and military aviation within the United States. Civil Aeronautics Act of 1938, Pub. L. No. 75-706, 52 Stat. 601 (1938); Federal Aviation Act of 1958, Pub. L. No. 85-726, 72 Stat. 737 (1958).

8. As such, it shows that expansion of the administrative state can occur just as readily out of concerns over the security and public safety implications of a new technol-

ogy as it can through the creation of a welfare state. For studies on the expansion of state power in response to "conservative" priorities, see Lisa McGirr, *The War on Alcohol: Prohibition and the Rise of the American State* (New York: W. W. Norton, 2016), and Theda Skocpol, *Protecting Soldiers and Mothers: The Political Origins of Social Policy in the United States* (Cambridge, MA: Harvard University Press, 1992).

9. By the time Congress passed aviation legislation, the need to regulate in the name of safety and security had become so accepted that a case challenging the validity of the 1926 Air Commerce Act or the Air Commerce Regulations never made it to the US Supreme Court.

Index

Page numbers in *italics* indicate figures or tables